THE ROUTLEDGE COMPANION TO THE AMERICAN LANDSCAPE

T0326201

The Routledge Companion to the American Landscape provides a comprehensive overview of the American landscape in a way fit for the twenty-first century, not only in its topical and regional scope but also in its methodological and disciplinary diversity.

Critically surveying the contemporary scholarship on the American landscape, this companion brings together scholars from the social sciences and humanities who focus their work on understanding the polyphonic evolution of the United States' landscape. It simultaneously assesses the development of the US landscape as well as the scholarly thought that has driven innovation and continued research about that landscape. Four broad sections focus on key areas of scholarship: environmental landscapes, social, cultural, and popular identities in the landscape, political landscapes, and urban/economic landscapes. A special essay, "American Landscapes Under Siege" and accompanying short case studies call attention to the legacies and realities of race in the American landscape, bridging the discussion of social and political landscapes.

This companion offers an invaluable and up-to-date guide for scholars and graduate students to current thinking across the range of disciplines which converge in the study of place, including Geography, Cultural Studies, and History as well as the interdisciplinary fields of American Studies, Environmental Studies, and Planning.

Chris W. Post is Professor of Geography at Kent State University at Stark.

Alyson L. Greiner is Professor and Head of Geography at Oklahoma State University.

Geoffrey L. Buckley is Professor of Geography and Interim Associate Dean, Ohio Honors Program, Honors Tutorial College, Ohio University.

THE ROUTLEDGE COMPANION TO THE AMERICAN LANDSCAPE

Edited by Chris W. Post, Alyson L. Greiner, and Geoffrey L. Buckley

Routledge
Taylor & Francis Group

LONDON AND NEW YORK

Cover image: © Photos by contributors

First published 2023
by Routledge
4 Park Square, Milton Park, Abingdon, Oxon OX14 4RN

and by Routledge
605 Third Avenue, New York, NY 10158

Routledge is an imprint of the Taylor & Francis Group, an informa business

British Library Cataloguing-in-Publication Data
A catalogue record for this book is available from the British Library

Library of Congress Cataloging-in-Publication Data
Names: Post, Chris W., editor. | Greiner, Alyson L., 1966– editor. |
Buckley, Geoffrey L., 1965– editor.
Title: The Routledge companion to the American landscape /
Edited by Chris W. Post, Alyson L. Greiner, and Geoffrey L. Buckley.
Description: Abingdon, Oxon ; New York, NY : Routledge, 2023. |
Series: Routledge international handbooks |
Includes bibliographical references and index.
Identifiers: LCCN 2022038475 (print) | LCCN 2022038476 (ebook) |
ISBN 9780367640156 (hardback) | ISBN 9780367640187 (paperback) |
ISBN 9781003121800 (ebook)
Subjects: LCSH: Cultural landscapes–United States. |
Geography–Social aspects–United States. |
Landscapes–Political aspects–United States. |
Environmental geography–United States. | United States–Historical geography.
Classification: LCC GF503 .R68 2023 (print) | LCC GF503 (ebook) |
DDC 304.20973–dc23/eng20230106
LC record available at https://lccn.loc.gov/2022038475
LC ebook record available at https://lccn.loc.gov/2022038476

ISBN: 9780367640156 (hbk)
ISBN: 9780367640187 (pbk)
ISBN: 9781003121800 (ebk)

DOI: 10.4324/9781003121800

Typeset in Bembo
by Newgen Publishing UK

CONTENTS

Contents

Contents

FIGURES

List of Figures

BOXES

TABLES

CONTRIBUTORS

Editors

Chris W. Post is Professor of Geography at Kent State University at Stark. He holds an MA and PhD, each in Geography, from the University of Kansas and a BS in Secondary Education from the University of Oklahoma. His research as a cultural and historical geographer focuses on the heritage of place, particularly as it becomes manifest through the cultural landscapes of commemoration, place naming, music, and sense of place. He has published in *Geoforum*; *Journal of Cultural Geography*; and *The Geographical Review*; and in books including, *Affective Architectures*; *Explorations of Place*; and *Social Memory and Heritage Tourism Methodologies*. He advises on the May 4 Visitors Center and annual commemorations at Kent State and currently serves as the University's inaugural May 4 Researcher in Residence.

Alyson L. Greiner is Professor and Head of Geography at Oklahoma State University. She is a broadly trained cultural and historical geographer, a textbook author, and for many years she served as the editor-in-chief of the *Journal of Cultural Geography*. Her diverse interests include necrogeography and landscapes of the dead, placemaking, and the history of geography, among others.

Geoffrey L. Buckley is Professor of Geography and Interim Associate Dean, Ohio Honors Program, Honors Tutorial College, at Ohio University. He holds a Master's degree from the University of Oregon and a Ph.D. from the University of Maryland. His research interests include environmental history and historical geography; environmental justice; mining landscapes; public lands, esp. national parks; urban environments; and urban sustainability. He is the author of five books including, most recently, *The American Environment Revisited: Environmental Historical Geographies of the United States* (co-edited with Yolonda Youngs, 2018) and *North American Odyssey: Historical Geographies for the 21st Century* (co-edited with Craig Colten, 2014).

Contributors

Derek H. Alderman is Professor of Geography at the University of Tennessee and a past President of the American Association of Geographers. He studies race, civil rights, and the politics of commemoration within the context of the landscapes of the southeastern USA.

Timothy G. Anderson is Associate Professor in the Department of Geography at Ohio University. His research focuses on the historical settlement geographies of the United States, and the production of space in Germanic diasporic cultural landscapes.

Samuel Avery-Quinn is University College Senior Lecturer at Appalachian State University in Boone, North Carolina, where he studies the intersections of religion, landscape, and material culture in America.

Thomas L. Bell is Professor Emeritus in the Department of Geography, University of Tennessee. In addition to music geography, his current research interests involve the craft beer industry and the geopolitical and economic aspects of postage stamps.

Lisa Benton-Short is Professor of Geography at the George Washington University. She is an urban geographer with an interest in the dynamics of the urban environment from many angles, including: urban sustainability, planning and public space, monuments and memorials, urban national parks, globalization, and immigration.

Brian C. Black is distinguished professor of Environmental Studies and History at Penn State Altoona, where he also currently serves as Head of the Division of Arts and Humanities. He is a specialist in energy history and is the author of several books, including: *Petrolia: The Landscape of America's First Oil Boom* (Johns Hopkins, 2003); *Crude Reality: Petroleum in World History* (Rowman & Littlefield, 2021); and *To Have and Have Not: Energy in World History* (Rowman & Littlefield, 2022).

Jordan Brasher is an Assistant Professor of Geography at Columbus State University in Columbus, Georgia and a Research Fellow for Tourism RESET, a collaborative initiative to identify and challenge racial inequity in the tourism industry. His areas of research specialization lie at the intersection of race, place, and public memory, and his work has been featured in popular news outlets such as *The Conversation, USA Today*, and *The Washington Post*.

Craig E. Colten is Professor Emeritus of Geography at Louisiana State University and is the author *of Southern Waters: The Limits to Abundance* (2014) and *State of Disaster* (2021). During his career, he developed an award-winning exhibit on the Illinois River, investigated groundwater pollution, studied efforts to manage water resources, and researched historical adaptations to flood and hurricane hazards.

Christina E. Dando is Professor of Geography and Chair of the Geography and Geology Department at the University of Nebraska at Omaha. Her research interests include landscape (specifically the Great Plains), gender, and media (newspapers, photographs, films, maps) and the various intersections of these elements, both historical and contemporary.

Fiona Davidson has been on the Geosciences faculty at the University of Arkansas since 1992. Her primary research interests are in the geographies of science fiction, specifically the geopolitics of the future, and in contemporary US political culture and elections.

Dydia DeLyser is a feminist cultural-historical geographer whose research is qualitative. Her co-authored book, *Neon: A Light History* draws from over a decade of archival research, interviews, participant observation, and autoethnography, and is a community-based collaboration that benefits the Museum of Neon Art.

Erika Doss is a Professor of American Studies at the University of Notre Dame.

Emily Fekete, PhD Geography, University of Kansas, is the Community Manager at the American Association of Geographers. Her research focuses on the relationships between social media and sites of consumption, both on and offline.

Aretina R. Hamilton received her Ph.D. in Geography from the University of Kentucky with a Certificate in Gender and Women's Studies. She is a proud graduate of Kentucky State University, an HBCU where she received her BA in Political Science. Dr. Hamilton is currently the Director of Equity and Inclusion for the City of Raleigh, North Carolina. As a Cultural Geographer her work explores the intersection of Anti-Blackness, racial trauma, white violence, and white supremacist placemaking.

Now a Professor Emeritus in the Department of Geography and Spatial Sciences at Kansas State University, **Lisa M. Butler Harrington** is now located in Washington State. Her research addresses rural topics, environmental change, natural resources, and land management.

Andrew Herod is Distinguished Research Professor of Geography at the University of Georgia. He writes frequently upon matters of labor and the global economy.

Ryan Holifield is an Associate Professor of Geography at the University of Wisconsin-Milwaukee. His research interests include environmental justice policy and practice, social and political dimensions of urban environmental change, and stakeholder participation in environmental governance.

Heather Hollen holds a master's degree from the University of Oklahoma College of Public Health and a science writer and editor with the Hudson College of Public Health at the University of Oklahoma Health Sciences Center.

Ellen Hostetter is an Associate Professor in the Norbert O. Schedler Honors College at the University of Central Arkansas. Her recent scholarly work sits at the intersection of cultural landscape studies and legal geographies and focuses specifically on the automobile, particularly as it relates to how automobile regulation and energy consumption (gas stations) has impacted the cultural landscape.

Andrew Husa, PhD, is a Lecturer, School of Global Integrative Studies, University of Nebraska-Lincoln. His interests include small town geography and rural identity, and attachment to rural landscape, especially on the American Great Plains.

Joshua Inwood is a Professor of Geography and a senior research scientist at The Rock Ethics Institute at The Pennsylvania State University. Broadly interested in the human condition, his work spans human geography with a specific focus on race and processes of racialization.

Melissa Keeley is Associate Professor of Geography at the George Washington University and serves as the Director of the Environmental Studies Program and the Environmental and Sustainability Science Program. Her research on urban sustainability examines the intersection of urban ecology, engineering, and environmental planning and is clustered around several themes including green infrastructure and transnational policy adaptation.

David J. Keeling is Distinguished University Professor Emeritus in the Department of Earth, Environmental, and Atmospheric Sciences, Western Kentucky University. In addition to music geography, his current research interests involve the transportation industry in Latin America and the built-environment impacts of global accessibility and mobility challenges.

Christopher R. Laingen is a Professor of Geography at Eastern Illinois University. His research centers on rural and agricultural land-use change in the United States, specifically the Midwest and Great Plains.

Maria Lane is a Professor of Geography at the University of New Mexico, where she also directs the R.H. Mallory Center for Community Geography. Her research publications focus on the historical geographies of science, environmental knowledge, and cartography.

John Lauermann is an Associate Professor at the Pratt Institute, where he also directs the Spatial Analysis and Visualization Initiative. An urban geographer, his research uses GIS to analyze socio-spatial inequality in US cities, on topics such as gentrification and mega-projects.

Chris Lukinbeal is a Professor of Geography at the University of Arizona. He has published books on *The Geography of Cinema, Mediated Geographies and Geographies of the Media, Place Television and the Real Orange County*, and *Media's Mapping Impulse*.

Kenneth D. Madsen is Associate Professor of Geography at The Ohio State University at Newark. His work in political and cultural geography considers borders and bordering processes with an emphasis on the US–Mexico boundary.

Martin V. Melosi is Cullen Professor Emeritus of History and Founding Director of the Center for Public History at the University of Houston. His most recent book is *Fresh Kills: A History of Consuming and Discarding in New York City* (2020).

Jacque Micieli-Voutsinas, PhD, is an Assistant Professor of Museum Studies at the University of Florida, Gainesville. Her research focuses on sites of difficult heritage and the affective politics of trauma and memory

Matthew L. Mitchelson is an Associate Professor of Geography at Kennesaw State University. His research centers on urban-economic and political geography, and the vast consequences of imprisonment.

David R. Montgomery is Professor of Geomorphology at the University of Washington where he studies the evolution of topography and the influence of geomorphological processes on ecological systems and human societies.

Cheryl Morse is Associate Professor of Geography and co-director of the Environmental Program at the University of Vermont. She specializes in rural geography, place and identity, and human–environment interactions.

Velvet Nelson is a Professor in the Department of Environmental and Geosciences at Sam Houston State University in Huntsville, Texas, USA. She is a human geographer with a specialization in the geography of tourism.

Martin J. Pasqualetti is Professor of Geography at Arizona State University. He has written extensively on the public acceptance of wind energy landscapes.

Mark Alan Rhodes II is an Assistant Professor of Geography at Michigan Technological University where he teaches in the Department of Social Sciences and advises Industrial Heritage and Archaeology MS and PhD students. His research critically examines the intersections of memory, identity, culture, and landscape, particularly in Wales.

Sarah Fayen Scarlett is Associate Professor of History at Michigan Technological University where her research in public history investigates the applications of GIS-based models of post-industrial landscapes for heritage management. Her book *Company Suburbs: Architecture, Power, and the Transformation of Michigan's Mining Frontier* (University of Tennessee Press 2021) focuses on the spatial, material, and experiential aspects of early twentieth-century domestic architecture and landscapes to examine relational social identities.

Richard Schein is Professor of Geography at the University of Kentucky. He teaches and writes about the place of landscape in everyday life.

Fred M. Shelley is Professor Emeritus of Geography and Environmental Sustainability at the University of Oklahoma. His research interests include the political, electoral, and cultural geography of the United States.

Emily Skop is Professor and Chair in the Department of Geography and Environment Studies and Founding Director of the Global Intercultural Research Center at the University of Colorado Colorado Springs. Her scholarship explores global migration and its consequences, particularly placemaking for refugees and migrants in the US.

Ted Steinberg is the Adeline Barry Davee Distinguished Professor of History at Case Western Reserve University in Cleveland.

Stephen Cho Suh is Assistant Professor of Asian American Studies at San Diego State University. His scholarly interests lie at the intersections of race, ethnicity, gender, migration, and culture, especially in relation to Asian Americans and the Korean diaspora. His current book project examines the culinary entrepreneurship of 1.5 and 2nd-generation Korean Americans in the US and South Korea.

Brandi T. Summers, PhD is Assistant Professor of Geography and Global Metropolitan Studies at the University of California, Berkeley. Her research examines the relationship between and function of race, space, urban infrastructure, and architecture.

Barney Warf is a Professor of Geography at the University of Kansas. His research concerns political economy, social theory, and globalization.

Randall K. Wilson is Professor of Environmental Studies at Gettysburg College. His research focuses on environmental history and politics with special attention given to public lands, collaborative management, and the political ecology of the American West.

Yolonda Youngs is an Associate Professor in the Department of Geography and Environmental Studies at California State University, San Bernardino. She specializes in cultural geography, environmental studies, national parks and protected areas, public lands, environmental justice, social science GIS, conservation of natural resources, environmental policy and land management, field methods, American West, Europe.

ACKNOWLEDGEMENTS

Many people came together to make this volume possible. First and foremost, it was a delight to work with our many contributors. We appreciate their commitment to this project through the disruptions of the pandemic as well as their valuable insights on the American landscape.

As a team we'd also like to thank our editors at Routledge, particularly Andrew Mould, who first approached Chris about developing this book. We'd also like to thank Claire Maloney for her help with the details of editing and submission to ensure this volume became the best possible resource on the US landscape. We know several other staff at Routledge were also involved in this book's evolution and we appreciate their labor, too. We would like to thank Deans Denise Seachrist and Rob Sturr at Kent State University's Stark Campus and the University Teaching Council at Kent State, Roxanne Male'-Brune and Carma West, Research Division, Faculty Research Support Program at Ohio University, and Oklahoma State for funding in support of this project. Thank you also to our indexer, Heidi Blough, for her timely and important help.

Chris thanks his wife, Amy, and their children, Kiera and Dylan, for their support and patience while he read, edited, and wrote this text. Chris also thanks his parents for their inspiration to take on big projects and see them to completion. Finally, to the "anonymous" colleague (Derek Alderman!) who proposed to Andrew Mould that I be the one to contact about quarterbacking this project, thank you so much for this wonderful opportunity!

Alyson appreciates her co-editor compadres, Chris and Geoff, whose perspectives and levity greatly enriched this experience. She remains deeply indebted to her life-partner, Luis, for his patience and support.

Geoff would like to thank his wife, Alexandra, for always being there when he's needed you (and he has needed you a lot over the past three years!), as well as their children Ingrid, Peter, and Owen. I am so fortunate to have all of you in my life! I have learned so much from you over the years and look forward to our next adventure together.

FOREWORD
READING THE LANDSCAPE

Richard H. Schein

Introduction

Driving along the eight-mile stretch of US 62 between the Woodford County seat of Versailles and the little town of Midway takes you through the classic Kentucky Bluegrass horse farm landscape (Figure F.1). If you are paying attention you might notice that farms on the east side of the road all display elements of the classic regional landscape of the superrich: white or brown four-board fences, manicured fields, barns often more elaborate than most houses in the area, a big house on the farm's interior whose roofline is just visible, and elegant and lithe Thoroughbred horses grazing in the fields. If you are really paying attention you will notice one farm on the other side that seems somehow out of place (Figure F.2). It is scruffier, even unkempt. The fence is wire, the grass is irregular and clumpy, there are pockets of woods in the low areas amongst the rolling fields displaying corn stalks and rolled hay bales, the barns are for tobacco, and the animals in the fields are beef cattle.

Whether you were conscious of the fact, you read the landscape on that short drive. Perhaps you noticed regularity in the horse farms or marveled at the amount of money it takes to get them looking that way. Perhaps you detected a difference between the horse farm and the cattle farm. At the very least you could not help but pay attention to the road and take a cue from the landscape to watch out for tractors or horse vans just to avoid an accident.

Donald Meinig (1979) wrote an essay many years ago in which he took the reader through ten versions of the same scene before their eyes, variously describing the landscape as nature, habitat, ideology, history, problem, and so on. His point was that we all read the landscape, and the manner in which we read—the things we notice, the questions we ask, the assumptions we make—derives in part from the perspective we bring to the scene, from what lies in our mind. That perspective might derive from our experiences, our training, and our emotional responses. It might not be confined to our mind and could also stem from the visceral experience of our being in the landscape. Meinig's point was to draw attention to the manner in which we read the landscape.

Returning to the Midway Road as an animal activist might mean you read the landscape as one of exploitation; where horses are valued solely for their ability to win races or produce foals and cattle for their place in an agro-industrial food commodity chain. If you are a political or cultural ecologist you might see two very different landscapes, both humanly contrived, but one

Figure F.1　A typical Bluegrass horse farm landscape. Photo by author.

Figure F.2　A Bluegrass cattle and tobacco farm directly across the road from Figure F.1. Photo by author.

more homogenous and sterile, with fields scraped and tiled and smoothed and planted in fescue so that million-dollar horses don't break their exceedingly thin legs. The other with hedgerows intact and home to a thriving diversity of wildlife, and a mixed farming regime which allows at least some nitrogen fixing without intensive applications of fertilizer. If you were a lifelong city dweller you might have simply been perplexed, indifferent, or even afraid of the scene passing before your eyes. In short, we already are conditioned to read landscapes, albeit often unconsciously and ephemerally. To move to the next step entails at least two realizations: first, if we are to move beyond our own, sometimes solipsistic engagement with the landscape we need to take landscape seriously as a social phenomenon; second, to move to the social requires a more disciplined attention to reading the landscape itself.

This essay pays attention to the idea of "reading the landscape." First, it addresses the tradition of reading in the historiography of landscape study, which opens myriad possibilities for conceptualizing the place of landscape in everyday life. Second, it reviews one method for beginning to systematically read the American cultural landscape. Finally, it builds upon the social theoretical foundations of landscape reading to propose some aphorisms for thinking about how landscapes work as part of the everyday processes of human social, cultural, political, and economic interaction.

Reading the American Landscape

Reading the landscape is a metaphor that likens the landscape to a book or a text. Indeed, the Geographer Peirce Lewis (1979, 12) wrote in the preface to his famous Axioms for Reading the Landscape that the "human landscape is our unwitting autobiography, reflecting our tastes, our values, our aspirations, and even our fears in tangible visible form." He proposed a set of axioms—purportedly self-evident propositions—that he saw as some basic rules for reading that autobiography.

The autobiographical aspect of Lewis's maxim captured scholarly imaginations intrigued with developments in literary theory at the time. On the one hand, the textual metaphor appealed to landscape scholars looking for more formalized interpretive frameworks. On the other hand, many unquestioned theoretical and social assumptions in Lewis's approach and its antecedents in Cultural Geography's "Berkeley School" were tackled as part of a conceptual renaissance regarding the landscape concept. Rethinking landscape theory was not an insulated academic exercise for the landscape tradition located in Geography. It coincided with Geography's empiricist critique and reached across academic boundaries, to engage debates in literary theory, in Anthropology's "crisis of representation," in Sociology and, perhaps most importantly, in the academy's engagement with French, British, and German social theory. It also intersected with critical geography and an increasing political commitment to acknowledging questions of power, class, race, and gender in the social sciences and humanities in general, and for landscape interpretation in particular.

That renaissance raised questions about how we looked at landscapes. What was the meaning of "culture"? Who was the "our" in "our unwitting autobiography"? How could/should we get at "meaning" associated with landscape? Was "authorial intention" primary or did we create landscape by reading? What was the relationship between landscape and power, between landscape and politics, between landscape and justice? The questions generated new literatures and important articles and books. For example, we took seriously the textual metaphor (Duncan and Duncan 1998). We saw landscape as both an epistemology and as a politically expedient mechanism for legitimating and facilitating the capitalist transformation of modern European social class structure (Cosgrove 1984). We discussed landscape's assumption of the masculinist universal subject in the distanced view or the "gaze" central to the reading of landscape (Rose 1992, 2001). We explored the critical links between landscape, labor, and class negotiations (Mitchell 1996). We came to see the place of landscape in subject formation and social intervention through its discursive constitutions (Schein 1997). The theoretical foment in the waning decades of the twentieth century reestablished landscape as a staple geographic idea in Anglophone literature, at least, and suggested many prospects for continued landscape study (Wylie 2007). We added to the landscape repertoire new concepts and topics and reworked old ones. Some of these included exploring landscape's political ecologies in the Anthropocene (Robbins 2007), the urban landscape's role in the carceral state (Derrickson 2016), Black Geographies' landscape stories (McKittrick 2006), engaging more-than-representational landscapes (Waterton 2019), and a reinvigoration of landscape, nature, and justice (Olwig 2019).

Landscape study in the twenty-first century emerged from that foment with at least four important outcomes. First was the renaissance of a strong, broadly interdisciplinary field of landscape studies. Second was to enliven the landscape: giving it a central place in constituting cultural meaning, understanding that landscapes work in the interest of political and social and economic (re)production; and engaging theories that took seriously more-than-representational questions of embodiment, affect, and emotion. Third was to realize that cultural landscapes are always becoming and always open to (re) and (multiple) interpretations, a point that coincided with and drew from the rising academic challenge to the idea of the master narrative. Finally, we sought methodological innovation and challenge in places like oral history, feminist methodologies, and Black Geographies in order to not only bring the margins to representation but to also celebrate the relational ontologies of landscape.

The cultural or human landscape moved from being simply the result of human activity (our unwitting autobiography) that we could read to also being a tangible, visible element important to constituting our everyday lives. We rethought our "reading" and reformulated our methodologies in a more critical vein.

A Methodology for Reading the American Landscape

A simple methodological framework for beginning to more critically read the American Landscape starts with four steps (Schein 2009).

Landscape History

The framework's foundation rests on the conviction that landscape study should be grounded in good historical-geographical description and that the palimpsest qualities of landscape and the lasting imprint of the tangible visible scene are a structural part of the worlds we inherit and inhabit. Returning to the eight-mile stretch of the Midway road that opened this chapter, we could begin to describe the historical development of the landscape moving past our windshield as the image of a regional landscape created first by native occupants who burned forests and so contributed to a forest and grassland landscape that suited the first imperial Euro-American colonizers who modified the landscape palimpsest with a mixed grain and livestock agricultural economy undergirded by enslaved labor. White Burley tobacco for cigarette production was added after the Civil War (Figure F.3), at the same time that changes in the structure of the national Thoroughbred racing industry led northern industrialists to invest in horse farms, bringing in landscape architects and capital to create a signature landscape of conspicuous consumption. Those farms today are doubly protected by the purchase of development right programs and publicly and privately funded conservation easement programs that maintain a continuous green space. These are a few facts of a regional landscape history that we can empirically document and trace and establish without much controversy. It is a statement about what a landscape is and, to an extent, when and how it came to be and these facts are important to contemporary interpretations and experiences with the horse farm landscape, for both the people who live there and those who valorize and fight over and revere and despise that landscape.

Paying attention to landscape history also entails knowing the histories of the landscape's constituent material objects. In this example that might mean familiarity with regional and

Figure F.3   White Burley tobacco being harvested and taken to the barn for air-drying. Photo by author.

national trends in house style and construction, technical aspects of urban and rural land use law and planning practices, the changing categorical uses of green space in American cities, or a history of roads and their private, community, local, state, and federal ordinances and specifications.

Landscape Meaning

Meaning can be derived from varied perspectives, scales, and intentions. In landscape studies more traditionally, meaning often was holistic, somewhat vaguely "cultural," and stemmed from the interpretive position of the expert viewer, "trained" to see and understand the landscape from the purportedly objective vantage point outside the scene. Critiques of that perspective allowed us to also ask about meaning in other manifestations and forms, including individual and group social dynamics and identity, questions of nationalism and regionalism, or in a different register or set of power axes, as imbricated with gendered, classed, or racialized social norms and values.

Ascribing meaning is not as easy as documenting the built environment's creation. People vest significant meaning into the landscape, especially in a nation founded on the principle of life, liberty, and the pursuit of happiness, where happiness is about freehold property tenure. And those meanings have consequences. The signature horse farm landscape might be an aesthetically pleasing scene for many, and so deemed worthy of public investment in private property through public subsidies that purchase development rights, effectively paying farmers to keep their land in agriculture and to visually maintain a rural aesthetic. The resulting landscape and its public subsidy might be supported by the local tourist industry, which does not have a vested interest in horses per se, but is happy to bundle horse farms with bourbon distilleries to promote a tourist destination as the now-symbolic landscape means economic development to the ancillary local service economy of tour guides, motels, hotels, restaurants, and shopping. To others, that regional landscape might reflect opulent and sybaritic capitalist excesses and appear simply as a playground for the rich and a symbol of the long-standing and widening gulf between the 1% and the rest of the country. For white landowners with a multigenerational ownership history, the landscape might be a source of individual pride and identity grounded in the land and its relation to their family history. To an African American of similar multi-generational occupancy, the landscape might hold mixed meanings, standing for both the oppressive system of white supremacy that enabled the landscape to be physically molded and built by enslaved labor; even as that labor over several generations might also be a source of pride as it now manifests in material culture including objects like the limestone rock fences that came to be associated with skilled African American labor and which are now venerated and celebrated for their origins. In this way, meaning might not only be interpretive or symbolic, but also might stem from our bodily, affective, and emotional engagement with the landscape.

Landscape as Mediator

Once we acknowledge meanings invested in landscapes, we can interrogate the role of the cultural landscape in facilitating and mediating political, social, economic, and cultural intention and debate. Design professionals take this as a given, as designed landscapes ostensibly are meant to facilitate meaning and intention. The vernacular, or everyday landscape, also can facilitate and mediate. The capacity of the landscape to naturalize and normalize social intention or values or ideologies can make the landscape normative, or a tangible, visible statement of what ought to be, which often carries with it moral or aesthetic understandings and expectations (Schein 2003). It is easy to imagine that some aspects of the perfectly groomed Bluegrass horse farm are explicitly designed to facilitate investment or to mark class status. No one wants to invest hundreds of thousands of dollars in Thoroughbred stud fees in a farm that is shoddy or unkempt. And the gated entry ways to many horse farms are meant to simultaneously keep out the hoi polloi even as they make a visible statement about the class position of those on the other side. Cultural landscapes are especially interesting in their mediating capacity, to act as the focal point or catalyst for ideas, meanings, ideologies, and actions. This is especially true when we engage the places where landscapes obviously mean so much to different people that there is conflict over the landscapes that people fight over. Those mediations—or even fights—generally transcend the actual landscape in question to engage broader social and cultural and ideological conflicts or tensions that exist beyond the landscape itself, but often depend upon the particular landscapes for their instantiations. The conflict is both over the landscape and works through the landscape.

The Bluegrass landscape itself mediates a debate over land use and suburban expansion. On the one hand, Thoroughbred horse farms are the focus of land and agricultural preservation efforts supported by: government and private purchase of development rights programs; strictly maintained urban service boundaries; rural zoning regulations mandating minimum lot sizes and land use; farm owners and breeders; preservationists; and the tourism industry, to mention only a few. On the other hand, there is an uneasy alliance between property developers, affordable housing advocates, mortgage bankers, anti-gentrification activists, and thousands of individual potential homeowners who feel they are operating in a market artificially limited by the privileging of horse farmland. The fight is over and mediated through the horse farm landscape.

Locally the fight might seem to pit preservationists against houses. But the fight/mediation also transcends the local to engage national actors, debates, and ideologies on several fronts, including long-standing American debates about property rights, the role of the state in the market, and the right to shelter.

Landscape as Discourse Materialized

The cultural landscape's capacity to naturalize and normalize social relations signals the landscape's role in social and cultural reproduction. We can think of the cultural landscape in this capacity as discourse materialized, where the landscape is both conceptual framing of the world and material presence in the world. The landscape disciplines, interpreting subjects in its articulation of a series of discourses. Those discourses can be thought of as shared meanings or commonsense assumptions or socially constituted ideologies. The landscape as discourse materialized also provides a space for human agency; moments where we can intervene in the discourse, where we can directly engage discursive formations, to negotiate and challenge them through the landscape (Schein 1997).

To follow with the example drawn from the Bluegrass horse farm landscape, we can envision that tangible, visible scene articulating any number of discourses that flow through and materialize in some form on any particular farm. The local zoning code, for instance, stipulates the range of land use practices on the farm. It is a legal discipline that can nevertheless be (hypothetically) changed through applications for a variance or lobbying for a zone change. An aesthetic discourse of "proper" horse farm appearance is part of a long tradition that disciplines individual farm owners (Raitz and Von Dommelen 1990). Owners can challenge it of course, but often not without social or cultural capital costs among their colleagues in the region or even the customers who seek out a particular farm for boarding or breeding mares, who might demand a certain aesthetic as indication of a farm's value. The discourse of contemporary immigration practice and policy materializes in the region in the bodies of farm workers, including those from Mexico who have been working in the region since the era of the Bracero program and now are subject to the legal and ideological disciplines of immigration debate and policy in the US at large. The discourse of land preservation is a powerful one in the region, and embroils farmers who agree to maintain "community" green space by taking their farms out of the development market. Each of these discourses materializes in some form on the farm, each serves to discipline the farm owners, workers, visitors, and others. Each provides a moment to accept or challenge the forms and practices they entail. The sum total of these (examples of) discursive formations materializing in one farm can make the landscape seem monolithically overdetermined in its form and function; and sometimes it is. But the point is that our reading of the landscape can decide which discourse or discourses to engage for a better understanding of the landscape's place in everyday life; and perhaps even as a site of intervention toward social change.

These methodological distinctions are heuristic, and the conceptual categories or "steps" presented here overlap of course. Anyone reading the landscape might variously fall into one of the methodological moments of this framework or engage all in a comprehensive study of a particular landscape. In the end we recognize that all landscapes are everywhere in the constant state of becoming and that our landscape reading is about asking how landscapes work. Once we get to this point—asking how landscapes work as part and parcel of our everyday lives—the question of reading is opened up to any manner of theoretical or conceptual framings, and the question of how a landscape works can focus our reading of the landscape in more directed and often more action-oriented ways.

Ten Aphorisms for Reading the Landscape

The idea that landscape works is an aphorism. The term is chosen specifically to follow from, yet challenge, Lewis's Axioms for Reading the Landscape which opened this chapter. Axioms generally are defined as self-evidently true statements, and ones that are accepted without challenge. But a generation (or two) of scholars has challenged the underlying assumptions and incontrovertibility of Lewis's Axioms (e.g., Mitchell 2008). Lewis sensed this, of course, and wrote so in his essay, stating that "what seems self-evident now was not obvious to me a few years ago" (Lewis 1979, 15), even as he apparently preferred to err on the side of overstatement in the interest of writing down some basic guidelines for reading the landscape.

The term aphorism signals an element of (theoretical) truth encapsulating both our own experience and a conceptual or theoretical understanding. One attraction of the aphorism is that, like the metaphor, we are given some leeway applying it. The match does not have to be perfect if it triggers the meaning. Some definitions of aphorism also include a sense of moral interrogation, and that too seems fitting. This chapter will close with an introduction—to some aphorisms drawn from the experience of our theoretical peregrinations over the past forty years (at least) in search of ways to make sense of the cultural landscape. It is a journey that should not have an end, if we believe in the importance and the utility of reading the everyday landscapes that, to borrow a phrase from Meinig (1979), are the mold and mirror of our lives.

There potentially are myriad aphorisms expressing truths and/or opinions about the place of landscape in a concise manner. The ideas that landscapes both mediate and serve as discourses materialized are themselves aphorisms, each of which holds in brief a conceptual or theoretical historiography that is captured in the statement. The utility of an aphorism as a general statement stems from the questions one asks of the landscape qua landscape (from the theoretical context held in the aphorism) and the questions one asks of a particular landscape, as the focus of both empirical documentation and examination and as the object of interrogation, interpretation, and analysis. Aphorisms also can be "shorthand" or mnemonic devices to keep in mind as we read and write about the American cultural landscape. They remind us that landscapes do work and to keep the question of how and to what ends at the forefront of our reading. Following are a few very brief examples of landscape aphorisms that draw upon the thematic place of landscape in racial formation; which are, in short, examples of racialized landscapes (Schein 1999, 2003, 2018).

The aphorism "landscape as a social wrong" draws on a concept of critical discourse analysis concerned with unequal power relations and their ideological processes that bear upon human wellbeing, especially as manifest in the semiotic aspects of landscape (Fairclough 2009, 163–164). At one end of the eight-mile stretch of US 62 that opened this chapter is the Sons and Daughters Relief Cemetery, a final resting place for primarily African American "inhabitants." One faded grave there marks the death of Richard James, whose lynching at the hands of a white mob in 1921 is materially manifest in the gravestone itself, and for many in the community still might "appear" in the remnants of the distillery he purportedly robbed and where he was accused of killing a man, and in the local field that held the tree displaying his gruesome and horrible death to the public and became a visible reminder for all. This clearly was a social wrong materialized in the landscape through a pervasive discourse of white supremacy, Jim Crow, and racist practice; a long-standing discourse we still are engaging and challenging, in part through landscape.

"Landscapes of belonging" acknowledges that the landscape is central to identity at any number of scales and can be seen as a marker of someone's (or some group's) own sense of belonging as well as the manner in which one's presence is accepted as belonging in a societal sense (Figure F.5).

Figure F.4 A typical Bluegrass dry-laid rock fence. Photo by author.

In the case of Richard James, the indications suggest that he somehow "did not belong," given that his rights and basic humanity were denied. On the other hand, in the same place, there is a movement in which the material contributions of African Americans are held out as points of pride for local communities of color. These include, for instance, the valorization of the regional form of limestone rock fences that line the road we first travelled in this chapter (Figure F.4). Those fences are recently the subject of historic preservation efforts, are heralded for the Black labor of enslaved and free people of color who built them, and are designated as integral to the internationally recognized Bluegrass landscape. Similarly, the role of African American labor in general, including as builders and architects, is increasingly celebrated locally, and especially in those parts of central Kentucky towns where Black-owned businesses predominated. Belonging can be ascribed and claimed through the landscape, and as a conscious act of addressing a social wrong not only as an individual act, but in its discursive or seemingly structural form.

As part of the material record of the nation, those rock fences and African American buildings can be envisioned as a heretofore unappreciated component of the landscape as (national) archive. Archives are a repository for the historical record of a parent organization and usually are thought of as paper materials governed by professional rules of collection, preservation, provenance, and control. But if the landscape is our unwitting autobiography, we can add to the American (US) national archive a material record found in the landscape, and one that allows a broader range in the collection including many people left out of the American national story as recorded in the paper archive or, more importantly, whose record in the paper archive seems to lock them into narrow categories. The National Memorial for Peace and Justice has

Figure F.5 A gravestone memorializing an individual sense of (African American) belonging in central Kentucky. Photo by author.

a Community Remembrance Program conceived as restorative truth telling that aims to place historical markers engaging and discussing past and present issues of racial justice focused on lynching sites that include that of Richard James (https://museumandmemorial.eji.org/memorial). Those markers, too, will become an important part of our national archive.

DeSilvey's (2007) work on archives reminds us that archives call into existence associative meanings, a point which might help us to see that landscapes assemble meaning. We hold the various material and visual elements of any landscape we call home in a relational framework that ultimately means something to us as individuals, as part of a group, or not. The distillery and Richard James' gravestone and the rock fences and the Bluegrass fields together assemble a picture of a regional landscape that means something to people, albeit in a relational framework that might vary significantly from one person to the next. But that is the place of landscape as symbol. This chapter's focus on a short drive down US 62 in central Kentucky has introduced a handful of material and visual components of an internationally famous agriculture landscape. Taken in toto that landscape symbolizes a regional identity that can no longer be regarded as monolithic but nevertheless stands for histories and ideologies of some aspects of American life. It is at once a social wrong, a landscape of belonging, a part of our national archive, and an assembled relational network of meaning and identity.

This last section has introduced the idea of landscape aphorisms under the proposition that landscapes work, which is itself the meta-aphorism serving as synecdoche for a landscape historiography documenting the place of landscape in everyday life. Examples here: identify the landscape as discourse materialized, the landscape as archive, and the landscape as social wrong; note the importance of landscapes of belonging; claim that landscapes mediate, that landscapes are normalizing and normative, that landscapes symbolize, and that landscapes assemble meaning. We might add to these by suggesting in these many examples that landscapes are not innocent and that we can find hope in the possibility of identifying landscapes as sites of intervention. That makes ten. There are undoubtedly more.

References

Cosgrove, D. 1984. *Social Formation and Symbolic Landscape*. London: Croom Helm.

Derickson, K. D. 2016. "Urban Geography in the Age of Ferguson." *Progress in Human Geography*. 31: 230–244.

DeSilvey, C. 2007. "Art and Archive; Memory-work on a Montana Homestead." *Journal of Historical Geography* 33: 878–900.

Duncan J. and N. Duncan. 1988 "(Re)reading the Landscape," *Environment and Planning D: Society and Space* 6: 117–126.

Fairclough, N. 2009. "A Dialectical-Relational Approach to Critical Discourse Analysis in Social Research." In *Methods of Critical Discourse Analysis*, edited by R. Wodak and M. Meyer, 162–186. Thousand Oaks: Sage.

Lewis, P. F. 1979. "Axioms for Reading the Landscape." In *The Interpretation of Ordinary Landscapes*, edited by D.W. Meinig, 11–32. NYC: Oxford University Press.

McKittrick, K. 2006. *Demonic Grounds*. Minneapolis: University of Minnesota Press.

Meinig, D.W. 1979. "The Beholding Eye." In *The Interpretation of Ordinary Landscapes*, edited by D.W. Meinig, 33–48. NYC: Oxford University Press,

Mitchell, D. 1996. *The Lie of the Land*. Minneapolis: University of Minnesota Press.

———. 2008. "New Axioms for Reading the Landscape: Paying Attention to Political Economy and Social Justice." In *Political Economies of Landscape Change*, edited by J.L. Wescoat, Jr. and D.M. Johnston, 29–50. Dordrecht: Springer.

Olwig, K. 2019. *The Meanings of Landscape*. NYC: Routledge.

Raitz, K. and D. Van Dommolen. 1990. "Creating the Landscape Symbol Vocabulary for a Regional Image." *Landscape Journal* 90: 109–121.

Robbins, P. 2007. *Lawn People*. Philadelphia: Temple University Press.

Rose, G. 1992. "Geography as a Science of Observation: The Landscape, the Gaze, and Masculinity," in *Nature and Science: Essays in the History of Knowledge*, edited by F. Driver and G. Rose, 8–18. Historical Geography Research Series, No. 28.

———. 2001. *Visual Methodologies*. London: Sage.

Wylie, J. 2007. *Landscape*. NYC: Routledge.

Schein, R.H. 1997. "The Place of Landscape: A Conceptual Framework for Interpreting an American Scene." *Annals of the Association of American Geographers* 87: 660–680.

———. 1999. "Teaching "Race" and the Cultural Landscape." *Journal of Geography*, 98: 188–190.

———. 2003. "The Normative Dimensions of Landscape." In *Everyday America: Cultural Landscape Study after J.B. Jackson*, edited by C. Wilson and P. Groth, 199–218. Berkeley: University of California Press.

———. 2009. "A Methodological Framework for Interpreting Ordinary Landscapes: Lexington, Kentucky's Courthouse Square." *The Geographical Review*, 99: 377–402.

———. 2018. "After Charlottesville: Reflections on Landscape, White Supremacy, and White Hegemony." *Southeastern Geographer*, 58: 10–13.

Waterton, E. 2019. "More-than-Representational Landscapes," In *The Routledge Companion to Landscape Studies*, edited by P. Howard, I. Thompson, E. Waterton and M. Atha, 123–133. London: Routledge.

INTRODUCTION

Twenty-first Century American Landscapes

Chris W. Post, Alyson L. Greiner, and Geoffrey L. Buckley

Welcome to our investigation of the American (US) landscape. Our primary mission for this volume is to update readers on two interrelated transformations over the past 25 years, which we present here as questions: First, how has the American landscape itself changed as a product of US social, economic, political, and environmental processes and activities? Second, how have our studies, perceptions, and understandings of these landscapes evolved over that same period? We should not, and do not, take these questions lightly since so much has changed in that span of time. Events including the attacks on September 11, 2001, the COVID-19 pandemic, the launch of cryptocurrency, the January 6, 2021 attack on the Capitol, and ecological deterioration due to climate change have significantly altered the ways we view the US and its role in the world. Additionally, the country continues to struggle to meet the needs and desires of all persons who consider this country home, with respect to their ethnic, religious, gender, and sexual identity, or ability, as the Black Lives Matter and #metoo movements attest. With these changes and challenges come shifting landscapes and places that, with frequent resistance from powerful interests, aim to be more inclusive for all Americans. This progress and pushback produce a discourse that forms our landscape at specific points in time. Such changes and tensions do not occur in a "zero-sum" fashion. We can all benefit from more just landscapes. And such changes have occurred in all forms—economic, social, political, environmental, urban, and rural—providing a template from which we can continue to develop that "more perfect union."

We present a fourfold rationale for this volume. First, we wish to bridge traditional and contemporary scholarship on the landscape of the United States. Second, we want readers to understand the development of the American landscape simultaneously with the development of that more critical perspective through a variety of approaches. Third, we use engaging and empirically rich case studies by a collection of recognized and rising scholars to highlight these developments. Fourth, this text explores topics and landscapes not traditionally covered in edited volumes on the American landscape, such as chapters on gender and sexuality, memory, and environmental justice. Together, these objectives enable us to create a unique volume notable for its inclusivity of topics and methodologies within four broad sections on key areas of scholarship—environmental, social and cultural, political, and urban and economic landscapes.

This introduction presents to readers 1) our inspiration for taking on this project, 2) how we understand the idea of landscape throughout this book and how the American landscape

DOI: 10.4324/9781003121800-1

and its study have changed with greater acceptance of critical methodologies over the past two-plus decades, 3) major themes emerging throughout this volume, and 4) how we structure this book, before sharing some concluding thoughts.

Inspirations

We present this book to readers looking for an update on how the American landscape is understood in multiple contexts and themes, and through multiple theoretical lenses. In short, things have gotten complicated. We see this book as an update that builds on the foundation constructed by scholars who have come before us and outlined their analyses of the American landscape. Our own careers are sprinkled in this transitional moment between the more traditional underpinnings of landscape and newer interpretations that urge equality for all those persons who call the country home. But where is the textual anchor that could bridge these eras?

To date, there is no single volume that connects these eras nor critically surveys the contemporary scholarship on the American landscape while focused on rich thematic examples. Several texts—such as Donald Meinig's *The Shaping of America* set (1986, 1993, 1998, 2004), or Thomas McIlwraith's and Edward Muller's edited volume (2001; see also Robert Mitchell and Paul Groves 1987) provide excellent canvases on which we all learned. Yet scant attention is paid to the removal of Indigenous peoples, or a deeper discussion of racialization and its role in US state formation. One recent work comes closer as Craig Colten and Geoffrey Buckley focus on the North American landscape from a particularly historical geography perspective with several more critical discussions (2014). Another volume (Conzen 2010) likewise focuses on the American landscape using a historical-evolutionary approach but also emphasizes the more traditional practice of "reading the landscape" for what it can tell us about settlement and occupancy. Absent from the literature is a work that approaches the American landscape from contemporary and interdisciplinary perspectives that also covers how our perspectives have changed over the past 25 years. How does our abstraction of water both enable our livelihoods and yet simultaneously put our settlements at risk? What does the prison economy mean for small towns and rural areas? What are the implications of transnationalism for ethnic and social identity? Arguably, the need to ask and answer these and other questions provides a compelling rationale for this book. More to the point, none of the extant books grapple with the rapid and successive changes wrought by expanding access to the internet, the events of 9/11, the #metoo and Black Lives Matter movements, and the COVID-19 pandemic. Concurrently, these developments have forever altered American politics, environments, economics, and society, and the resulting landscape. With this in mind, we approach this book as an update to the US landscape and its study over the past 25 years. We asked our contributors to cast wide nets such that all chapters also include a literature review that offers up-to-date appraisals of substantive debates and developments, and to bring a combination of voices from venerable established scholars and rising stars in geography and beyond.

Landscape

While a conventional—or general public—approach to the landscape often understands it as the "lay of the land" or environmental scene, we draw on a twofold conceptualization of landscape that builds upon the work of Denis Cosgrove (1984), Peter Jackson (1989), Don Mitchell (2000, 2008), and Richard Schein (1997, 2009), among others. On the one hand, we recognize that the landscape is the material expression of our actions, preferences, and desires. Four broad and seemingly innocuous systems—environmental, social, economic, and political—produce

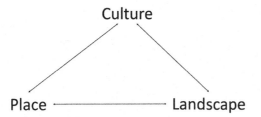

Figure I.1 Conceptual relationship between culture, place, and landscape.

a variety of material forms and landscapes—memorials, farms, skyscrapers, freeways, schools, removed mountaintops, pilfer-proof trashcans, and gated communities. On the other hand, we also acknowledge that the landscape is neither accidental nor apolitical. Thus, the landscape not only reflects our actions, but it also encodes social values which shape power relations and have significant consequences for equity, access, and justice—and the lack thereof. This volume applies this understanding of landscape as an appeal to our contributors and our readers. What is more, it also highlights the continuing debate over how to define American culture as it is manifested through the collective landscape of a nation.

To illustrate this concept best for a wide audience, we start with culture broadly defined as everything we do that means something to us. These activities, singularly and in combination, produce two things—places and landscapes. Landscapes are the collective material manifestations of those otherwise immaterial cultural, social, economic, and political actions. Through our experiences we produce meaningful places out of these locations and among those material manifestations of landscape. As a co-productive system, the landscape works back on our place making and cultural identities (and place impacts culture over time, too). Thus, conceptually, we envision a triangle consisting of culture, place, and landscape where each impacts the others (Figure I.1).

To further develop the landscape idea over the past two decades, we asked Rich Schein to contribute his latest work on what the landscape means, how our understanding of it has changed, and how we can attempt to understand it. Known for his work that inhabits the complex transitional space between established conceptions of landscape and newer developments, Schein often presents the idea of landscape as a materialized discourse (1997, 2009) that constantly adjusts to power structures (particular those related to race and racialization; 2006) and our struggle to be a more inclusive and conscientious society. He builds on these past works to further develop the landscape idea with several aphorisms.

We also provide a series of more methodological essays that allow readers to step back and understand the application, or nature and variety of perspectives, that our authors employ in their chosen topic. Such essays appear as boxed readings to highlight a particular case study or the use of a particular research or analytical technique. Combined, these methodological essays complement their chapters by providing additional insights on the changing American landscape and the study of it.

Emerging Themes

We edited this volume from 2020 through the first quarter of 2022, years that will go down as among the most unique in world history (let alone US history). Finishing a presidential campaign during a pandemic in a politically divided country, citizens dissatisfied with Donald Trump's failed reelection bid attempted a violent coup at the US Capitol on January 6, 2021, in order to reinstate (unconstitutionally) the former president. Trump's domestic policies of family separation along the border and environmental destruction, along with his corroboration of racial inequality and other matters, greatly altered the American landscape.

Throughout that campaign year and beyond, the United States witnessed the deaths of over 900,000 citizens due to COVID-19, the respiratory disease that develops from infection of the novel coronavirus, SARS-Cov-2. Trump's general disregard for science throughout the pandemic only made things worse. Many Americans refused to wear masks for the safety of others or resisted vaccination for various reasons—many falsely contrived. The pandemic thus continued and worsened in the US beyond what many other countries experienced (or continue to). Still, and despite a number of challenges, by the time we came to the one-year anniversary of the COVID-19 pandemic in March 2021, local health departments and pharmacies were vaccinating citizens at the rate of up to two million Americans each day. This pandemic also changed the American landscape, in ways that we are just starting to understand, such as stay-at-home policies that altered patterns of labor and child care, the downfall of music venues and other entertainment industries, an invigorated push for online consumerism, and an epic backlog of shipping from a bottlenecked supply chain that for months had to prioritize the transportation of health care equipment, masks, vaccines, and a plethora of home goods, even toilet paper.

During the pandemic, the US witnessed the unjustified killings of black Americans by police officers and regular citizens. In Minneapolis, Minnesota, a former police officer knelt on George Floyd's neck for nearly ten minutes, killing him. In Louisville, Kentucky officers killed Breonna Taylor while using a no-knock warrant for the wrong apartment. Kenosha, Wisconsin officers shot Jacob Blake seven times in front of his three children. Ordinary citizens chased down Amaud Arbury and killed him for exploring a house under construction. Protests over these murders filled the streets of the country with citizens calling for legal justice and equality. While several chapters connect to this theme, Joshua Inwood and Derek Alderman make a particularly important and timely contribution. Focused on landscape change and contrapuntal political action, Inwood and Alderman deftly navigate both US politics and social identities in a call to action. Two shorter essays support this chapter, which readers will find between our sections on social and political landscapes. Aretina Hamilton offers a short history on how race has always been evident in the American landscape and Brandi Summers provides a deeper look at the role of race in the landscape of Washington, DC. This discussion aligns nicely with the points made by Schein that start the volume and together they underscore the critical relevance of landscapes in our everyday lives as agents of change that may bring about a better more humane experience for future generations of all Americans.

All the while, in 2020, the Gulf Coast experienced two tropical cyclones, including Hurricane Laura, within days of one another as part of an Atlantic storm system that produced so many hurricanes that the National Oceanic and Atmospheric Administration turned to the Greek alphabet as an emergency naming method. Meanwhile, fires killed more than two dozen people and burned millions of acres in the American West. These events—focused on race, economic consumption, climate change, public health, politics, free speech—spurred place name changes, monument toppling, protests, mass burial sites, empty cities and freeways, and bottlenecked ports that occurred during the production of this volume. However, these all provide ground truth and hard physical evidence that is the landscape. In other words, the American landscape—its social, economic, and political manifestations, took center stage the last three years—just as it always does, but under some new terms and conditions. We hope this volume connects people to these events and their larger contexts, particularly as an understanding of landscape and its power in American society.

Structure

We divide this book into four parts and a special essay contribution. After this introductory section comes the heart of this volume—four sections on a theme. These sections focus on, in order: Environmental Landscapes, Social, Cultural, and Popular Identities, Political Landscapes, and Urban/Economic Landscapes. Of course, any topic possesses many opportunities to overlap with several of our artificially produced themes. Our distillation is by no means dogmatic. An introductory essay initiates each section with previews of each chapter and the major themes that emerge from that collection. Our volume ends with an afterword by William Wyckoff, who reflects on the text and main points of focus by our contributors.

Concluding Thoughts

Against this backdrop, and with this introduction, we present *The Routledge Companion to the American Landscape*. We believe the collection of topics, and the contributing scholars, represent some of the most pressing issues to explore from leading voices engaged in understanding the American landscape. As a collective, our contributors come from a variety of disciplines. And we are excited about assembling those voices for you. The last 25 years have taught us that the world constantly provides challenges to our idealistic hopes for a more just human existence in all its forms—a healthy environment, peace, quality livelihoods, just cities and the freedom to contribute to humanity's well-being that builds on our strengths as individuals, but also as part of a shared network of humans. High hopes perhaps, in a world that still experiences terrorism, climate change, a pandemic, an attempted domestic coup, and a war in Ukraine. Our argument, in the end, and to echo Don Mitchell (2010) is that the landscape is the force by which a lack of justice is recognized. But, it also provides us the opportunity to push back and continue that long arc toward justice.

References

Colten, C. E. and G. L. Buckley, eds. 2014. *North American Odyssey: Historical Geographies of the Twenty-first Century*. Lanham, MD: Rowman and Littlefield.

Conzen, M. P., ed. 2010. *The Making of the American Landscape, 2nd ed.* London: Routledge.

Cosgrove, D. E. 1984. *Social Formation and Symbolic Landscape*. Madison: University of Wisconsin Press.

Jackson, P. 1989. *Maps of Meaning*. London: Routledge.

McIlwraith, T. F. and E. K. Muller, eds. 2001. *North America: The Historical Geography of a Changing Continent, 2nd ed.* Lanham, MD: Rowman and Littlefield.

Meinig, D. 1986–2004. *The Shaping of America: A Geographical Perspective on 500 Years of History*. Vol. 1, *Atlantic America, 1492–1800* (1986); vol. 2, *Continental America, 1800–1867* (1993); vol. 3, *Transcontinental America, 1850–1915* (1998); vol. 4, *Global America, 1915–2000* (2000). New Haven, CT: Yale University Press.

Mitchell, R. D. and P. A. Groves., eds. 1987. *North America: The Historical Geography of a Changing Continent*. Lanham, MD: Rowman and Littlefield.

Mitchell, D. 2000. *Cultural Geography: A Critical Introduction*. Oxford: Blackwell.

———. 2008. New axioms for reading the landscape: Paying Attention to Political Economy and Social Justice, in *Political economies of landscape change*, edited by J. L. Wescoat, Jr. and D. M. Johnson, 29–50. New York: Springer.

Schein. R. 1997 The Place of Landscape: A Conceptual Framework for Interpreting an American Scene. *Annals of the Association of American Geographers* 87 (4): 660–680.

———. 2006. *Landscape and Race in the United States*. New York: Routledge.

———. 2009. A Methodological Framework for Interpreting Ordinary Landscapes. Lexington, Kentucky's courthouse square. *The Geographical Review* 99 (3): 377–402.

PART I

Environmental Landscapes

The American Environment

Geoffrey L. Buckley

In *Travels with Charley*, published in 1962, John Steinbeck famously embarks on a cross country road trip to "rediscover" America. With his French poodle by his side, he sets out from his home on Long Island, pointing his camper in the direction of California and sticking as much as possible to secondary roads. His descriptions of people and places have resonated with generations of readers. At one point in the journey, however, he finds himself on U.S. 90, "a wide gash of super-highway" (Steinbeck 1986, 89). The experience causes him to reflect on the future of travel in the US: "When we get these thruways across the whole country, as we will and must, it will be possible to drive from New York to California without seeing a single thing" (p. 90). Fast forward 60 years and we readily acknowledge how much the Interstate Highway System has transformed the way we view what Steinbeck once called this "monster land" of ours and, likewise, how difficult it is to interpret its myriad landscapes while hurtling along at 75 miles per hour (p. 6). And yet, as the authors in this section contend, there is much to see, learn, and appreciate if we slow down just a bit. Moreover, much has changed since Steinbeck's time.

Let's begin our examination of the American environment with a short quiz. Think about the area where you live or some of the places you have visited recently as you formulate your responses. Also, consider the forces that produce the environmental landscapes we see today. Okay, here goes. Can you identify the crop you are seeing through the windshield of your car? Why might a farmer—or, more likely, a large corporation—favor this grain over another? Where does any given community derive its water supply? Why does water quality vary—sometimes significantly—from one town to the next or even from one neighborhood to the next? What about the transmission lines overhead? How is this electricity generated, where is it coming from, and who is consuming it? And what of the "hidden" landscapes of pollution and waste that create the illusion that our consumption of material goods has no social or environmental impacts? These are just a few of the questions driving the research that unfolds in the following pages.

In the first chapter, Lisa Butler Harrington and Christopher Laingen offer us a primer on agriculture, which is, perhaps, the ultimate marker of rural America. Americans know a lot less about farms and farming than they used to—their obsessive quest for perfect turf notwithstanding (see Ted Steinberg's box on suburban lawns)—but after perusing this chapter, readers will have a much better understanding of how physical conditions, shifting economics, and new

DOI: 10.4324/9781003121800-2

technologies influence agricultural decisions and patterns, affecting everything from farm size and crop choice to water, chemical, and labor inputs. The second chapter shifts the focus to water landscapes. As dependent as we are on this fluid to meet our domestic, industrial, energy, and transportation needs, Craig Colten prompts us to recognize that water landscapes also reflect political and economic power. This includes power through naming, with all its cultural and colonial implications, but also the power exercised by entities like the federal government. Of course, rivers and other water bodies provide habitat for fish. How we view major projects such as dam construction—and removal—tells us much about how we value non-human living creatures and how these perceptions change over time (See David Montgomery's box on dams and salmon).

In the third chapter, Brian Black trains a spotlight on our energy landscapes. Everywhere we turn, we see evidence of our insatiable appetite for energy resources. Oil and gas pipelines, coal barges, solar panels, wind farms, hydroelectric dams, power lines, gas stations, suburban lawns, highways, and, of course, motor vehicles of all shapes and sizes. There is no questioning the centrality of energy in American life. Indeed, human lives are organized around the production, distribution, and consumption of energy resources. While many of these landscapes are obscured from view, others are not. In his box on wind power, Martin Pasqualetti argues that the presence of wind turbines on the land is discomfiting for many of us because the spinning blades serve as a constant reminder of our consumption. If there is an aspect of life in modern America that we deliberately seek to keep hidden—even more so than our landscapes of energy—it is waste. This is the subject of Martin Melosi's chapter. According to Melosi, "While humans are not alone in waste making, they may be the only species which passes judgment on it and also produces outsized amounts of it." But wastescapes are not simply dumping grounds for our discards. They provide a running commentary on our social and cultural values.

The last two chapters in this section address public lands and environmental justice, respectively. As one might imagine, the indicators associated with these landscapes are often more subtle and difficult to detect. In Chapter 5, Randall Wilson and Yolonda Youngs explore the origins and meaning of America's system of public lands. Often touted as symbols of shared national values, public lands—their creation and management—never fail to spark controversy and debate. Focusing on federal lands, the authors also highlight three of the most important challenges confronting resource managers today: the efficacy of ecosystem-based collaborative resource management, issues revolving around access and visitation, and the impacts of climate change. Ryan Holifield's chapter on environmental justice provides a fitting bookend to this section. As Holifield points out, environmental justice refers not only to the inequitable distribution of environmental amenities and hazards, but also to historical processes that produce distinctive landscapes. In addition to providing a "tour" of landscapes of environmental injustice in our cities, as well as three broadly defined sections of the US—the predominantly African American "Black Belt"; Indigenous territories of the Midwest, West, and island territories; and the Spanish-speaking Southwest—Holifield suggests a conceptual framework for future use.

Reference

Steinbeck, J. 1986. *Travels with Charley; in Search of America.* New York: Penguin Books.

1

CHANGES ACROSS RURAL AMERICA

Agricultural Landscapes

Lisa M. Butler Harrington and Christopher R. Laingen

> Agriculture may well be the most comprehensive of geographical topics. It involves modification of both the biological (plants) and physical (soil-landforms) components of the environment, and it incorporates social and economic components with distinctive spatial manifestations. It is also dynamic, playing a major role in recent human history.
>
> *(Doolittle 1992, 386)*

Agriculture continues to be the dominant land use in much of the United States—1.18 billion acres, or 52.5% of total US land area (Bigelow and Borchers 2017)—and farming is the activity most often brought to mind as a marker of rural areas. As the US grew in territory and population, and as technology developed and economic conditions shifted, agriculture has changed. Much of this change has taken the forms of specialization and extensification, but social shifts also have affected versions of agriculture across the country. Patterns of agriculture follow environmental conditions like topography and climate to a large extent, but market access, government policy, and anthropogenic landscape alterations that support agricultural production are also important. This chapter describes the variety of significant agricultural types and patterns in the US and how they have changed, and concludes with a short critique of modern, large-scale agricultural production.

Vegetable Production and Peri-Urban Farming

Vegetable and fruit production are generally considered to be "intensive" agriculture, as compared to grains and some root vegetables. Intensive crops require more labor, and sometimes other inputs like agrichemicals depending on whether production is "organic," "low input," or "conventional." Intensive crops are produced on smaller plots of land, and command higher prices per bushel or per pound than "extensive" crops. Most vegetable and fruit crops have shorter shelf lives than grains and are more likely to be consumed fresh. Due to these characteristics and the costs of transport—with greater need for care in packaging and shorter time to market—production has historically taken place nearer to cities. This historic agricultural location pattern was described by Johann Heinrich von Thünen in the early 1800s; while

DOI: 10.4324/9781003121800-3

it is no longer as fitting as it once was, the pattern of more intensive crop and milk production nearer to cities still is observable in many locations (O'Kelly and Bryan 1996; Block and DuPuis 2001).

Technological changes, crop breeding, and processing has meant that some vegetable farming now takes place in broad areas beyond urban fringe or peri-urban areas. Much of the agricultural production in California is representative of this—and largely dependent on high inputs of irrigation water. Today there is significant international trade in vegetables and fruit, although local sourcing is desirable to reduce greenhouse gas emissions associated with transport.

Fruit and vegetable production takes place on large corporate farms to small farms and gardens. It may be "conventional," reliant on fairly high agrichemical (fertilizer and pesticide) inputs; may rely on reduced chemical inputs as in the case of "low input" or "low input sustainable agriculture"; or may be organic, without use of manufactured chemicals. Although household gardening has taken place for centuries to help feed families, garden-based production fell off in the US during the last century or more but has seen a recent resurgence (Tornaghi 2014). This kind of urban agriculture and peri-urban production may be pursued as backyard gardens on private property and as cooperatives, community gardens, and allotments. Some efforts attempt to address concerns with food security, food deserts, and social equity.

Orchards and Vineyards

Many fruit and nut crops are grown in orchards, with trees of the same type planted in rows. Grapes also are produced in rows of woody plants and with long-term production expectations, though vineyards do not count as orchards. Olmstead (1956, 193) noted that into the early 1900s a typical farm in the eastern part of the country would include a small "kitchen orchard" of less than an acre to produce for family use; larger orchards contributed to farm sales. Concentration of production on "commercial fruit farms" took place in the first half of the twentieth century (Olmstead 1956).

As with many other crops, tree fruits, nuts, and grapes are grown where the environment is most suitable. Since these are long-lived plants rather than annual crops, year-round climatic suitability is even more important for orchards and viticulture (grape-growing). Fruit orchards include apples, pears, peaches, plums, cherries, lemons, limes, oranges, grapefruit, and other crops. Probably the oldest orchard crop in the US would be apples. Apple orchards are most concentrated in New York, south-central Pennsylvania, Michigan, and a wide swath of central Washington (USDA NASS n.d.), but are found in most states. In contrast to apples, citrus crops are found in the warmest locations, including much of Florida, southern California, southern Texas, and along the Gulf Coast.

Besides fruit crops, tree nuts like pecans, almonds, and English walnuts are produced in orchards. Pecans are mainly found from the Carolinas to Arizona, and from Texas to parts of Missouri and southern Illinois. In western areas, production is dependent on irrigation. California is the focus of almonds and much of the English walnut production.

Vineyards have been established across the US for juice, table grape, and wine grape production. There are many different varieties, with wines named for the type of grape used in their production. Wine grapes were mainly produced in California, where the climate is closest to that of the Mediterranean, until the latter part of the twentieth century. California is still dominant, but production has spread around the country as the popularity of wines and wine-making has grown and grapes appropriate to other regions have been identified.

Dairy Regions

The traditional US Dairy Belt extended from New York state to the upper Midwest (Hartshorne 1935; Durand 1951) (Figure 1.1). In the 1800s and through the first decades of the 1900s most farms in the US had enough dairy cattle to meet the needs of the family, and small dairy farms served local communities across much of the country until at least 1970. Dairy farms remained concentrated in the Dairy Belt until about 1990 (Harrington, Lu, and Kromm 2010). Wisconsin was a particular focus of dairying and by the 1920s was labeled "America's Dairyland" (Cross 2001), with New York and Minnesota following in importance. The relatively cool climate of this area lent itself to dairy farming, which also was favored because poor soils allowed little more than the production of forage crops to feed livestock during the winter (Hudson and Laingen 2016). Additionally, this region served earlier metropolitan areas, given milk's tendency to spoil quickly and more limited transportation distances before the advent of refrigeration.

In the second half of the twentieth century, the geographical pattern of dairy farming experienced major change as production increased rapidly in the western US, starting in California. By the early 1990s, California had become the largest milk-producing state, with massive, highly automated milking parlors in the Central Valley (Cross 2001, 2006; Harrington, Lu, and Kromm 2010). In the latter twentieth century, there also were several push factors—especially urban sprawl and land costs—that caused movement of some California operations to other parts of the western US, including Idaho, New Mexico, and Kansas. While the old Dairy Belt farms may have milked as many as a few hundred cows, new western facilities have the capacity to milk several thousand and are operated as feedlots or CAFOs. Spatial reorganization was driven by shifts in economies of scale (larger farms allow for lower production costs), population shifts, and land and other resource availability. The traditional Dairy Belt still produces large volumes of milk, although it has lost its dominance.

Feedlots and other Concentrated Feeding Operations (CAFOs)

Livestock of all types were common on most American farms up until the middle part of the twentieth century, but livestock production has become quite specialized over the past half-century. Specialization has led to an implosion in the number and proportion of US farms that continue to raise livestock and an explosion in the number of overall animals produced. Concentration of production was driven, in part, by reorganization and consolidation of agricultural production and meat-packing industries in response to increased demand for pre-packaged, pre-processed, inexpensive, and easy-to-purchase meat.

Hogs and Beef Cattle

Fattening hogs and cattle on grain as part of a mixed grain and livestock farming system originated in Virginia and was brought west over the Appalachians into the Midwest in the early 1800s (Clampitt 2018). Today, Iowa and North Carolina are first and second, respectively, in hog production, with Minnesota ranked third (Figure 1.2). These three states produce and sell 52% of all US hogs.

Beef cattle continue to be the most widespread type of livestock; smaller herds are common in the eastern US, while larger herds dominate the west. Most are born and raised in areas of the country that are too arid, too topographically rough, or that have soils too poor to grow high-value crops (Hart 2003, 40). The cattle are sold and transferred to feedlots for fattening before slaughter; a

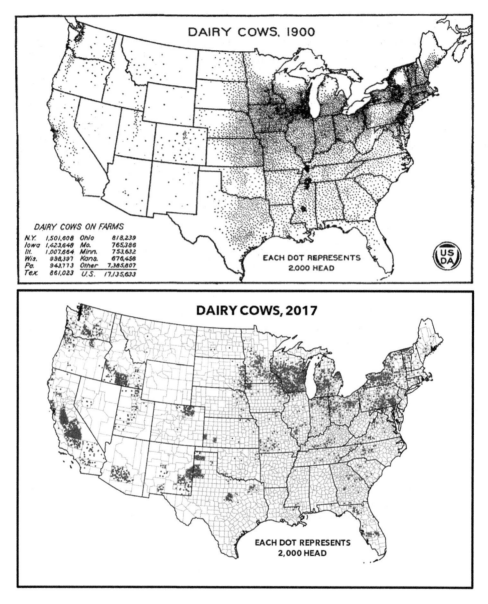

Figure 1.1 Dairy cow distribution 1900 (top) and 2017 (bottom). Top map from USDA *Yearbook 1922*, bottom map by C. Laingen from QuickStats (USDA NASS 2020) data.

minority are fattened on pasturage and sold as "grass-fed beef." While this system is not without its flaws, it has created a highly efficient and year-round supply of cheap beef for American consumers.

Poultry

No other country in the world consumes more poultry products on a per capita basis than the United States (USDA ERS 2016). Broiler and egg chicken production is concentrated across the

Figure 1.2 Hog CAFOs in Martin County, Minnesota. Each of these two barns contains about 1,000 hogs; each barn will see 2-3 batches of hogs that are brought in and fattened for slaughter each year. Photo by C. Laingen.

southern and southeastern part of the US from Pennsylvania through Texas. CAFO-based broiler and egg production began in Delaware in the 1920s. Production in the South grew as former cotton farmers, needing a replacement for the crop devasted by the boll weevil, began raising chickens large-scale in confined barns. Cheap feed, mostly corn and soybeans, is delivered from Cornbelt states and processed at feed mills owned by large poultry companies. Like the beef and pork industries, chicken and egg packing/packaging facilities are found near the clusters of production.

Corn and Soybeans

Corn and soybeans are two of the most important crops in the US. When westward-migrating farmers crossed the Appalachians and moved into southwestern Ohio, they encountered flat expanses of land covered with fertile soils in the Scioto and Miami River valleys (Hudson 1994). Here, Midwestern Cornbelt agriculture originated as a mixed-grain and livestock farming system where corn was grown primarily to feed and fatten hogs and beef cattle.

Soybeans were not widely planted on the landscape until well into the 1920s and 1930s and did not reach most parts of the region until after World War II when its industrial uses and nitrogen-fixing capabilities were realized (Hudson and Laingen 2016). Farmers also raised small grains, hay, and oats—the latter used to fuel the original *horse*power of farms.

During the twentieth century, the Midwest shifted from a mixed-grain and livestock system to one focused on cash-grain production where farmers' almost singular goal was to increase per-acre crop production for immediate sale (Hart 2003). This transition began earliest in the dual cores of the region: central Iowa/southern Minnesota and east-central Illinois/west-central Indiana. Here, soil fertility and return-on-investment for focusing on these two crops were greatest, and from there the pattern of specialization diffused outwards (Napton 2007).

The contemporary American Cornbelt remains centered on western Ohio, Indiana, Illinois, Iowa, southwest Minnesota, and the eastern portions of South Dakota and Nebraska (Hunt et al. 2020). What continues to set this region apart is the overwhelming amount of cropland growing only corn or soybeans. States in the Midwest typically produce over 80% of the nation's

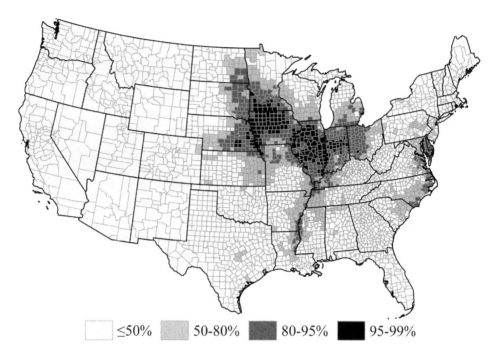

Figure 1.3 Corn and soybeans as a proportion of harvested cropland in 2017. Map by C. Laingen from QuickStats (USDA NASS 2020) data.

corn and soybean acreage (USDA NASS n.d.); a good indicator of a "Cornbelt county" is one where corn and soybeans are grown on at least 80% of a county's cropland area (Figure 1.3).

The Cornbelt's boundary has changed over the years due to technological innovations related to irrigation and agrichemical use, changing scales of production, seed hybridization and genetic modification, and changes to federal Farm Bill policies that once limited where corn and soybeans could be planted (Anderson 2009; Laingen 2017). This is especially noticeable in the northwest part of the region where corn and soybeans have replaced small grains (e.g., wheat, barley, rye, and flax, whose demand has decreased) and pastureland (as livestock disappear from regions focused on crop production) in both Dakotas. Beef cattle feedlots atop the southern and central Great Plains, overlapping the Ogallala aquifer, have increased the area of cropland devoted to irrigated corn production, nearly all of which is used to fatten cattle (Harrington and Lu 2002).

Small Grains

The British and French first introduced wheat to the Americas. By the early part of the eighteenth century, wheat production was greatest in northern Virginia, western Maryland, southeastern Pennsylvania, western New York, and eastern and southern Ohio (Hudson and Laingen 2016). Following population as it moved west and transitions in dominant modes of transportation (rivers/canals to rail), wheat milling centered on cities like Minneapolis and St. Louis. Wheat followed migration into the Great Plains, and throughout much of the late nineteenth and early twentieth centuries was increasingly grown there rather than in the Midwest.

By the beginning of the 1900s two separate zones of wheat production had emerged, with winter wheat in the central and southern Great Plains and eastern Washington, and spring

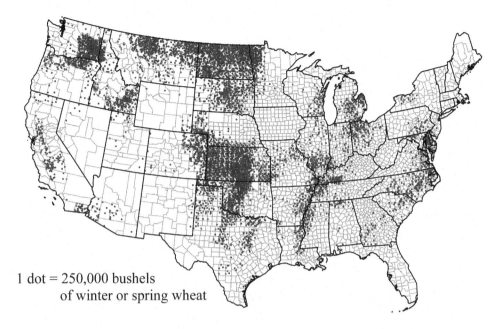

Figure 1.4   Bushels of wheat harvested, 2017. Map by C. Laingen from QuickStats (USDA NASS 2020) data.

wheat in the northern Plains (Figure 1.4). These zones are largely a product of climate. Winter wheat is planted in the fall and harvested in early summer, going dormant during relatively short and mild winter months. In many locations this timing allows for the planting and harvesting of a second crop such as soybeans before the end of the growing season. In Montana and the Dakotas, the long and often severe winters necessitate spring planting and fall harvest.

Another story of wheat on the American agricultural landscape is one of stagnation. Wheat, barley, oats, flax, and rye once occupied over 140 million acres of US cropland; by 2020 that area had been reduced to 50 million acres (USDA NASS n.d.). The US produces only 10% of the global wheat crop and 40–50% is exported overseas (USDA ERS 2020c). Wheat is widely grown in other regions of the world, usually in surplus amounts, lessening even more the demand for US farmers to grow more of it. Unlike corn and soybeans, there are no genetically modified varieties of wheat approved for sale by US farmers, indicating little interest in trying to find ways to grow more wheat (Hudson and Laingen 2016). Corn varieties have been developed for starch-centric industrial uses (feed, energy, and food derivatives). Nationwide corn yields over the past century have increased 507% (from 27 to 164 bu/acre) (Kucharik and Ramankutty 2005); wheat, used mostly to produce flour, has increased 235% (14 to 43 bu/acre) (USDA NASS n.d.).

Over 45 million acres of oats were once planted on the American agricultural landscape; by 2020, that area had shrunk to only 2 million. In the early part of the twentieth century, on a typical 160-acre farm, about 25 cropland acres grew the oats that fed the average of four horses found on each farm. As horses disappeared when tractors became more widespread, so did oats (Hoffman and Livezey 1987). Most oats milled and consumed in the US are now imported from Canada (from Saskatoon to Winnipeg) because of the specific quality of oat produced from that region's climate and soils. Aside from wheat and oats, barley is the next most plentiful grain and is grown mostly in North Dakota, Montana, and Idaho. The majority is used in the production of beer.

Pasture and Rangelands

Over the past half-century, the share of the American landscape occupied by agricultural uses has declined from 63% in 1950 to around 55% in 2017 (Bigelow and Borchers 2017); cropland was just under 400 million acres and grassland pasture/range was 650 million acres. The regions with the greatest amounts of pastureland are the non-glaciated central and northern Great Plains, the southern Plains, and the intermountain West. Additionally, central Florida has a sizeable amount of rangeland, as does the Southeast between the Mississippi River and the coastal plains (Hart 2007).

While most beef cattle are "finished" in the massive feedlots of the central and southern Great Plains, those cattle are born and raised on pasture or rangeland found mainly in the central and northern Great Plains, and to a lesser extent in the intermountain west. After calves are weaned, they enter the backgrounding stage as steers or heifers where they graze on pasturage until they are sold to feedlots where they are "finished."

The number of farms raising sheep in the US decreased from 538,593 in 1920 to 101,387 in 2017; inventory also declined significantly (USDA NASS n.d.). It is likely that meat preferences and lessened use of US-grown and processed wool have played a role in the reduction. Roughly 25% of all sheep are found in Texas and California, with another quarter in Colorado, Wyoming, Utah, and South Dakota (USDA ERS 2020b). The number of farms raising goats has increased since 1969, while the inventory has fluctuated between 1.2 and 3.1 million. Goats are raised for meat, milk and cheese, and mohair fiber/textiles. Central Texas is the core area for Angora (mohair) goats.

Organic Agriculture

One of the fastest-growing agricultural sectors is organic production. Labeling indicating "organic" or "natural" is widely used, but to be considered "certified organic" farmers must avoid the use of synthetic chemicals, GMO seed, adhere to feed/housing/breeding requirements for livestock, and allow on-site inspections by USDA officials (Duram 2005). Since the first widespread USDA organic survey was completed in 2008, the number and size of organic farms has increased (USDA NASS 2020). California sold four times as much organic food in 2017 than any other state, accounting for 36% of total US production. However, organic farming's footprint remains very small. In 2017, organic farms represented 0.9% of all US farms; 1.9% of farm sales; 0.6% of farmland; and 1.1% of all US cropland (USDA NASS n.d.). The most productive organic farms, aside from adhering to certified organic regulations, operate much like large non-organic farms. In 2017, the largest organic farms (those with sales of over $500,000) accounted for only 17% of all organic farms but were responsible for 84% of all certified organic production. Large farms, organic or not, tend to get bigger and accumulate a larger share of overall production—an overarching theme of modern agricultural systems (Hart 2003).

The Case of the Ogallala-High Plains

The Ogallala-High Plains aquifer system underlies parts of the Great Plains, running from eastern New Mexico and adjoining Texas into southern South Dakota. The groundwater of this system has been central to agricultural development where it has been most available and where other water sources have been less available, particularly in southwestern Kansas, and the Texas panhandle.

In southwestern Kansas, early agricultural development was reliant on the low levels of precipitation and the few streams. Use of groundwater for irrigation began significant expansion in the region in the 1950s, and combining modern pumps and center pivot irrigation led to transformation of production and the agricultural landscape in the 1960s–70s (Harrington, Lu, and Harrington 2009). Crops requiring more water than would naturally be available, like corn, and those that often are grown with lower rainfall, but are more productive with irrigation (like wheat), are grown in the area.

Although the groundwater here was once considered to be nearly unlimited, heavy use has proved otherwise. In Kansas, there are large areas where aquifer saturated thickness has been reduced by over 45% of pre-development conditions, and much of this by over 60% (KGS n.d.). As depth to water increases, pumping costs increase. In some locations, there has been abandonment of irrigation.

Agriculture Critiques

Criticism of agriculture comes easily. It has caused soil erosion (the Dust Bowl), habitat loss (plowing the prairies, wetland drainage), and degraded water quality/quantity (CAFO manure spills, aquifer depletion). Genetic modification has led to the questioning of the safety of some of the foods we consume and of pollinators and ecosystems, as has the use of GMO-enabled chemicals. Inadvertent contamination of neighboring (and sometimes organic-certified) farms with genetic material from GMOs also has been a concern.

Modern agriculture has its flaws but provides the country with relatively inexpensive food. Americans spent an average of 17% of disposable personal income on food in 1960; by 2019, food spending had been reduced to 9.5% (USDA ERS 2020a). At the same time, farmers' knowledge of best practices and the effects of various farming methods has led to improvements, including adoption of no-till practices, cover crops, and precision agriculture.

Today, a majority of Americans lack much understanding of production. Although this was once a nation of farmers, most Americans now have no connection to farmland, people, or agricultural activities. Only 1% of Americans are farmers, and fewer than 20% live in rural areas. For many suburbanites, the quest for the perfect lawn is the closest they come to the cultivation of a crop.

Box 1.1 The Rise of the Perfect Lawn

Ted Steinberg

Nothing is more American about the American landscape than the perfect lawn, the weed-free, supergreen, monoculture that has transformed the suburban yard into a cross between an outdoor living room and a sweeping chemical experiment.

Lawns started out as an expression of class privilege among the British aristocracy. Likewise, George Washington and Thomas Jefferson had lawns but because the lawn mower had yet to be invented the American lawn remained until well into the nineteenth century a rich man's affair or, more precisely, an ecological conceit available to slaveholders who could dispatch their slaves, scythes in hand, to cut the grass. These lawns, however, were not *perfect* lawns because no pesticides or specially bred grass seeds were available.

Only after the Second World War did the perfect lawn slowly take root as housing starts, mass consumption, and suburban development marched in lockstep. Some 17,000 lawns were planted

---

in Levittown, New York, between 1947 and 1951 in what is arguably one of the most famous exercises in suburban expansion. The Levitt family, who built the development, insisted that homeowners mow the lawn on a weekly basis and wrote the requirement into the covenants found in deeds. But beyond requiring that the grounds be kept neat, the Levitts did not require residents to tend a monoculture; plant species such as clover and plantain remained prolific occupants in suburban yards.

But a turning point had been reached with the Scotts Company playing a leading role as it cleverly used advertising and marketing to sell Americans on the need for lawn perfection. One 1950s advertisement pictured a young couple in their car gazing at an immaculately groomed green lawn. The text read: "Your lawn is the first thing they see," language that preyed on people's social anxieties about how others perceived them. Scotts also successfully tapped into postwar trends in brightly colored consumer products which many in the postwar era embraced to project an image of themselves as forward-looking people who had left behind the black-and-white world of urban life for the bright kaleidoscope of modern suburbia.

In 1948, to aid in the quest for the perfect lawn, Scotts began selling a combination fertilizer and herbicide called Weed and Feed. From an ecological perspective the product was one of the worst things to happen to the suburban landscape because it encouraged overtreatment with chemical inputs—there is no need to put down a herbicide as a matter of course—and eliminated clover, a plant that had evolved with lawn grass to capture nitrogen from the air and add it to the soil and thus helped fertilize the grass plants for free. Before Weed and Feed, Scotts even sold something called Clovex that *helped* homeowners sow clover in the yard. Now Scotts helped eliminate a plant that added nutrients to the soil so the company could sell homeowners more artificial fertilizer. It was a great business model but a terrible ecological one. Weed-and-feed products remain ubiquitous in the United States. They contribute to water pollution and, because they contain toxic herbicides, they represent a public health threat, which is why Canada's national health agency banned weed-and-feed products in 2010.

Reference

Steinberg, T. *American Green: The Obsessive Quest for the Perfect Lawn*. New York: Norton, 2006.

Concluding Thoughts

Agricultural landscapes of the US have evolved over time (Figure 1.5). Native American hunting-gathering techniques, EuroAmerican subsistence farming, and more recent local/regional-scale mixed-grain and livestock farming have been replaced by large-scale, highly sophisticated crop and animal production systems to a very large extent. Throughout, technological changes have been important, from the invention of the cotton gin and John Deere's plow to refrigeration, development of agrichemicals, and genetic modifications to crops. At times, economic pressures have pushed people out of farming, sometimes in concert with the economics and reduced labor needs associated with adopting newer technologies. With time, there also has been improved understanding of impacts and benefits of agricultural practices, retirement of marginal croplands, and alternative approaches.

Figure 1.5 The changing agricultural landscape of eastern South Dakota: in the foreground a small herd of beef cattle graze; further in the distance bales of hay signify land being used for harvesting forage; while off in the distance sits a newly constructed loop-rail grain terminal just east of Andover, SD that ships corn and soybeans to a port at Longview, Washington, for export to East Asia. Photo by C. Laingen.

Often, political and social changes resulted in the displacement of Native American peoples, and government policies favored EuroAmerican farmers over African Americans. Social changes also have created a variety of approaches to farming, including the proliferation of large-scale production to keep food prices low, while other approaches focus on smaller-scale, organic and low-input agriculture, that attempt to integrate food production with other eco-system services in the form of agroecosystems (Foley et al. 2005). Varying attitudes have thus resulted in a breadth of approaches to agriculture.

Throughout most of the late nineteenth and early twentieth centuries, farmers and ranchers determined which specific types of agricultural production were most suitable for different land areas. This evolving specialization and extensification resulted in distinct regional landscapes that produce our country's food, fuel, and fiber.

References

Anderson, J.L. 2009. *Industrializing the Corn Belt: Agriculture, Technology, and Environment, 1945–1972.* DeKalb, IL: Northern Illinois University Press.

Bigelow, D.P., and A. Borchers. 2017. *Major Uses of Land in the United States, 2012,* EIB-178, US Department of Agriculture, Economic Research Service, August.

Block, D., and E.M. DuPuis. 2001. "Making the Country Work for the City: Von Thünen's Ideas in Geography, Agricultural Economics and the Sociology of Agriculture." *American Journal of Economics and Sociology* 60 (1): 79–98.

Clampitt, C. 2018. *Pigs, Pork, and Heartland Hogs: From Wild Boar to Baconfest.* Lanham, MD: Rowman and Littlefield.

Cross, J.A. 2001. "Change in America's Dairyland." *The Geographical Review* 91 (4): 702–714.

———. 2006. "Restructuring America's Dairy Farms." *The Geographical Review* 96 (1): 1–23.

Doolittle, W.E. 1992. "Agriculture on the Eve of Contact: A Reassessment." *Annals of the Association of American Geographers* 82 (3): 366–401.

Duram, L.A. 2005. *Good Growing: Why Organic Farming Works.* Lincoln, NE: University of Nebraska Press.

Durand, L., Jr. 1951. "The Lower Peninsula of Michigan and the Western Michigan Dairy Region: A Segment of the American Dairy Region." *Economic Geography* 27 (2): 163–183.

Foley, J.A., R. DeFries, G.P. Asner, C. Barford, G. Bonan, S.R. Carpenter, F.S. Chapin, M.T. Coe, G.C. Daily, H.K. Gibbs, J.H. Helkowski, T. Holloway, E.A. Howard, C.J. Kucharik, C. Monfreda, J.A. Patz, I.C. Prentice, N. Ramankutty, and P.K. Snyder. 2005. "Global Consequences of Land Use." *Science* 309: 570–574.

Harrington, L.M.B., and M. Lu. 2002. "Beef Feedlots in Southwestern Kansas: Local Change, Perceptions, and the Global Change Context." *Global Environmental Change* 12 (4): 273–282.

Harrington, L.M.B., M. Lu, and J.A. Harrington, Jr. 2009. "Fossil Water and Agriculture in Southwestern Kansas." In *Sustainable Communities on a Sustainable Planet: The Human-Environment Regional Observatory Project*, edited by B. Yarnal, C. Polsky, and J. O'Brien, 269–291. New York: Cambridge University Press.

Harrington, L.M.B., M. Lu, and D.E. Kromm. 2010. "Milking the Plains: Movement of Large Dairy Operations into Southwestern Kansas." *The Geographical Review* 100 (4): 538–558.

Hart, J.F. 2003. *The Changing Scale of American Agriculture*. Charlottesville: University of Virginia Press.

———. 2007. "Bovotopia." *The Geographical Review* 97 (4): 542–549.

Hartshorne, R. 1935. "A New Map of the Dairy Areas of the United States." *Economic Geography* 11 (4): 347–355.

Hoffman, L.A., and J. Livezey. 1987. *The US Oats Industry*. USDA ERS Agricultural Economic Report No. 573.

Hudson, J.C. 1994. *Making the Corn Belt*. Bloomington: Indiana University Press.

Hudson, J.C., and C.R. Laingen. 2016. *American Farms, American Food: A Geography of Agriculture and Food Production in the United States*. Lanham, MD: Lexington Books.

Hunt, E.D., H.E. Birge, C.R. Laingen, M.A. Licht, J. McMechan, W. Baule, and T. Conner. 2020. "A perspective on changes across the US Corn Belt." *Environmental Research Letters* 15, doi.org/10.1088/1748-9326/ab9333.

Kansas Geological Survey (KGS). n.d. "Water Levels." *Kansas High Plains Aquifer Atlas*. Accessed February 21, 2021. www.kgs.ku.edu/HighPlains/HPA_Atlas/Water%20Levels/index.html.

Kucharik, C., and N. Ramankutty. 2005. "Trends and Variability in US Corn Yields Over the Twentieth Century." *Earth Interactions* 9: 1–9. doi.org/10.1175/EI098.1.

Laingen, C.R. 2017. "Creating a Dynamic Regional Model of the US Corn Belt." *International Journal of Applied Geospatial Research* 8 (4): 19–29.

Napton, D.E. 2007. "Agriculture." In *The American Midwest: An Interpretive Encyclopedia,* edited by R. Sisson, C. Zacher, and A. Cayton, 142–145. Bloomington: Indiana University Press.

O'Kelly, M., and D. Bryan. 1996. "Agricultural Location Theory: Von Thünen's Contribution to Economic Geography." *Progress in Human Geography* 20 (4): 457–475.

Olmstead, C. 1956. "American Orchard and Vineyard Regions." *Economic Geography* 32 (3): 189–236.

Tornaghi, C. 2014. "Critical Geography of Urban Agriculture." *Progress in Human Geography* 38 (4): 551–567.

US Department of Agriculture, Economic Research Service (USDA ERS). 2016. "Chicken's Popularity Makes It the Most Consumed US Meat." Last modified January 25, 2016. www.ers.usda.gov/data-products/chart-gallery/gallery/chart-detail/?chartId=78715.

US Department of Agriculture, Economic Research Service (USDA ERS). 2020a. "Americans' budget share for total food was at a historical low of 9.5 percent in 2019." Accessed February 21, 2021. www.ers.usda.gov/data-products/chart-gallery/gallery/chart-detail/?chartId=76967.

US Department of Agriculture, Economic Research Service (USDA ERS). 2020b. "Sheep, Lamb, and Mutton–Sector at a Glance." Last modified June 24, 2020. www.ers.usda.gov/topics/animal-products/sheep-lamb-mutton/sector-at-a-glance/.

US Department of Agriculture, Economic Research Service (USDA ERS). 2020c. "Wheat." Last modified June 24, 2020. www.ers.usda.gov/topics/crops/wheat/.

US Department of Agriculture, National Agricultural Statistics Service (USDA NASS). n.d. "Quick Stats" (multiple years). Accessed February 12, 2021. doi: quickstats.nass.usda.gov/.

US Department of Agriculture, National Agricultural Statistics Service (USDA NASS). 2020. "Organic Farming: Results from the 2019 Organic Survey (ACH17-20)." www.nass.usda.gov/Publications/Highlights/2020/census-organics.pdf.

2

FLUID LANDSCAPES

Craig E. Colten

Water is fluid. As rainfall, it can rinse other elements of the landscape. A portion of the cleansing liquid filters into the roots of countless trees, shrubs, crops, and weeds. Where water is abundant, so are plants. Vegetation in turn offers shelter and sustenance for many creatures—wild and tame. Water reshapes the physical landscape, incrementally moving mountains toward the sea, realigning streams and shorelines in the process, and it can corrode or damage structures. Thus, the physical and cultural landscapes are a reflection of water—whether in surplus or deficit. Water also inspires conflicts among people who require it. Arising from those water needs are myriad human landscape features that seek to capture, control, regulate, or utilize water. Its fluidity and its submission to gravity powers its movement toward the oceans and inspires interventions to delay, reroute, or intercept its inevitable migrations. Water is itself a prominent landscape feature, but the many ways we view water and the features that societies have erected to manipulate its utility along with the underlying social and cultural impulses that motivated their construction are the focus of this chapter.

William Least Heat-Moon undertook a novel, transcontinental journey by water in 1995. In his inimitable style, he reintroduces the reader to past explorers who followed waterways, points out the re-engineering of stream channels, both grand and modest, and directs our attention to an America viewed from the water's surface, or "those secret parts hidden from the road traveler" (Heat-Moon 1999, 5). Least Heat-Moon exposes the evolving landscapes at the intersection of land and water. In the interest of approaching the landscape as what lies before our eyes and as shaped in part by our perspective, this chapter considers the water-oriented landscapes that reflect power and authority, motivate social action, and stand as cultural symbols. Water's fluidity enables it to serve many purposes for diverse societies across America. This quality allows water to convey many ideas, designs, aspirations, and memories suspended in its turbulent flow.

Landscapes Viewed from Water

European explorers and colonizers may have viewed water through a different lens than Indigenous Peoples, but they shared many concerns. The French explorer Jean Ribaut recorded observations that reveal colonial intentions. He noted coastal features advantageous to European

DOI: 10.4324/9781003121800-4

designs such as protected harbors, freshwater supplies, food resources, and the potential for riches (Ribaut 1563). Reporting on the Virginia coast, John Smith wrote about opportunities for harboring ships, and essential supplies such as freshwater springs, timber, and fish (Smith 1907, 32–39). He affixed European names to prominent physical features that would allow English navigators to orient themselves, while adopting mostly Indigenous names for inland streams (Figure 2.1). The view from sailing ships guided initial European assessments of waterfronts.

Europeans probing deeper into the North American continent reported on inland waterways. In his explorations of the lower Mississippi, Iberville comments on the wildlife along the river's banks and fish caught from the waterway which were essential for provisioning his expedition. He observes dense stands of cane, large grapes, numerous Indigenous settlements, and the susceptibility of the riverside to inundation (d'Iberville 1699). In sum, the riverside was a narrow linear resource base already inhabited by a string of villages.

Numerous expeditions along waterways during the early national period further document the orientation of Indigenous settlements to waterways. Lewis and Clark note the presence of mounds along the Missouri and describe a Mandan village about fifty feet above the river consisting of several round houses shared by several families (DeVoto 1953, 58). Their expedition (1804–1806) expanded the formal geographic knowledge of the western waterways and confirmed that there was no transcontinental water route. Their accounts of storms and capsizing boats along with the arduous efforts to complete the upstream legs of their journey add an adventurous appeal to the western waters that continue to inspire whitewater adventurers to the present.

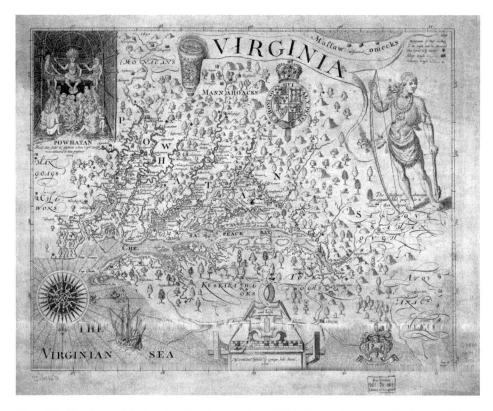

Figure 2.1 Detail from Virginia, Smith, John. London, 1629. Richmond, Va., Franklin Press, 1819. Library of Congress.

Set in the populated northeast, Henry David Thoreau's trip down the Merrimack and Concord (1839) is less a journey of geographic exploration as one of personal self-discovery. Thoreau recounts the sensations of steering his craft down the river and observes that his thoughts make sudden bends just as the river does (Thoreau 2014, 339). It was a temporal journey from the present to the future, as much as a geographic undertaking (Thoreau 1839, 20). He moves from the cerebral to the physical landscape and comments on the flora and fauna along the way, more as a poet than a natural scientist. He notes the presence of fishermen, bridges, and silent villages that he passed. This journey was a far cry from that of the Lewis and Clark expedition and underscores that the landscape is as much a product of the observer as the observed.

A utilitarian guide for riverboat pilots honed-in on visible navigational details (Cummings 1845). The author provides a portrait of the Ohio River as an unrivaled and still wild waterway, with some florid passages. The mainstays of the volume are river charts and descriptions that are attuned to the navigational landmarks a steamboat pilot needed to recognize.

John Wesley Powell explored the Colorado in 1869 and provided a harrowing account of his journey through the Grand Canyon. Wild rapids in the deeply entrenched river challenged this expedition. His narrative focuses on the towering rock formations and minerals of this immense gorge and an inventory of the canyon's geological resources (Powell 1895).

Landscapes of Power and Authority

Water has carved the physical North American landscape and provided colonial authorities with handy political boundaries. Naming those waterways was an act of imperial authority (Cohen and Kilot 1992; Rose-Redwood et al. 2010). Along the eastern seaboard and the Gulf of Mexico, explorers and settlers affixed European names to rivers such as the St. Lawrence, Charles, James, St. Johns, and Rio Grande. Yet, many Indigenous words persisted for watercourses such as the Penobscot, Potomac, Apalachicola, and Mississippi, challenging the notion of toponymic hegemony (Herman 1999). Some rivers or their watersheds served as colonial boundaries assigning political authority over a basin. Lord Baltimore's charter specified the southern boundary of his grant would follow the Potomac River (Report 1873). French explorer LaSalle claimed the entire Mississippi River basin for France in 1682. Tiny Bayou Manchac in Louisiana, the northern edge of the "Isle" of Orleans, served as an international border between colonial powers and became the site of military outposts which stood as explicit landscape markers of the cartographic delineation (Figure 2.2). When France transferred its holdings to Spain through the Treaty of Paris in 1763, it specified the Mississippi River as the eastern boundary for that cession (US Senate 1879). The Mississippi River remains a broad delineation between states. Its meanderings, however, have created jurisdictional headaches. Towns like Kaskaskia, Illinois's territorial capital, are separated from the state by the Mississippi River which changed course in 1881. Its residents vote in Illinois but must drive about 15 miles to enter the main territory of their state. Names reflect the naming society, and rivers demarcate the limits of political power, a delineation that has been subject to change.

As Don Mitchell (2008) cautions, landscapes are more than material forms. They reflect political and economic power with deep historical roots. Political authority extends legal rights attached to rivers, lakes, and oceans which give shape to riparian landscapes. Europeans drew on Roman law which categorized navigable rivers as commons. This means that riverfront property owners may not obstruct a water course in such a way that impedes free passage of boats or interferes with the normal quantity of downstream flow. Industrialization in nineteenth-century New England changed expectations for river use and challenged traditional legal concepts. Soon after Thoreau paddled down the Concord and Merrimack, mill owners launched an

Figure 2.2 Detail from A plan of the Coast of part of West Florida & Louisiana: Including the River Yazous, Gauld, George and J.E. Hilgard. 1778. Library of Congress.

environmental re-ordering of the region's legal hydrology. Seeking to ensure a reliable water supply New England industrialists built an expanding hydrologic network of dams and canals to collect and distribute water as needed. This network was the visible manifestation of a profound landscape change that challenged riparian common law. Transformation of water regimes with distinct seasonal patterns into reliable year-round power sources required a reach far upstream and into neighboring states that accelerated between 1820 and1850. This regional re-plumbing removed the mill pond from the side of the mill and created a string of upstream impoundments that unleashed impacts across the agricultural countryside. They flooded crop and pastureland and prompted attacks on the dams by farmers, along with legal challenges (Steinberg 1991). The courts tended to side with the manufacturers which solidified the legal basis for a landscape of company-controlled dams and ponds, canals, and gritty manufacturing towns such as Lowell and Lawrence (Steinberg 1991). New England's reworked hydrology powered the dependable spinning of thread, weaving of fabric, and assembly of garments.

The river-bound mill towns and factory cities with handsome stone or brick mills adjacent to myriad canals, worker dormitories, and support structures became the dominant landscape that eclipsed the traditional meetinghouse-centered New England village. This urban form spread throughout New England, westward across the Mohawk Valley, and as far west as Rockford, Illinois's waterpower district. As dependence on direct waterpower declined in the twentieth century, many working mills migrated southward. The sturdy, 2–3 story brick or stone structures, built to support the weight of heavy machines, were not easily converted to modern manufacturing uses or even the tech industry that has grown up around Boston. Following the evacuation of most textile activity, some of the former factories underwent adaptive reuse as museums, shops, and apartments, while some succumbed to dereliction or the wrecking ball (Figure 2.3).

Figure 2.3   Converted mill on the Mohawk River, Little Falls, New York. Photo by the author.

Natural waterways served colonial and early national-era shipping needs, but nineteenth-century commercial ambitions demanded a transportation network that aligned with notions of industrial efficiency. This motivation compelled a brief phase of landscape re-arrangement along canals which provided a more dependable means to move bulky cargos. Completion of the Erie Canal in 1825 sparked the construction of feeder canals across upstate New York, and competing canals from the seaboard cities of Philadelphia, Baltimore, and Washington towards the Ohio Valley. Eminent domain allowed the taking of private land in the pursuit of a public good. The swift transfer of title to private canal companies made this landscape transformation possible. New York City captured the greatest benefit from its position as the terminus of a canal. In the interior, improved access to markets endowed small towns with advantages that concentrated populations and agricultural processing industries, and augmented growth of their distinctive main street business districts. Along the Erie Canal, Syracuse, Rochester, and Buffalo became important commercial centers, and based in part on their early development, grew into twentieth-century industrial cities eventually served by rail. Canals stretched across Ohio, linking Lake Erie with the Ohio River. Costly and less successful canals crossed the Appalachian Mountains in Pennsylvania and Maryland. Illinois even participated in the canal fever with a connection between Lake Michigan and the upper Illinois River. None of the original canals function presently. Portions of most of these once important routes now host recreational or historical institutions. Interpretive exhibits provide a glimpse into their former purpose and function, and the tow paths now allow visitors to view segments of these former economic king makers. Historical warehouses, derelict industrial structures, or street names identify the route of the former canal.

In the South, conflicts also arose among competing waterway users that shaped the landscape, but with different results. From the colonial period, mills for grinding wheat or corn and sawing timber hugged the banks of innumerable small streams. Before the revolution, mills and associated commercial activities clustered along the fall zone—the seaward edge of the Piedmont where rapids offered excellent hydropower potential (Mitchell and Hofstra 1995). Towns along the fall zone from Virginia to Georgia exemplified the urbanization that accompanied the processing of farm and forest products using waterpower. Conflicts between mill operators and timbermen seeking to float log rafts downstream erupted in the backcountry by the 1820s. Territorial and state legislatures tended to side with the rights of navigation and insisted that mill operators provide locks around mill dams (Colten 2014, 126–127). The legal priority given to navigation favored agriculture and forest trades over manufacturing and contributed to the dominance of rural economies in the South. There were efforts to emulate waterpower enterprises along southern Piedmont rivers, but poor connections to the coast, however, stifled the ambitions to rival New England industries (Hunter 1949). No industrial-scale cities emerged along southern rivers during the antebellum period.

Rural agriculture indirectly contributed to fluvial refashioning of the landscape. Clearing inland forests for agriculture allowed the transfer of huge amounts of sediment from the uplands to lowlands along the region's rivers. Analysis of soils in the Neuse River valley of North Carolina documents significant transfer of Piedmont soils to mill ponds and low-country wetlands. Erosion removed as much as eight inches of soil from the North Carolina uplands and deposited a large portion in rivers, estuaries, and on floodplains (Phillips 1997). Streams transferred so much sediment from the Georgia Piedmont that it formed swamplands along the Alcovy River during the 1850s and prompted farm abandonment (Trimble 1970; Phillips 1997). Across the region, rivers became turbid after rainfall, sedimentation clogged rivers, and prompted Southern states to appeal to the Army Engineers to clear channels following the Civil War. Dredging opened up the lower course of numerous waterways and aided timber transport and also the delivery of cotton to coastal ports. These "improvements" fortified the status of the planter society and contributed to the continuing prominence of plantation agriculture across the region and the development of numerous small coastal ports from Virginia to Alabama (Colten 2014, 135–136).

Waterpower development followed a different course from New England (Swain 1885). Private power companies began erecting dams for the generation of hydroelectric power in the late nineteenth century. Electricity was destined for towns and cities and an emerging textile industry across the Piedmont. Beginning in the 1930s, the Tennessee Valley Authority (TVA) created a massive collection of 49 federally funded dams of which 29 generate electrical power to lift an economically disadvantaged basin out of poverty. In addition to power, the multipurpose dams mitigated flooding, facilitated navigation, and offered recreation (Foresta 2013; Manganiello 2015). The reservoirs flooded hundreds of small farmsteads, scores of communities and cemeteries, and in doing so largely obscured a traditional agricultural landscape. Touted as a recreation benefit, the Land Between the Lakes, although embraced by local boosters as an economic windfall, offered little in the way of memorable landscape vistas so prized in national parks and initially failed to lure visitors (Foresta 2013). The large-scale hydrological transformation brought a new generation of industry as well during and after World War II. Chemical manufacture near Muscle Shoals has contributed to the creation of hazardous waste sites and nuclear industries near Knoxville added non-agricultural industries to the landscape.

Casting a federal stamp on the country's fluvial landscapes was not limited to the TVA. Federal appropriations underwrote nineteenth-century snag clearing and the construction of locks and dams throughout the Mississippi River basin that reworked the physical and human landscape. The Mississippi and its tributaries were once a free-flowing system of waterways that carried spring floods born of snow melt and spring rains. These pulses were relatively short lived and drier conditions in the summer and fall caused great variation in river stages and navigation safety. Locks and dams created a controllable water stairway which enabled the synchronization of an industrial society with a seasonal and undependable natural waterway (Scarpino 1985; Paskoff 2007). Greater reliability and agricultural development of the Midwest and Great Plains prompted the growth of factories processing grain, cattle, and hogs, and the construction of massive grain elevators adjacent to the waterways. Refineries occupied the floodplain near St. Louis to take advantage of the crude shipped from the Gulf Coast.

In the American West, there were fewer rivers that could be improved for navigation, but that did not minimize revamping the fluvial geography of the region. Historians have discussed the revaluation of water from a "divinely appointed means for survival" (Worster 1985, 52) to a negotiable commodity. Federal ambitions to reclaim the arid West and redirect meager water supplies to agricultural lands and cities produced an unrelenting drive to remake the watery landscape. Federal programs enabled the construction of a massive set of dams along the Colorado River and rivers feeding into the Pacific Ocean (Worster 1985; Pisani 1992). These dams altered the natural flow of rivers while storing water for use by farmers and urban residents. A massive pipeline moves some of the water from the Colorado basin to southern California. The allocation of Colorado's limited flow during a time of relative abundance is now under serious stress as drier conditions strain the ability of the river to meet an expanding demand with diminished supply.

In the Pacific Northwest, federal efforts led to the transformation of the Columbia River into what Richard White (1995) refers to as an organic machine. It is a natural waterway, subject to seasonal fluctuations, but regulated with locks and dams that facilitate navigation to the interior and enables hydropower production and irrigation (Figure 2.4). The abundance of

Figure 2.4 Roza Irrigation Canal in Yakima County, Washington. Photographer Dorothea Lange, 1939. Library of Congress.

cheap hydroelectric power prompted the development of a major nuclear weapons complex in western Washington. Irrigated fields in the Palouse region of eastern Washington yield substantial harvests of wheat while the dams that capture that water impede the migration of salmon. Competing demands for water contribute to conflicts among shippers, fishers, and farmers. Vast fields of wheat, the sprawling nuclear complex at Richmond, the aluminum smelters and lumber mills along the lower river, port facilities at Portland, and the fishing stations on Native American reservations reflect the multiple uses of the Columbia.

Box 2.1 Salmon and Dams: Changing Views of Rivers

David R. Montgomery

Americans have seen and used rivers in many different ways, from source of life and sustenance, to avenues for exploration and commerce, and for irrigation, power, and recreational opportunities. And as people altered their habitat native salmon populations in New England and the Pacific Northwest declined dramatically, with populations now down to less than a tenth of their historical abundance. How people viewed rivers drove changes to river systems that altered access to and the character of habitat, playing major roles in salmon declines, and potential recovery.

Natural rivers are beautiful—and messy, a kind of organized chaos. They move, erode their banks, and periodically flood. Logs and logjams that locally block flow and trap sediment were perceived in the late nineteenth century as impediments to commerce and in the Pacific Northwest thousands of logs and logjams were removed in de-snagging efforts that continued into the mid-twentieth century. Today, however, these same features are seen as critical salmon habitat, scouring deep pools in riverbeds and providing shelter from predators (like bears, eagles, and fishermen). River restoration efforts now commonly reintroduce engineered logjams to rivers from which natural ones were previously removed.

Views of dams also changed dramatically over time. From colonial days to the Civil War waterpower fueled American industry, beginning with dams along small rivers and expanding to larger rivers instrumental to powering the industrial revolution in New England. Thousands of milldams reshaped the region's hydrological plumbing—and blocked salmon from making the journey from completing their natural life cycle. By the close of the nineteenth century salmon were blocked from roughly 90% of New England's rivers. Today, the plight of wild Atlantic salmon is motivating efforts to remove obsolete salmon-blocking dams across the region.

A similar story played out in the Pacific Northwest. The era of dam construction from the 1930s to 1960s harnessed the power of the Columbia River, turning it from a swift flowing highway that moved juvenile salmon to the sea into a series of predator-filled lakes and a gauntlet of deadly turbines. Other dams around the region also blocked salmon from completing their life journey, including those on the Elwha River in Washington State. Built in the early twentieth century to provide hydroelectric power, these dams blocked the river's salmon from 90% of their potential habitat, most of which lay pristine inside the Olympic National Park. In 1992 Congress authorized removal of the Elwha dams. After almost two decades of political wrangling, the dams were removed between 2011 and 2014, paving the way for salmon recovery and serving as a model for a new view of the relationship between salmon, rivers, and dams. In recent decades more than a thousand dams have been removed in the United States. With an estimated 84,000 dams in the continental United States river restoration and dam removal efforts are likely to continue for decades to come.

Substantial investment in redistributing meager water supplies in the West was matched by federal spending on flood control—much of it in the more humid east. Flooding was no stranger to residents in the lower Mississippi River. Colonial authorities required private landowners to build levees to fend off the nearly annual floods. Parishes and later the state assumed the cost of constructing these safety features. Low earthen berms paralleled the lower Mississippi by the start of the Civil War, but travelers aboard paddle wheelers could still observe plantation "big houses" from their upper decks. Neglect during the conflict led to levee deterioration and appeals for federal funding during the post-war period. Reluctantly in the late nineteenth century, the federal government assumed primary responsibility and instituted a program that delivered more consistently designed levees. Still, the great flood of 1927 overwhelmed the massive protection system, causing tragic damage and deaths, and prompted a shift to a revised arrangement. The Corps of Engineers built a pair of giant outlets to redirect flood waters out of the main channel and into the Gulf of Mexico, thereby reducing the strain on levees near New Orleans. Federal funding for levees in California, Florida, and on the upper Mississippi expanded in the post-1927 period (O'Neill 2006). Some 28,000 miles of levees line US rivers, about half of that managed by the Corps of Engineers (US Army Corps of Engineers 2020). The most extensive federal levees line the lower Mississippi (below the mouth of the Ohio River), in areas of Florida, and California (Colten 2014) where they contribute to the "levee effect." Urban and industrial development huddle in the shadow of the levees. Protected from river crests that do not exceed the levees' design, these areas remain susceptible to more extreme floods. When such events occur, the damages to commercial and personal property are more costly as became evident in the aftermath of Hurricane Katrina in 2005. It left nearly 80% of the levee-encircled city of New Orleans under water and caused billions of dollars in damage (Figure 2.5).

Figure 2.5 Houses ripped from their foundations and cars tossed asunder by flooding in New Orleans, Louisiana during Hurricane Katrina, 2005. Photo by the author.

Landscapes of Pollution

Urban and industrial concentration along waterways tends to affect water quality. Offensive smells, oil slicks, and flotsam from packing plants attracted the attention of civic officials in the nineteenth century. Riparian common law protected downstream users from diminished water quality caused by upstream polluters. Legal action was not always able to attach responsibility to the specific source of foul conditions, and some pollution was short-lived which undermined recourse. Nonetheless, by the twentieth century, accounts of degraded waterways became more common. Bubbly Creek that drained Chicago's packing district was a particularly offensive eyesore and the target of outrage. In the 1950s, islands of suds on the Ohio River reflected the development of new detergents without comparable development of pollution control and prompted legislation to control this emerging problem. The release of the pesticide Endrin into the Mississippi River during the winter of 1963–1964 left millions of fish carcasses floating on the water's surface and inspired efforts to regulate industrial waste releases to waterways. Fires erupted on the Cuyahoga River in the 1960s due to oily wastes that covered its surface providing further impetus for pollution control. These ephemeral landscapes represent the reliance on waterways to transport wastes. There was a visceral response to such visual spectacles.

Many cities and factories had installed some basic pollution abatement equipment from the early decades of the twentieth century. Yet, public outrage, sparked in part by the visible impacts of pollution, spawned a renewed effort for policies that demanded treatment facilities. Since the 1970s, sewage and industrial waste treatment infrastructure (Figure 2.6) has appeared along rivers in an attempt to meet new laws and satisfy public demands for water that is safe for consumption and fish life. These massive public works are now one of the more prominent elements of riverfront landscapes.

Figure 2.6 Baton Rouge sewage treatment works on the banks of the Mississippi River. Photo by the author.

Concluding Remarks

Water is a fluid that is often in motion and seldom considered as a landscape feature unless stationary as a lake or tumbling over a precipice as a waterfall. It flows and with that it has potential power and acts to shape many landscapes elements. It can transport cargo and generate electricity, but as a flood can devastate a riverside community. Thus, it can be both a resource and a hazard. Also, while society has constructed visible landscapes attached to waterbodies and writers have commented on the view from the water, many of water's qualities are invisible. The riverboat pilot needs a nine-foot channel, which is maintained by dredges and obscured by the turbid flow. A municipal water treatment plant reduces harmful chemicals to safe levels before sending the liquid to consumers. Most contaminants are not visible to the naked eye. Our perceptions of water are as varied as our uses. Arid land farmers fear dropping reservoir levels. Recreational paddlers revel in the rise of rapids. While those perceptions change from place to place and over time, humans will remain tightly attached to water and the landscapes they have created that solidify that relationship.

References

Cohen, S. B., and N. Kliot. 1992. "Place-names in Israel's Ideological Struggle over the Administered Territories." *Annals of the Association of American Geographers* 82 (4): 653–680.

Colten, C. E. 2014. *Southern Waters: The Limits to Abundance*. Baton Rouge: Louisiana State University Press.

Cummings, S. 1845. *The Western Pilot; Containing Charts of the Ohio and the Mississippi*. Cincinnati: G. Conclin.

DeVoto, B. 1953. *The Journals of Lewis and Clark*. Boston: Houghton Mifflin.

d'Iberville, Pierre Le Moyne. [1661–1706] 1981. *Iberville's Gulf Journals*. Translated by R.G. McWilliams. Tuscaloosa: University of Alabama Press.

Foresta, R. A. 2013. *The Land Between the Lakes: A Geography of the Forgotten Future*. Knoxville: University of Tennessee Press.

Heat-Moon, W. L. 1999. *River Horse: A Voyage across America*. New York: Houghton Mifflin.

Herman, R. D. K. 1999. "The Aloha State: Place Names and the Anti-conquest of Hawai'i. *Annals of the Association of American Geographers* 89 (3): 76–102.

Hunter, L. C. 1949. *Steamboats on the Western Rivers: An Economic and Technological History*. Cambridge: Harvard University Press.

Manganiello, C. J. 2015. *Southern Water, Southern Power: How the Politics of Cheap Energy and Water Scarcity Shaped a Region*. Chapel Hill: University of North Carolina Press.

Mitchell D. 2008. "New Axioms for Reading the Landscape: Paying Attention to Political Economy and Social Justice." In *Political Economies of Landscape Change*. The GeoJournal Library, vol. 89, edited by J.L. Wescoat and D. M. Johnston, 29–50. Dordrecht: Springer.

Mitchell, R. D., and W. R. Hofstra 1995. "How Do Settlement Systems Evolve? The Virginia Backcountry during the Eighteenth Century." *Journal of Historical Geography* 21 (2): 123–147.

O'Neill, K. 2006. *Rivers by Design: State Power and the Origins of US Flood Control*. Durham: Duke University Press.

Paskoff, P. F. 2007. *Troubled Waters: Steamboat Disasters, River Improvements, and American Public Policy*. Baton Rouge: Louisiana State University Press.

Phillips, J. D. 1997. "A Short History of a Flat Place: Three Centuries of Geomorphic Change in the Croatan National Forest." *Annals of the Association of American Geographers* 87 (2): 197–216.

Pisani, D. J. 1992. *To Reclaim a Divided West: Water, Law and Public Policy, 1848–1902*. Albuquerque: University of New Mexico Press.

Powell, J. W. [1895] 1961. Reprinted. *The Exploration of the Colorado River and Its Canyons*. New York: Dover Publications.

Report and Accompanying Documents of the Virginia Commissioners Appointed to Ascertain the Boundary Line between Maryland and Virginia. 1873. Richmond: R.F. Walker.

Rose-Redwood, R., D. Alderman, and M. Azaryahu. 2010. "Geographies of Toponymic Inscription: New Direction in Critical Place-Name Studies." *Progress in Human Geography* 34 (4): 453–470.

Rubin, J. 2016. A *Negotiated Landscape: The Transformation of San Francisco's Waterfront Since 1950*. Pittsburgh: University of Pittsburgh Press.

Scarpino, P. V. 1985. *Great River: An Environmental History of the Upper Mississippi, 1890–1950*. Columbia: University of Missouri Press.

Smith, J. [1608] 1907. "A True Relation by Captain John Smith." In *Narratives of Early Virginia, 1606–1625*, edited by L.G. Tyler, 27–71. New York: Scribner's Sons.

Steinberg, T. 1991. *Nature Incorporated: Industrialization and the Waters of New England*. Amherst: University of Massachusetts Press.

Swain, G. F. 1885. *Report on the Water Power of the Streams of the United States, Parts 1&2*. Washington: US Department of the Interior, Census Bureau.

Thoreau, H. D. [1839] 2014. Reprinted. *The Illustrated Week on the Concord and Merrimack Rivers (1839)*. Princeton, NJ: Princeton University Press.

Trimble, S. W. 1970. "The Alcovy River Swamps: A Result of Culturally Accelerated Sedimentation." *Bulletin of the Georgia Academy of Sciences* 28: 131–141.

US Army Corps of Engineers. 2020. National Levee Data Base. doi://levees.sec.usace.army.mil/#/ [accessed July 22, 2020]

White, R. 1995. *The Organic Machine*. New York: Hill and Wang.

Worster, D. 1985. *Rivers of Empire: Water, Aridity and the Growth of the American West*. New York: Pantheon Books.

3

ENERGY LANDSCAPES

Brian C. Black

The layers of our energy story are evident in a landscape such as mine in Central Pennsylvania (Figure 3.1). Today, one finds wind turbines topping the mountain ridges that earlier were mined extensively for minerals such as coal that powered the nation's era of industrialization in the nineteenth century. Prior to that, the valleys presented opportunity for agriculture and forests were felled to power early, proto-industrial endeavors such as furnaces for iron manufacture. And wherever they were found, the Commonwealth's rivers hosted water wheels to enable milling and, ultimately, very often the waters to cool the residue from nuclear power plants. In many regions, these landscapes enabled and extended human labor in important ways that applied energy to human endeavors—to our varied efforts to work and to produce.

In one of our era's most unique developments, across disciplines ranging from geology to philosophy, a growing number of scholars now refer to our imperfect moment as the *Anthropocene epoch*: a moment marked by Earth being perilously near to various cataclysmic tipping points—largely due to the discernible causality of human activities such as burning fossil fuels—and yet recognizably salvageable, with the help of human ingenuity and insight. While scholars debate certain aspects of the merit of the Anthropocene term, it is indisputable that just as the term portends a common forecast for the future of all humans, it also affords a language for organizing our past (Malm 2016).

Often this narrative of development and expansion is concealed in forms and lifestyles that humans deem normal or acceptable, particularly in the form of landscapes of productivity that we take for granted and overlook (LeMenager 2016). In a vicious feedback loop, humans' ability to overlook the systems that enable their lifestyle has often led them to consume even more aggressively (Marks 2007). In early stages of energy consumption, writes historian Chris Jones, "The direct connection between labor and energy gave Americans an embodied experience of their energy choices and provided strong incentives to limit their consumption" (Jones 2017, 2). In the later energy exchanges that defined the Anthropocene, delivery systems and landscapes helped to conceal energy transactions—including costs—from consumers. Scholars consistently expose this veneer of consumption to understand the implications of modern life.

While many characteristics define human life in the Anthropocene, energy consumption is clearly the prime component. With energy a critical example, historians now have the opportunity to apply macroscopic organizational concepts to human history in order to better overcome global borders and to bring to the fore elements of global ecology that are held in common. As theorist Vaclav Smil writes, "Energy is the only universal currency; one of its many forms must be transformed to get anything done" (Smil 2017, 12–14).

Previous scholars, such as David R. Meyer, have focused on the industrial landscape as one composed of "specialized activity and mechanical integration, of growth and decline, and of

DOI: 10.4324/9781003121800-5

Figure 3.1 Layers of energy history show in Central Pennsylvania where coalfields now give way to ridgeline wind generation, Hollidaysburg, PA. Photo by the author.

abandoned and reused relics" (Meyer 1990, 249). In the approach that was often referred to as industrial archaeology, built forms of production were emphasized while scholars offered some consideration of the systems of industrial distribution and the "byproducts of production" (such as pollution) that were most often tied to local or regional industries. Building on this scholarly foundation, this essay refuses to take for granted the power that enabled each industrial under-taking to occur—indeed, it will emphasize the energy exchanges that define them. Each site of production must be revealed for the web of connections that come together in order for it to function. In these landscapes of energy, we can discern the systems at work behind much of our familiar life.

This essay organizes energy landscapes into three classifications: production, distribution, and consumption. By emphasizing landscape, this essay demonstrates how energy is often camouflaged behind a veneer of familiar forms that compose our everyday lives. Clearly per-ceiving the energy landscapes around us might serve as the single-best demonstration of modern humans' high-energy existence—distinct from that seen at any time in the past.

Landscapes of Energy Production

Energy exchanges form the basic foundation of humans' life in any environment in any era. At this essential level, the management of harnessing sunlight to grow crops and animals shapes the agricultural revolution that ultimately leads to farms as one of the most basic energy landscapes. In American history, the exchange is made more complicated by increasing scale and scope of

Energy consumption in the United States (1776-2040)
quadrillion Btu

Figure 3.2 Energy transitions often occur gradually. The fluctuation in reliance on coal, for instance, has defined the last century and a half. Renewables have clearly emerged as Americans have better understood the limitations of fossil fuels. U.S. Energy Information Administration, 2015.

the undertaking, normally by variations in supplementary power input, including additional human labor or mechanization (McShane and Tarr 2011). In the American South, for instance, the plantation form was created to structure the system of adding additional human labor. As mechanization increased in the twentieth century, the same undertaking expanded Westward and required the systematized landscape known as the "grid." And, finally, this expansion grew in scale and scope through the incorporation of the coal-powered railroad that allowed the same energy exchange of photosynthesis to create and access markets that connected producers and consumers in a fashion that allows us to view the entire continent as a landscape of energy production originating from photosynthesis (Figure 3.2).

Similarly, wind power was another energy exchange that reaches back to pre-industrial eras of human life. In the American experience, of course, the power of the wind carried European settlers to North America and provided colonists with the clearest system for organizing their settlements. Creating seaports, particularly along the East Coast, the young Republic made the sailing ship the essential link to foreign trade and travel. These vessels utilized an energy technology that had been perfected by Chinese sailors and others as early as the 1300s; now, however, organization such as the "Atlantic system" used a general seaport design, equipped with wharfs, warehouses, and roads to the interior, to stabilize and routinize sea-born trade, whether for sugar, cotton, or human slaves.

Upon this foundation and often within this general system of trade, new energy sources emerged in the 1800s that fueled the expansion of industrial endeavors and enabled the production of energy to become a commodity that existed outside of the actual undertakings to which it might be put. The basic cultural desire for more flexible and reliable (at least in the short term) sources of power defines the age of industry and a selection of the productive landscapes that enabled it to grow in the 1800s (Wallace 2005).

Energy Source: Rivers

Richard White's *Organic Machine* (2011) uses the Columbia River to connect generations of human enterprise with the sun's energy. It provides a template for how a landscape analysis might be applied to understand any river's energy story through time and changing use.

The river as prime mover is explored with the most clarity by Theodore Steinberg in *Nature Incorporated* (1991), which explores the Industrial Revolution in New England as it caused alterations to gender and class relations but also in the way the natural world was handled. Focusing on the legendary Waltham-Lowell style mills, *Nature Incorporated* examines the legal, economic, and social methods by which these textile factories brought water under their exclusive control. Steinberg offers a reinterpretation of industrialization that centers on the struggle to control and master nature. In *Fish Versus Power* (2004), Matthew Evenden applies Steinberg's general approach to debate and conflict over energy development to the Fraser River in British Columbia, Canada in the twentieth century. Evenden writes:

> From the role of state systems and geopolitics to the importance of monumentalism and modernism in dam design, environmental historians have sought to understand the forces that have made dam development politically possible, economically feasible, and ideologically acceptable. Complementing a number of important studies focusing on fisheries, they have also catalogued a host of environmental consequences of river development, not only on fish, but also on changed flow regimes and on human settlements. Relatively few national and international studies have sought to resurrect and understand protest movements against dam development.
>
> (Evenden 2004, 15)

In his book, Evenden analyzes the way such protest movements defined this powerful waterway (which, incidentally, never was dammed). Most helpful, though, might be a variety of sources that explore dams and their design as material artifacts of larger political and social ideas; in particular, see works by Donald C. Jackson, including *Big Dams of the New Deal Era* (2006). Although primarily about flood control, Martin Reuss' *Designing the Bayous* (2004) stands as the preeminent history of the Corps of Engineers and the management and control of the Atchafalaya Basin. Finally, Jeff Stine's fine study of the conflict over the Endangered Species Act of the 1970s, *Mixing the Waters* (1993), demonstrates one of the basic conflicts facing hydroelectric development in the era of environmental policy.

Energy Source: Coal

As the predominant energy source behind the industrial revolution, coal and its effects on society have interested historians—particularly labor, social, and economic scholars—for generations (Wallace 1987). Many of these sources, such as Anthony F.C. Wallace's *St. Clair* (1985), contain useful insights about the social implications of coal mining as they play out on the landscape. Early in his career, Martin Melosi wrote *Coping with Abundance* (1985), which provides a useful introduction to energy transitions in the nineteenth century. In describing the adoption of coal in the early 1800s, he ties the energy transition from biomass to coal as one of available supply, particularly due to the War of 1812. Growing reliance on coal would fuel America's industrial revolution, creating landscapes and land uses unlike those of previous eras. To demonstrate these transitions in nineteenth-century industry, landscape historians should draw from the rich work of industrial archaeologists including Robert Gordon and Patrick Malone's *The Texture of Industry* (1994).

Richard Francaviglia used industrial archaeology to study such locales of extraction. In *Hard Places* (1997), he makes clear that actual places in which people live and work have taken form from extractive beginnings, first in the mid-Atlantic area of the US and later in Western States. In *Extracting Appalachia* (2004), geographer Geoffrey Buckley uses photographs to delve into the landscape and culture of Appalachian coal during the early twentieth century. Buckley's research centers on a set of approximately four thousand photos collected in the archive of the Consolidation Coal Company, which were taken primarily in Kentucky, West Virginia, and Maryland. Using approaches from history and geography, Buckley carefully manages the limits of his source material to yield a "collective glimpse of life and labor in central Appalachia's coalfields" (Buckley 2004, xvii). He writes that the photos also "allow us to follow closely the construction and development of mines and company towns; inspect the work of miners; to track the technological advances…; to observe conditions in and around the mines…; and to speculate on social and cultural aspects of coal town life" (xvii).

Mining is at the center of current environmental issues that influence regional life after coal extraction has discontinued. Environmental historian Chad Montrie emphasizes the social implications of mining in *To Save the Land and People* (2003), which examines the twentieth-century movement to outlaw surface coal mining in Appalachia and in *Killing for Coal*, Thomas Andrews (2008) connects labor uprisings to locales in the American West. Finally, for a general survey of coal in world history, historians might also use Barbara Freese's *Coal: A Human History* (2003). In her brief overview, Freese covers the global rise of industrialization through the application of coal burning technology. Most important, her account finishes the cycle by connecting the various implications of burning coal. In particular, her chapter "A Burning Legacy" takes the global story of industrialization and matches it with the global story of the science clarifying the impact on the global environment, including climate change.

Energy Source: Petroleum

Petroleum is not like other sources of energy. With similarities to valuable minerals such as gold, petroleum's utility was realized in very different landscape patterns from coal. In recent years, many environmental historians have come to investigate the culture formed around the extraction of petroleum. In *Petrolia: The Landscape of America's First Oil Boom*, Brian C. Black (2003) used landscape history and cultural geography to create a portrait of the industrial ethics taking shape in the earliest development of petroleum in the fields of Pennsylvania from 1859–1873. Running outside the bounds of typical property law, petroleum required a unique landscape of extraction. The trial and error of the earliest fields in Pennsylvania made the landscape one unlike any seen before. Black primarily uses the oil fields as a case study of changing ethics in all of American industry. Similar usage patterns can be seen throughout industrial development after the 1880s; however, the story that Black tells is based in the subtleties that are peculiar to petroleum. These details continue to help define human use of crude today, when petroleum has become one of the world's most important commodities.

Similarly, one of the best accounts of the culture of oil on existing societies is the *Ecology of Oil* (2006) in which Myrna Santiago provides a critically important new source in the historical consideration of petroleum as well as in the social and cultural impacts of massive industrial change. For specific consideration of offshore drilling, see Tyler Priest, *The Offshore Imperative* (2009) and Teresa Spezio, *Slick Policy* (2019). For overviews of oil in human living patterns more broadly speaking, see Brian C. Black, *Crude Reality* (2020) and Matthew Huber, *Lifeblood* (2013).

Energy Source: Nuclear

Inquiry into nuclear power's history builds on Thomas Wellock's *Critical Masses* (1998), which discusses California's public movement against nuclear power between 1958 and 1978.

Additionally, William Kovarick's *Radium Girls* (online: www.runet.edu/~wkovarik/envhist/rad ium.html) as well as sources on Native American uranium mining are a promising start (*If You Poison Us*, by Peter H. Eichstaedt [1994] and Valerie Kuletz, *The Tainted Desert* [1998]). Finally, Kate Brown's *Plutopia* (2013) offers a fascinating comparative analysis of American and Soviet plutonium mining communities. For a very textured, cultural approach to the implications of nuclear development, readers should also consider Terry Tempest Williams' *Refuge* (1992).

Landscapes of Production in the Era of Extreme Energy

New landscapes of energy production have emerged in the twenty-first century, particularly due to shifts in public attitude and in the remaining supplies of fossil fuels. When traditional reserves of coal, oil, and natural gas waned over recent decades, energy producers embraced new methods that are classified as "extreme" due to their reliance on additional technology or expense for extraction. These more intense practices formed new landscapes, including: off-shore drilling platforms for petroleum miles below the ocean's surface in locales such as the Gulf of Mexico; tar sand removal that first required aggressively stripping all of Earth's surface in stretches of Alberta, Canada; and horizontal drilling to access underground seams of shale that could then be fractured (fracked) with hydraulic fluids to transform it into oil or gas where geology allows, including the Bartlett Shale of Texas and Marcellus Shale of Pennsylvania.

During this same era, growing public awareness of environmental costs as well as the increased expense of fossil fuels stimulated new energy landscapes that applied renewable harvesting techniques, including solar and wind, on a scale that could begin to compete with fossil fuels. Founded in awareness of climate change and carbon pollution, these new developments must also be classified as evidence of our moment of "extreme" energy even though they promise future energy production in a more sustainable manner.

Box 3.1 Wind Energy Landscapes are a Good Thing

Martin J. Pasqualetti

Until recently, America's energy landscapes were localized creations that were largely out of sight and mostly out of mind. For example, coal recovery areas did not impinge on Baltimore, Philadelphia, New York, or Boston. While these cities might have hosted coal transport terminals, coal users rarely crossed paths with the landscapes that mining activities were ravaging many miles away.

Emerging decades later, Pennsylvania, Texas, and California's oil landscapes took on a different form, but with a similarly muted reaction. As intrusive as oil activities were, the landscapes they created reflected a wealth that went hand in hand with the prideful conceit of human dominion over nature that was common at the time.

Similar disinterest prevailed in the Great Plains when wind energy landscapes first appeared on the American scene. Here, tens of thousands of primitive but effective wind-powered water pumps marked family farms and helped early settlers convert the untamed grasslands into a grid-controlled breadbasket. These wind machines attracted no public resistance because of their utility and because—like coal and oil—they symbolized progress.

Eventually, however, such wind pumps gradually disappeared, much like the bison that had preceded them. Well before the end of the twentieth century, they had become little more than iconic echoes of a simpler time in American landscape evolution (Pasqualetti 2012).

Today, the development of the Great Plains' ample wind power has made a strong comeback, if in a different form. While the new machines are of the same genus, they are of a separate species. Like a Basset Hound and a Cheetah are both four-footed, new wind machines both spin and rotate but differ in every other way. They are taller than the Statue of Liberty, their blades move faster than a jetliner, the diameter of their swept area exceeds the Wright brothers' first flight, they are erected in clusters rather than singly, they generate electricity rather than suction, they are made of composite materials instead of wood, and corporations rather than families finance them. In sum, our age's wind turbines are quite different from their ancestors, including that they are certainly no longer quaint.

Modern wind turbines differ in another way; they change landscapes to a degree unimaginable 150 years ago, and there is little that can be done to avoid it. They resist compact positioning, and they allow no siting flexibility. You either install the turbines where the winds are suitable or not at all. Such genetic characteristics plant the seeds of conflict wherever development coincides with where people live, work, drive, or recreate. As a result, the expansion of wind power frequently provokes strong public resistance to the novel landscapes it creates. It is one of the great ironies of modern times. A resource we cannot see is producing landscapes we cannot ignore.

But, when considered in the broader context, such visibility is a good thing. Conceding that wind turbines should never be allowed to degrade high-value landscape treasures, their no-compromise visibility serves a higher purpose. Wherever we encounter them, they remind us that our electricity must come from somewhere. Once we accept that reality, we have to admit that the costs of wind energy landscapes are counterbalanced in large part by the benefits that accompany them. Wind turbines require no water, produce no greenhouse gases, poison no soil, while typically boosting land values wherever they appear. In this context, wind energy landscapes are a good thing. We can live with them. At least until something better comes along.

Landscapes of Energy Connection and Distribution

The true mastery of the American high-energy existence relied on hidden systems that distributed the power acquired from coal mines, oil wells, and nuclear power plants to consumers. The single most important element in this process was electricity, which could be generated from a variety of sources and moved almost invisibly across the landscape to tasks ranging from lighting Times Square to recharging an iPhone. Canals, railroads, and tanker trucks began the process by moving coal or oil from their points of harvest and they remain vital cogs today in our energy system. And, finally, pipelines crisscross the American landscape to consistently distribute power sources far from their site of extraction. The additional layer in this system, though, that is most taken for granted is electricity made from such energy sources and the rigid grid that allows it to be organized and dispersed while operating under the guise of flexibility.

In this type of system's level analysis of the landscape, two larger works can be read with energy at their core: William Cronon's *Nature's Metropolis* (1992), which applies central-place theory to the role of Chicago's development in the American West, and Richard White's *Railroaded* (2012). In each case, the railroad becomes a mechanism for organizing the landscape and general lifestyle for the American West. This meta-level analysis of energy systems is also related to the industrial infrastructure that is discussed in Jones' *Routes of Power*, which was discussed above, and more specifically in titles such as Julie Cohn's *The*

Grid (2017), an analysis of the system of electricity distribution, Chris Turner's *The Patch* (2017), concerning Oil Sands and pipeline development, and Adam Mandelman's *The Place with No Edge* (2019), which studies the connection between Bayou regions and pipelines. On the larger patterns, particularly of electricity, see David Nye's *Electrifying America* (1991) for a seminal exploration of the connection between social and technological history through the implications of electricity in American life and his *Consuming Power* (1984). In each case, an energy technology becomes transformative through the landscapes that take shape to disperse or utilize it.

Landscapes of Energy Consumption

Making energy available created the raw opportunity for transformational living patterns in the US, but it was landscapes of consumption that created the ubiquity and standardization that actually allowed a new American standard of living to take physical form. Transportation corridors offer the clearest pattern on the landscape of increasing energy intensity. In *The Horse in the City* (2011), Clay McShane and Joel A. Tarr tell the story of animal centrality. As the authors put it, their story is a symbiotic one: the lifecycle of the horse was mirrored in the city. For all the other aspects of the organic city that Tarr and others have explored in other works, in this work McShane and Tarr see the nineteenth century city as the "climax of human exploitation of horse power" (McShane and Tarr 2011, 1).

The modern era of transportation is the focus of Thomas McCarthy's *Auto Mania* (2007) in which he creates a history of technological innovation with a particular emphasis on the environmental impact of the automobile, which John R. McNeill has called one of the twentieth century's most socially and environmentally "consequential" technologies (McNeill 2001, 308–311). The American automobile is no typical consumer good, as "All of the automobile's environmental impacts occurred in a larger context shaped by consumers as much as (and sometimes more than) producers" (McNeill 2016, xiv). Other related aspects on the topic of transportation, include: Shane Hamilton's *Trucking Country* (2008), a social history of long-haul trucking; Peter D. Norton's *Fighting Traffic* (2008), which describes the landscape and spatial impact of automobile use in the twentieth century city; Thomas Zeller's *Driving Germany* (2007), which unpacks the social meaning of Hitler's Autobahn as a technology with serious social and environmental implications; and *The World Beyond the Windshield* (Mauch and Zellner 2008), which is a comparative analysis of the impact of auto use on the landscape in twentieth-century Central Europe and the United States.

From gasoline filling stations to Coney Island, landscapes of consumption are defined by available cheap energy. Whether in cities or suburbs, though, similar patterns can be drawn out by scholars by tracing the energy systems that make them possible. For instance, Adam Rome's *Bulldozer in the Countryside* (2001) most specifically considers the larger systems that enable American suburban development and expansion. With only a touch more specificity, Theodore Steinberg's *American Green* (2006) investigates energy concerns in the American fetish with lawns, a truly unique American landscape. His careful study of this pseudo-nature that covers 40 million acres of the US is engaging and humorous. Just having a lawn was insufficient: the ideal quest was to have the *perfect* lawn. "Perfection is elusive," says Steinberg. "And it constantly creates the need for people to return to the hardware store to buy more chemical inputs in the quest for the perfect yard" (Steinberg 2006, 11). This landscape of conformity was made achievable and perpetuated through technology (chemicals, biotechnology, and mower technology) and marketed through sporting activities (sports fields and Astro-turf as well as golf and the sales of Scotts Turfbuilder), even when host environments were too arid for it.

Finally, all American cities rely on energy and transportation networks. Platt's *Electric City* (1991) looks at the expansion of electricity in Chicago and others, including John Findlay's *Magic Lands* (1993), have connected the expansion of electricity to the development of urban areas of the American West. Finally, in their collection on Houston, Texas, *Energy Metropolis*, Martin V. Melosi and Joseph A. Pratt write: "Cities are by their very nature energy intensive. A key challenge in urban and environmental history is to identify and analyze the central impacts of energy production and use on the evolution of cities" (Melosi and Pratt 2007, 12). Although it represents the energy intensiveness of all urban life, as a city defined by the creation of energy, Houston becomes an entirely unique city at once creating itself but always also fulfilling the expectations and needs of the nation's insatiable need for cheap energy in the twentieth and early twenty-first centuries.

Conclusion: Tracing Energy History through the Landscape

As a system of energy exchanges, the natural environment hosts each species and provides what is needed for it to satisfy its basic needs—to survive. Although our more modern history distinguishes us from other species, our distant past associates us with all else. Restoring the continuity of this story is a compelling purpose for scholars to study energy. Our story has become more complicated; however, the point of origin is undeniable. Humans first strove for survival just like all other living things and it was unthinkable that we or any other member of Earth's system could disrupt the scale and scope of its innate patterns.

As its point of origin, Earth's energy patterns begin with the Sun that radiates from the center of our planet's universe. In short, its energy makes possible our very existence (Crosby 2007, 2). From this point of origin, changes in human living can be traced through methods of acquiring and mastering energy. Indeed, access to the systems and supplies predicts and determines issues of energy justice. Societies also need to choose what to do with energy or how much of a value to put on utilizing sustainable resources. Each society's responses to such questions regarding new regimes of energy have fueled them to develop in dramatically different ways from one another. Properly placing the American human in this pattern, of course, presents the portrait of a modern culture entirely wedded to cheap energy and mastery of its concealment within a twentieth-century standard of living organized by effortless consumption. Our landscapes reflect and perpetuate the centrality of energy in American life.

References

Andrews, T. 2008. *Killing for Coal: America's Deadliest Labor War*. Cambridge: Harvard University Press.

Black, B.C. 2003. *Petrolia: The Landscape of America's First Oil Boom*. Baltimore: Johns Hopkins University Press.

———. 2020. *Crude Reality: Petroleum in World History*. New York: Rowman & Littlefield.

Brown, K. 2013. *Plutopia: Nuclear Families, Atomic Cities, and the Great Soviet and American Plutonium Disasters*. New York: Oxford University Press.

Buckley, G.L. 2004. *Extracting Appalachia: Images of the Consolidation Coal Company, 1910–1945*. Athens, OH: Ohio University Press.

Cohn, J. 2017. *The Grid: Biography of an American Technology*. Cambridge: MIT Press.

Cronon, W. 1992. *Nature's Metropolis: Chicago and the Great West*. New York: W.W. Norton & Company.

Crosby, A. 2007. *Children of the Sun*. New York: W.W. Norton & Company.

Eichstaedt, P.H. 1994. *If You Poison Us: Uranium and Native Americans*. Albuquerque: Red Crane Books.

Evenden, M. 2007. *Fish Versus Power: An Environmental History of the Fraser River*. New York: Cambridge University Press.

Findlay, J. 1993. *Magic Lands: Western Cityscapes and American Culture after 1940.* Berkeley: University of California Press.

Francaviglia, R. 1997. *Hard Places: Reading the Landscapes of America's Historic Mining Districts.* Iowa City: University of Iowa Press.

Freese, B. 2003 *Coal: A Human History.* New York: Basic Books.

Gordon, R., and P. Malone. 1994. *The Texture of Industry: An Archaeological View of the Industrialization of North America.* New York: Oxford University Press.

Hamilton, S. 2008. *Trucking Country: The Road to America's Wal-Mart Economy.* Princeton: Princeton University Press.

Jackson, D.C., and D. Billington. 2006. *Big Dams of the New Deal Era: Confluence of Engineering and Politics.* Norman: University of Oklahoma Press.

Jones, C.F. 2017. *Routes of Power: Energy and Modern America.* Cambridge: Harvard University Press.

Kuletz, V. 1998. *The Tainted Desert: Environment and Social Ruin in the American West.* London: Routledge.

LeMenager, S. 2016. *Living Oil: Petroleum Culture in the American Century.* New York: Oxford University Press.

Malm, A. 2016. *Fossil Capital: The Rise of Steam Power and the Roots of Global Warming.* New York: Verso.

Mandelman, A. 2019. *The Place with No Edge: An Intimate History of People, Technology and the Mississippi River Delta.* Baton Rouge: LSU Press.

Marks, R. 2007. *The Origins of the Modern World: A Global and Environmental Narrative from the Fifteenth to the Twenty-First Century.* Lanham: Rowman & Littlefield Publishers.

McNeill, J.R. 2001. *Something New Under the Sun: An Environmental History of the Twentieth-Century World.* New York: Norton.

McNeil, J.R., and P. Engelke. 2016. *The Great Acceleration: An Environmental History of the Anthropocene since 1945.* Cambridge: Belknap Press.

McShane C., and J. Tarr. 2011. *The Horse in the City: Living Machines in the Nineteenth Century.* Baltimore: Johns Hopkins University Press.

Melosi, M.V. 1985. *Coping with Abundance: Energy and Environment in Industrial America.* New York: Knopf Publishers.

Melosi, M.V., and J.A. Pratt. 2007. *Energy Metropolis: An Environmental History of Houston and the Gulf Coast.* Pittsburgh: University of Pittsburgh Press.

Meyer, D.R. 1990. "The New Industrial Order." In *The Making of the American Landscape*, edited by M. Conzen, 249–268. Boston: Unwin Hyman.

Montrie, C. 2003. *To Save the Land and People: A History of Opposition to Surface Coal Mining in Appalachia.* Chapel Hill: University of North Carolina Press.

Norton, P.D. 2008. *Fighting Traffic: The Dawn of the Motor Age in the American City.* Cambridge: MIT Press.

Nye, D. 1991. *Electrifying America: Social Meaning of a New Technology, 1880–1940.* Cambridge: MIT Press.

———. 1984. *Consuming Power: A Social History of American Energies.* Cambridge: MIT Press.

Pasqualetti, M.J. 2012. *Wind Power in View.* New York: Academic Press.

Priest, T. 2009. *The Offshore Imperative: Shell Oil's Search for Petroleum in Postwar America.* College Station: Texas A&M University Press.

Rome, A. 2001. *The Bulldozer in the Countryside: Suburban Sprawl and the Rise of American Environmentalism.* New York: Cambridge University Press.

Santiago, M. 2006. *Ecology of Oil: Environment, Labor, and the Mexican Revolution, 1900–1938.* New York: Cambridge University Press.

Smil, V. 2017. *Energy and Civilization: A History.* Cambridge: MIT Press.

Spezio, T. 2019. *Slick Policy: Environmental and Science Policy in the Aftermath of the Santa Barbara Oil Spill.* Pittsburgh: University of Pittsburgh Press.

Stilgoe, J. 1994. *Alongshore.* New Haven: Yale University Press.

Steinberg, T. 1991. *Nature Incorporated: Industrialization and the Waters of New England.* New York: Cambridge University Press.

———. 2006. *American Green: The Obsessive Quest for the Perfect Lawn.* New York: Norton.

Stine, J. 1993. *Mixing the Waters: Environment, Politics, and the Building of the Tennessee-Tombigee Waterway.* Akron: University of Akron Press.

Turner, C. 2017. *The Patch: The People, Pipelines, and Politics of the Oil Sands.* New York: Simon & Schuster.

Wallace, A.F.C. 1987. *St. Clair: A Nineteenth-Century Coal Town's Experience with a Disaster-Prone Industry.* New York: Knopf Publishers.

———. 2005. *Rockdale: The Growth of an American Village in the Early Industrial Revolution*. Lincoln: University of Nebraska Press.

Wellock, T. 1998. *Critical Masses: Opposition to Nuclear Power in California, 1958–1978*. Madison: University of Wisconsin Press.

White, R. 2011. *The Organic Machine: The Remaking of the Columbia River*. New York: Hill & Wang Publishers.

———. 2012. *Railroaded: The Transcontinentals and the Making of Modern America*. New York: Norton.

Williams, T.T. 1992. *Refuge: An Unnatural History of Family and Place*. New York: Vintage Publishers.

Zeller, T. 2007. *Driving Germany: The Landscape of the German Autobahn, 1930–1970*. Berlin: Berghahn Books.

Zeller, T., and C. Mauch. 2008. *The World Beyond the Windshield: Roads and Landscapes in the United States and Europe*. Athens: Ohio University Press.

4

WASTESCAPES

Martin V. Melosi

Wastescapes are landscapes on the margins, wounded spaces, garbage graveyards, or derelict sites (Figure 4.1). Sociologist Kevin Hetherington pointed out: "Disposal … is not primarily about waste but about placing. It is as much a spatial as a temporal category" (Hetherington 2004, 159). Whether we are talking about landfills, brownfields, various toxic sites, or nuclear wastelands, human intent most often determines what and where waste will impact the landscape.

The Process of Wasting

Wasting itself is an inevitable part of life, and an integral part of consuming. Scholars have expended much energy trying to explain exactly what waste is and if it has value—materially or symbolically. Garbage, trash, rubbish, offal, refuse, junk, debris, clutter, litter, rejecta-menta, radioactive material, toxic chemicals, sewage, and middens. Waste goes by many names. In the natural world, waste is simply part of the life cycle; it is a change of substance returned to the physical environment merely in a different form. In human civilization, waste has an entirely different role and connotation. While humans are not alone in waste making, they may be the only species which passes judgment on it and also produces outsized amounts of it.

Social scientist Michael Thompson, in his book *Rubbish Theory*, is probably best-known for his effort to define waste. Thompson's basic premise is that there are two cultural categories that are "socially imposed" on objects—the durable and the transient. The former (antique urns) have "increasing value and infinite expected lifespans," while the latter (milk cartons) have "decreasing value and finite expected lifespans." The transient, however, has the possibility of entering the durable category (an old postcard perhaps?). Rubbish are those items that fall neither into the durable or transient category, but rather than reaching zero value and zero expected life, usually "continues to exist in a timeless and valueless limbo" where at some future time "it has the chance of being discovered… and successfully transferred to durability." Thompson concludes that "in order to study the social control of value, we have to study rubbish" (Thompson 2017, 27).

In its typical context, "to consume" gives the impression of finality, that is, completing the process of acquiring some material thing and using it up. In many cases, the lifespan of waste is longer than its useful phase. Sociologist John Scanlan is right when he states that "what we

DOI: 10.4324/9781003121800-6

Figure 4.1 Fire at a large tire dump in southeastern Ohio in the early 1990s. Photo by Tom O'Grady, Southeast Ohio History Center, Athens, Ohio.

consume never really disappears" (Scanlan 2005, 153). There is a certain ambiguity about waste, however. Philosopher Greg Kennedy argued, "Anything and everything can become waste... [W]hat one person discards, some other person likely covets" (Kennedy 2007, 1).

Wastescape Defined

A big part of our material legacy is the wastescape. Yet, in the case of wastescapes, scale matters. The collective accumulation of spent or cast-off material in most cases reduces the ambiguity of what constitutes a new landscape. Architect Silvia Dalzero has a good grasp of the nature of wastescapes, suggesting that landscape changes in the presence of waste, "reality that invades the territory in many different surprising ways in time and space." Essentially "waste becomes a place" (Dalzero 2016, 201). Yet, wastescapes do not occupy new or empty spaces, but transform places overlaying or replacing what came before, such as marshland or industrial sites. They acquire a new identity because of the waste itself.

In the last twenty to twenty-five years several book-length studies have contributed significantly to our understanding of wastescapes and their significance. Leading the list is Mira Engler's *Designing America's Waste Landscapes* (2004). A landscape architect, Engler intended to "open up waste landscapes and set free their contradictory powers" (Engler 2004, ix). Waste orders and shapes environments, she argued, and "creates geography." The landscapes of waste exist along both physical and symbolic margins (Engler 2004, xiii–xiv). Such an assessment is meant to bring these marginal landscapes within a conversation of landscape in general, rather than cordoning them off as perpetually separate.

The book discusses a variety of waste landscapes, such as private landscapes (e.g., plumbing), garbage dumps, sewage treatment plants, and more. Engler makes a common claim that these sites mirror our culture, and that waste—which offers paradoxes—forces us "to rethink ethics and aesthetics." There is also a pragmatic side to Engler's assessment of waste landscapes: while hoping we avoid problems created by waste, she is interested in how environmental scientists, engineers, artists, and designers deal with them and create new spaces by freeing the spirit "from the limitations of professional conventions and the chains of social taboos" (Engler 2004, 232). Thus, Engler's book is most important for raising the value of wastescapes as a central theme of landscape studies.

Several studies have broadened and deepened our understanding of wastescapes through defining or refining key ideas. Urban design professor Alan Berger makes an argument in *Drosscape* that:

> Landscape architecture tends to dwell on the traditional areas of landscape history— site engineering, construction detailing, and project-based design studio education. But beyond and behind these topics is an issue so huge we tend not to see it at all— what I call the drosscape, which implies that dross, or waste, is 'scaped,' or resurfaced, and reprogrammed for adaptive reuse.
>
> *(Berger 2006, 12)*

The book examines a wide range of urban landscapes across the country from Atlanta to Houston, from Boston to Chicago, seeking to develop "a new aesthetic and vocabulary" for these sites. He argues that drosscapes emerged out of urban sprawl or as "leftovers of previous economic and production regimes" (Berger 2006, 12). He also views drosscapes as organic phenomena growing naturally out of industrial growth. In this sense, he does not imply that the wastescapes are aberrations, but almost an inevitable outcome of economic transformation resulting from the Industrial Revolution. In a companion study, *Reclaiming the American West* (2002), he examined lands altered by extraction.

Drosscape, as a book within the field of urban design, depends upon dozens of visual images— photographs, maps, and graphs—that offer a set of striking examples of Berger's main ideas, stressing how the American landscape has been rapidly urbanizing. Illustrations of the varying types of drosscapes across the country—waste landscapes of dwelling, transition, infrastructure, obsolescence, and contamination—allow the reader to draw more general conclusions about these derelict (at least for the moment) sites. It also suggests a new paradigm where designers consider "working in the margins rather than at the center" (Berger 2006, 239, 241).

Wastes as Information

Professor of art, design, and public policy Dietmar Offenhuber's *Waste is Information* expands our understanding of wastescapes by introducing more current ways of grasping the nature of waste and how it transmits information to us. He is not the first, however, to suggest that "waste is just another aspect of societal rejection" and that dumpsites can offer "unprecedented snapshots of past civilization" (Offenhuber 2017, viii). For instance, artist Mierle Laderman Ukeles viewed Fresh Kills Landfill on Staten Island as "a social structure." Garbologist William Rathje reminded us that our "garbage heritage" is worth remembering (quoted in Melosi 2020, 491). Journalist Heather Rogers stated that "Through waste we can read the logic of industrial society's relationship to nature and human labor" (Rogers 2005, 3).

Offenhuber considers waste systems from the perspective of information. He argues that "waste is, above all, a designation. Waste is whatever is labeled as waste, and nothing exists that cannot become waste at some point" (Offenhuber 2017, 2). Offenhuber does not deny the materiality of waste but uses a postmodern vantage point to view waste as information. As such, our understanding of waste is not static since it can generate new information. Waste systems, therefore, capture data stored in a physical form. It is possible to "read" infrastructure by relying on the concept of "legibility" (used in the fields of design and urban planning). Legibility "does not depend on awareness," he argues, "but on the ability to differentiate features and the capacity to detect and assemble a coherent picture from them" (Offenhuber 2017, 42). While infrastructure legibility can be, among other things, a process for understanding waste and wastescapes, Offenhuber concludes that waste systems are "heterogenous, illegible, and contested infrastructures" (Offenhuber 2017, 225). That being the case, there is a great deal yet to be learned by exploring the idea of waste as information. For example, can a wastescape give us a sense of its extant timeline by examining its contents?

Anti-Landscapes and Sacrifice Zones

Contrary views about wastescapes also offer ways to rethink their meaning. In the Introduction to *The Anti-Landscape*, historian David E. Nye states that "The anti-landscape is a new term, but the phenomena of spaces that have become unlivable was already described in [George Perkins] Marsh's *Man and Nature*." As a point of departure, Nye refers to a definition of landscape by the renowned writer on landscape, J.B. Jackson: landscape is "A composition of man-made or man-modified spaces to serve as infrastructure or background for our collective existence." By contrast, Nye argues that anti-landscapes "are spaces that do not sustain life," and although they can derive from natural causes, they are usually the result of human action. In some instances, he added, anti-landscapes are capable of being improved (Nye and Elkind 2014, 11).

While Nye goes into some detail about the characteristics of anti-landscapes, the definition may be too sweeping. Some derelict landscapes—especially landfills—do not fit neatly into the definition of anti-landscapes. These landscapes usually carry heavy cultural baggage which impacts their value and use, and often are transient spaces that are part of the evolution of a particular place (to be discussed in more detail later). With respect to the former, one's first impressions of a landfill may not tell its story.

In Mexico City, for example, many landfills are pushed to the edges of the urban area rising high as banished structures beyond the gaze of city dwellers. But for a peasant class not welcomed as residents in the core of Mexico's capital, such places provide life-sustaining resources for those denied a viable existence in the central city and for people who must seek food and other needed goods by mining the landfills. Such a role for landfills is quite different from their American counterparts, which are essentially land sinks for items that people no longer need or want. Even in the United States, there was a time—in nineteenth-century New York City for example—when Italian immigrants were garbage trimmers who sold discarded materials buried in the refuse mounds of scows to second-hand stores or to salvage companies (Melosi 2005, 59).

The Anti-Landscape does raise several valuable insights that can be, or have been, applied to wastescapes in general. Nye suggests that some anti-landscapes came about without human intervention and pre-date humans, such as the surface of the moon. Intentionality thus depends on human action in creating anti-landscapes, and is an issue requiring further study. Anti-landscapes can be vulnerable to change, and that also is worth exploring further. Questions concerning the temporality and physical location of anti-landscapes—and wastescapes in general—are likewise important.

Broadening out the themes of wastescape and community is Steve Lerner's *Sacrifice Zones*. He stated that "The label *sacrifice zones* comes from 'National Sacrifice Zones,' an Orwellian term coined by government officials to designate areas dangerously contaminated as a result of the mining and processing of uranium into nuclear weapons" (Lerner 2010, 2). However, the term has come to be used in many contexts in a variety of publications. Lerner usefully stresses the relationship between sacrifice zones and the inequality in environmental contamination as expressed in the environmental justice movement. His focus is on the impact of chemical exposure, especially for those living "on the frontlines of toxic chemical exposure in the United States" (Lerner 2010, 2). He sees sacrifice zones as "the result of many deeply rooted inequities in our society," one of which being "unwise (or biased) land use decisions" (Lerner 2010, 6). His concluding remarks are powerful: "Sacrifice zones are a blight on the land. They bear witness to an ongoing and pernicious form of racial and class discrimination that this nation has yet to address" (Lerner 2010, 297). The book is important for fusing wastescapes with social injustice, thus developing a critique that goes beyond the environmental impacts of them.

Graphic expressions are also significant in visualizing wastescapes. Photographer David Hanson has accomplished that goal in his pictorial study, *Waste Land*. Based on fifteen years of photographs and mixed-media installations, he has "investigated the contemporary American landscape as it reflects our culture and its most constructive and destructive energies" (Hanson 1997, 5). A great many of the images of hazardous waste sites—"ravaged landscapes"—in the book are aerial photographs beginning with his hometown of Colstrip, Montana and then various locations across the country. In the Preface, writer and activist Wendell Berry summed up what the book presents: "It is unfortunately supposable that some people will account for these photographic images as 'abstract art.' Or will see them as 'beautiful shapes.' But anybody who troubles to identify in these pictures the things that are readily identifiable... will see that nothing in them is abstract and that their common subject is a monstrous ugliness" (Hanson 1997, 3).

Nuclear Landscapes

Nuclear landscapes are a unique form of wastescape because what is hidden from view is often more lethal than what is visible. In *The Tainted Desert*, Valerie Kuletz concentrates on the "social and environmental impact of nuclearism," illuminating "some of the hidden and unacknowledged costs of the Cold War—both environmentally and socially—in the interdesert West" (Kuletz 1998, xiii).

Land that had been sacred landscape and homeland to Indigenous people and contrarily viewed by others as wasteland with little value, became the sites for nuclear experimentation—above-ground weapons testing—and ultimately sacrificed to "national interests." The nuclear landscapes began in secrecy, behind electrified fences, and produced a radioactive terrain with a very long toxic life which also ignored the populations at risk. She begins by attempting to map the "invisible nuclear landscape," having "emerged piece by piece over the last fifty years, the nuclear landscape constitutes as much a social and political geography as it does an environmental region" (Kuletz 1998, 9). Thus, she also takes up the social justice themes, but for a very unique wastescape—a sacrificial geography in the most extreme sense. She concludes on a somber note: "The pursuit of nuclear power in both its militaristic and economic forms has had and will continue to have tragic consequences for life in the American desert and beyond" (Kuletz 1998, 290).

John Beck's *Dirty Wars* explores several of the same themes as Kultez, but through an examination of Western American literature. He asserted that the West was changed by nuclear arsenals, proving grounds, and disposal sites as reflected "metaphorically and literally" in the works of writers such as Cormac McCarthy, Don DeLillo, and Leslie Marmon Silko (Beck 2009, 4). The interpretation of "dirty wars" in the context of the Cold War's militarized West (due to nuclear weapons development) places in opposition what is "clean or just" versus what is "dirty or excessive." Most obviously, the disruption of the West's landscape was dirty and excessive (Beck 2009, 14).

Toxic Sites

Brownfields and many landfills are toxic sites. Sanitary landfills in the late 1930s and 1940s went from being sites for organic waste, rubbish, and junk, to including a variety of toxic chemicals, hospital wastes, and even radioactive materials. Landfills of various types produced methane gas and leachate that often made its way to sources of water. What have been defined as toxic or hazardous sites, however, have expanded greatly since the late twentieth century under legislation such as the Superfund law. Toxic and hazardous sites in the United States, while including brownfields and landfills, also have come to be associated with chemical wastescapes such as Love Canal and nuclear waste sites.

Love Canal, an incomplete canal dug near Niagara Falls in New York State, was the burial site for tons of hazardous materials from the manufacture of pesticides and other chemicals between the early 1940s and 1953. In the 1970s, however, residents in the area complained about the odors and ultimately about the severe health risks coming from the site. Love Canal became a *cause célèbre* for the growing toxic and hazardous waste controversy that arose at the time. Literature on Love Canal and its aftermath has been substantial, but little has focused on the larger picture of the site as a wastescape. In the most recent study, *Love Canal: A Toxic History from Colonial Times to the Present*, historian Richard Newman devotes modest, but useful attention to the Love Canal landscape: "By the late 20th century, Love Canal became associated with a different definition of 'landscape.' Now it was also a place where people had fundamentally altered the environment via human landmarks, industrial development, and/or daily living arrangements" (Newman 2016, 11). Connecting and melding the hazardous industrial site with residential neighborhoods, infrastructure, and commercial establishments, Newman broadened our understanding of Love Canal as a wastescape—one that encompassed not just a detached or distant disposal site, but a community. Unfortunately, that discussion was fleeting.

Fresh Kills and Landfilling

Probably the most recognizable form of wastescape is the landfill. The landfill is a tangible reminder of human habits and societal behaviors caught between material wants and valueless remnants. Why should we care what a landfill tells us about our history? Most obviously, consumption and waste are inextricably connected, but that relationship is complex and broadly significant. Consumption of goods rarely, if ever, results in nothing left over. Material goods generate residue, which must be reused in some form or disposed of in some way. Historically, the dilemma of consuming needs to be understood as a triangular relationship among consumption of goods, creation of waste, and disposal (often meaning place).

Understanding consumption patterns is an important starting point. Like Europe, the United States was confronted with huge amounts of refuse in its industrial era with two distinct dimensions. One was tethered to the physical distress caused by overcrowding, poor sanitation, and primitive methods of collection and disposal; the other was linked to the rising affluence of the middle class, abundance of resources, and consumerism, which continued into America's postindustrial era. Over time, the goods produced in American factories and imported from abroad would become more diverse and technically more intricate, and the discards more plentiful and complex.

Anthropologists and archaeologists have played a significant role in recognizing the need to understand the places where landfills arise. In *Archaeologies of Waste*, several scholars examine the value and social practices related to waste, but also the importance of space and time. Archaeologist Sabine Wolfram asserted that "we have to know about place/location first before we can understand space, i.e., how space was used." For example, this suggests that the archaeologist in studying a landfill must understand "the temporalities" the feature may possess (such as original function) and where the waste materials originated (Sosna and Brunclikova 2017, 124).

Fresh Kills: A History of Consuming and Discarding in New York City (Melosi 2020) centers on the mundane, yet acute problem of solid waste disposal in the modern metropolis. New York City's story is told through the history of a monumental structure, the 2,200-acre Fresh Kills Landfill on Staten Island, which is one of the largest human-engineered formations in the world. The landfill was opened in 1948 and closed in 2001. Fresh Kills reopened briefly in late 2001 through June 2002 to provide a receiving point for human remains and building rubble from the destroyed Twin Towers of the September 11 attack. What had been a notorious disposal site of unwanted trash, now also became a cemetery for the remains of those who lost their lives in the nation's worst terrorist tragedy. Today the landfill is the heart of a mammoth reclamation project to construct an expansive parkland known as Freshkills Park, three times the size of Central Park.

The backstory of Staten Island before the landfill also is essential to understanding Fresh Kills as a wastescape. The land upon which it was built was a salt marsh utilized for many generations by Lenape Indians and European immigrants for growing salt hay, and for hunting and fishing. About 100 years before the landfill opened, the city placed a quarantine station and a supporting medical complex on Staten Island to distance infected sailors and others from New York's main population. In World War I, the city sited an unwanted waste disposal plant on the very spot where the landfill would be built several decades later. Staten Island, therefore, had long been viewed as a place to deposit the infirm or store waste.

The overall history of the place is one of transition and various forms of exploitation. New York City's hunt for an ultimate disposal practice led it to a variety of options over the years, including ocean dumping, incineration, recycling, and landfilling. In all cases, timing, circumstance, impulse, and accident played a role. Sanitation had always been and will always be political, and politics as much as anything else led to Fresh Kills. The landfill also is a supreme example of massive environmental transformation with serious consequences extending beyond the borders of Staten Island, that is, a transformed landscape and social scape impacting the whole city and beyond. The space which the landfill occupies was transformed over the years from an apparent marginal landscape of salt marsh into a wastescape, a cemetery, and then into an ecoscape or reimagined park space.

Sites like Fresh Kills Landfill are spaces within spaces, trash footprints, and areas of waste displacement. Even as a landfill it faced constant change because of the ceaseless need to deal not only with great volumes of refuse, but with a waste stream frequently altered in composition, and one presenting an array of environmental challenges.

Landfills operate as long-term storage facilities, but also consumption junctions, where the remnants of consuming merely find a way station, not a final resting place. They are part of a provisional process because they frequently do not remain a landfill forever. The landfill is characterized as an infrastructure having a mediating role between nature and city in dealing with waste, but its permanence or impermanence as a land use is decided by a variety of objectives, desires, and future aspirations for the space itself. Landfills like Fresh Kills also are archives of material and memories—collectively the accumulation of real and imagined pieces of, in this case, every borough in New York City.

Landfills as disposal sites and as wastescapes demonstrate that place matters. Heather Rogers hits on a critical point:

> There's a reason landfills are tucked away on the edge of town, in otherwise, un-traveled terrain, camouflaged by hydroseeded, neatly tiered slopes. If people saw what happened to their waste, lived with the stench, witnessed the scale of destruction, they might start asking difficult questions.
>
> *(Rogers 2005, 16)*

But Rogers' interpretation of a landfill's location is only part of a bigger issue about landfills in the landscape. As a highly engineered space, it bears the mark of a building and design tradition which more than hints at the lengths that city leaders go to dispose of the city's waste. Landfills, incinerators, and reduction plants are elaborate and intricate capital-intensive technologies created to hide, destroy, or remake wastes with little concern about how that waste came to be. They are marvelous inventions, but not developed to stem the tide of consumption, only reminders that the residuals from consumption merely need to be tolerated. In many ways, recovering value from landfills required turning the wastescape into something else. As Engler optimistically stated, turning it into a park and golf course, "the dump was redeemed, the repugnance covered, and the bad memory erased" (Engler 2004, 94).

In the case of Fresh Kills, at least, the recovery value of constructing a park neither eliminated the memory of the landfill, nor restored the marshland's ecological integrity. The specters of 9/11 continued to haunt it as well. Fresh Kills as a landfill was one particular iteration for that particular piece of Staten Island real estate, but it hardly seemed so at the time. The dump would always be the dump, but the site's meaning and value would not so much change as be modified by the powerful forces of history represented by the Twin Tower disaster and the building of Freshkills Park. The landfill was never really an isolated space, hiding the discards of a great city. Its recognition grew out of its relationship not simply to Staten Island, but to Greater New York and the surrounding area—and also to the world of waste and wastefulness. Staten Island paid the price for its location, but eventually the forgotten borough could reap the benefit of the promised Freshkills Park. The experience at Fresh Kills, of course, is only one example of how a landfill might only be a phase in the evolution of a particular landscape. The transience of the space is an important element in understanding this type of wastescape.

In exploring landfilling in West Germany, historian Heike Weber points to the long-term utilitarian aspect of landfills. They were, she argued, a means of urbanizing nature for human use, and valued for ultimately offering landscaping opportunities and reclamation to provide, among other things, recreational green spaces (Weber 2019, 261). Until science proved otherwise, places like marshes had been viewed as worthless land made valuable by development. The justification for the siting of landfills in marshes was to construct something useful that might someday be reclaimed for "productive" purposes. There is a long history of converting landfills into parkland, recreational areas, and other useable space. Conversion of landfills into parks goes back at least to 1916 in the United States when Seattle built Rainier Playfield on Rainier Dump. While no one has counted all of the parks and recreational sites built on old landfills, the number could be anywhere from 250 to 1,000 or more (Melosi 2020, 493).

Landfills provide much insight about the nature of waste, wastefulness, and wastescapes. However, book-length treatments of important landfills are few. Beyond *Fresh Kills* and Robert Sullivan's *The Meadowlands* (1998) there is a great deal of opportunity for more such studies. Brownfields as a historical topic also needs more historical analysis, since most of what is available is largely technical in nature. Most brownfields are associated with derelict industrial facilities, but also vehicle repair shops, gas stations, and dry cleaners. The Environmental Protection Agency defined these sites as "real property, the expansion, redevelopment, or reuse of which may be complicated by the presence of a hazardous substance, pollutant, or contaminant" (DeSousa and Spiess 2019). There may be as many as 450,000 brownfield sites in the United States, not to mention greyfield areas, such as decaying regional shopping centers and malls.

Concluding Thoughts

As the various books under review suggest, wastescapes vary in type, scale, and impact. Some are transient, some are not. From a material standpoint they can be strikingly ominous. The notorious dump in Corona, Queens composed of furnace ash, cinder, and household waste, was described in F. Scott Fitzgerald's 1925 classic *The Great Gatsby* as "a fantastic farm where ashes grow like wheat into ridges and hills and grotesque gardens" (Fitzgerald 1986, 23). Toxic chemicals can hide below the surface at places like Love Canal, and atomic test sites can exist as "hidden realities." Whether resting on the surface, buried below ground, or invisible to the eye, wastescapes are landscape artefacts often holding a wide variety of meanings.

Given the variety of wastescapes, future research needs to focus more comprehensively on the collective forms, their common features, and their differences. Several of the studies discussed suggest how wastescapes transform particular landscapes, but historians and geographers in particular need to expand the research on placing wastescapes in longer timelines, which may help to understand them in the context of something like Fernand Braudel's *longue durée*.

References

Beck, J. 2009. *Dirty Wars: Landscape, Power, and Waste in Western American Literature*. Lincoln: University of Nebraska Press.

Berger, A. 2006. *Drosscape: Wasting Land in Urban America*. New York: Princeton Architecture Press.

Dalzero, S. 2016. "Rejected Landscapes-Recycled Landscapes." *Procedia Engineering* 161: 201–206.

DeSousa, C., and T. Spiess. 2019. "Brownfields." *Oxford Bibliographies*. www.oxfordbibliographies.com/view/document/obo-9780199874002/obo-9780199874002-0048.xml.

Engler, M. 2004. *Designing America's Waste Landscapes*. Baltimore: Johns Hopkins University Press.

Fitzgerald, F.S. 1986. *The Great Gatsby*. New York: Scribners.

Hanson, D.T. 1997. *Waste Land: Meditations on a Ravaged Landscape*. New York: Aperture.

Hetherington, K. 2004. "Secondhandedness: Consumption, Disposal, and Absent Presence." *Environment and Planning D: Society and Space* 22: 157–173.

Kennedy, G. 2007. *An Ontology of Trash: The Disposable and Its Problematic Nature*. Albany: State University of New York Press.

Kuletz, V.L. 1998. *The Tainted Desert: Environmental and Social Ruin in the American West*. New York: Routledge.

Lerner, S. 2010. *Sacrifice Zones: The Front Lines of Toxic Chemical Exposure in the United States*. Cambridge: MIT Press.

Melosi, M.V. 2020. *Fresh Kills: A History of Consuming and Discarding in New York City*. New York: Columbia University Press.

———. 2005. *Garbage in the Cities: Refuse, Reform, and the Environment*. Pittsburgh: University of Pittsburgh Press.

Newman, R.S. 2016. *Love Canal: A Toxic History from Colonial Times to the Present*. New York: Oxford University Press.

Nye, D.E., and S. Elkind, eds. 2014. *The Anti-Landscape*. Amsterdam: Brill-Rodopi.

Offenhuber, D. 2017. *Waste Is Information: Infrastructure Legibility and Governance*. Cambridge: MIT Press.

Rogers, H. 2005. *Gone Tomorrow: The Hidden Life of Garbage*. New York: The New Press.

Scanlan, J. 2005. *On Garbage*. London: Reaktion Books.

Sosna, D., and L. Brunclikova, eds. 2017. *Archaeologies of Waste: Encounters with the Unwanted*. Oxford: Oxbow Books.

Sullivan, R. 1998. *The Meadowlands: Wilderness Adventures at the Edge of a City*. New York: Anchor Books.

Thompson, M. 2017. *Rubbish Theory: The Creation and Destruction of Value*. London: Pluto Press.

Weber, H. 2019. "20th Century Wastescapes: Cities, Consumers, and Their Dumping Grounds." In *Urbanizing Nature: Actors and Agency (Dis)Connecting Cities and Nature Since 1500,* edited by T. Soens et al., 261–289. New York: Routledge.

5

PUBLIC LANDSCAPES

Randall K. Wilson and Yolonda Youngs

The division between private property and public land constitutes a defining characteristic in the history and evolution of the American landscape. Whereas private property is typically held by individuals that claim exclusive use and access, public lands tend to be collectively owned and managed by governmental bodies for the public good. While such lands may be managed by state or municipal governments for a multitude of purposes, in this chapter, we focus on federal public lands managed primarily for purposes of environmental protection.

Defined in this way, the public land system is comprised of many different types of public lands, ranging from national parks and monuments, to national forests, national wildlife refuges, and more (Figure 5.1). Each type is managed by different federal agencies located in various departments within the executive branch of the national government (Table 5.1).

Experiencing the public/private divide on the American landscape can be a fairly straightforward affair, marked by a national park entrance or signage identifying the boundary of a national forest, monument, or wildlife refuge (Figure 5.2). Such markers suggest a clear separation between public and private. A glance at a map can strengthen this impression, with public lands bounded by solid lines that seemingly demarcate a stark border with private property.

A closer look, however, reveals two important truths. First, the public/private divide exists not only on the physical landscape, but also on a conceptual level, marking differences in social, political, and economic values that have long served as subjects of debate in American society (Wilson 2020). While it is true that public lands, and especially national parks, frequently serve as symbols of shared national values, they can also serve as focal points for political discord and even violence (Walker 2018).

Secondly, the public/private boundary is much more porous than it may first appear. This idea is perhaps most profoundly experienced on the landscape itself insofar as the vast majority of public lands are not fenced in. Water, air, and wildlife of all kinds—including human visitors—may pass freely back and forth across the public/private divide. While it is true that automobiles on major roadways must stop at official entrances of some national parks and wildlife refuges, one can always just wander in on foot just about anywhere else along the border. And national forests, Bureau of Land Management lands, and other types of federal lands generally lack monitored entrances of any kind.

DOI: 10.4324/9781003121800-7

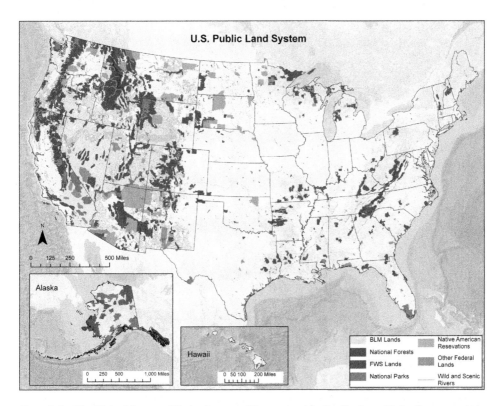

Figure 5.1 The United States public land system. The geographic distribution of federally managed lands illustrates a profound western regional bias in the public lands system. Map by Natalie Kisak.

Table 5.1 Federal Land Management Agencies

Agency	Public Land Unit/Type	Federal Department
National Park Service	National Parks, National Historic Sites, Battlefields, Recreation Areas, Sea and Lake Shores, etc.	Dept. of the Interior
National Forest Service	National Forests, Grasslands	Dept. of Agriculture
US Fish & Wildlife Service	National Wildlife Refuges	Dept. of the Interior
Bureau of Land Management	BLM (Public) Lands	Dept. of the Interior
All agencies (to various degrees)	Wilderness Areas, National Monuments, National Scenic Trails, National Wild & Scenic Rivers	Varies

This physical permeability also plays out conceptually. The guiding mantra of resource management on many public lands is multiple use. Under this approach (discussed further below), individuals can use natural resources on public lands for private economic gain. Activities like logging and mining are overseen by federal managers, but the end result is a hybrid public/ private model. Questions remain over how to balance resource use with protection, as well as the degree to which public lands should be managed to benefit private interests versus the public good.

Figure 5.2 The public/private divide is especially pronounced at entry points such as this one just outside Laramie, Wyoming. Photo by Geoff Buckley.

In this essay, we explore these issues through the lens of three major and interrelated challenges currently facing the public land system in the United States. They include the rise of ecosystem-based collaborative resource management, the issue of public land access and visitation, and the impacts of climate change.

In recent years, collaboration has provided federal officials with a tool for working across jurisdictional boundaries to manage resources at an ecosystem scale. But how successful have such efforts been? Meanwhile, questions over visitation and access to public lands have ignited impassioned movements for change across the political spectrum. Finally, how should federal managers address climate change dynamics that threaten the very flora and fauna for which many federal public lands were established to protect?

Before proceeding, it is worth considering two additional insights from the map of US federal lands. First, given the high value American society places on private property, the fact that approximately one-third of US land is publicly owned underscores the fact that the public/private divide reflects ongoing national discussions over competing values. Second, the uneven geography of public lands on the landscape, favoring the American West, reminds us of the colonial past that colors the US environmental movement. Therefore, before examining the three challenges identified above, we begin with a brief history of the US public lands system to better contextualize its political, economic, and social evolution.

History and Evolution of the US Public Land System

The starting point for understanding public lands in the United States is the US Constitution. Specifically, Article 4, Section 3 established federal authority over all territory within the nation's borders, including all lands previously claimed by the original thirteen colonies and all future acquisitions. Under this authority, Congress used the public domain to carve out new states, but land within state boundaries did not transfer to state governments. Instead, the federal government retained sovereignty, including the right to sell the land at auction, designate areas as military forts, establish Native American reservations, or provide land grants to states, private industry, or even military veterans as payment for their service (Gates 1968).

Designating portions of the public domain for environmental protection purposes would not emerge until the late nineteenth century, largely in response to the excesses of unregulated resource use. The decades-long transformation of the North American landscape via Euro-American settlement, urbanization, and resource development, generated great wealth for certain sectors of American society. But it also fueled an ideological backlash. The rise of American Romanticism articulated an alternative vision for the relationship between society and nature (Nash 2014). Flipping the dominant narrative on its head, Romantics contended that it was wilderness—not human progress or civilization—which offered the most direct path to moral and spiritual well-being. Rather than exploit nature as a commodity, they argued, wild lands should be respected, celebrated, and preserved.

Acting in part on these ideals, Congress began to designate portions of the public domain for environmental protection. Initially, this was done by transferring land to individual states as state parks, as was the case for Yosemite, ceded to California in 1864. In contrast, the first *national* park, Yellowstone, was established in 1872, but not necessarily by design, nor due solely to the influence of Romantic ideals. The case for establishing Yellowstone, much like Yosemite, hinged on several somewhat contradictory arguments. On one hand, protection was intended to avoid the embarrassing commercial exploitation that had denigrated Niagara Falls. On the other hand, proponents also viewed the park as an opportunity to promote the economic fortunes of the Northern Pacific Railroad and Montana Territory (Haines 1996). Because Wyoming, Montana, and Idaho were not yet states, Congress designated Yellowstone as a national, rather than a state park, thereby setting a precedent for a new form of federal lands that would eventually become a system of national parks.

In 1891, Congress established a new form of public land with passage of the General Revision Act. A last-minute amendment to this law gave the president power to create national forest reserves from the public domain. In 1905, with the establishment of the US Forest Service (USFS) in the Department of Agriculture under the leadership of Gifford Pinchot, these forest reserves evolved into a system of national forests with their own managing agency (Steen 1992).

Notably, the management of national forests followed the principles of progressive conservation (Hays 1959). In contrast to the ideals of Romanticism preservation, which held sway in national parks, this approach advocated for continued resource development through the idea of multiple use. Guided by scientific rationality, managers would extract resources such as timber, water, or minerals in a sustainable manner for the benefit of society writ large. The famous oft-repeated phrase "the greatest good to the greatest number in the long run" summarized what would serve as the guiding managing principle for most types of federal lands.

In contrast, management priorities for national parks emphasized environmental preservation and human recreation opportunities. In 1916, the National Park Service Act formally established the National Park Service (NPS) in the Department of Interior as the agency dedicated to managing the growing system of national parks. Significantly, the law ordered managers to preserve the land, historic objects, and wildlife in "unimpaired" form, while at the same time promoting tourism and recreation (Dilsaver 1997). The tensions embedded in this contradictory mandate continue to challenge park managers today.

As the twentieth century progressed, new forms of public lands emerged upon the American landscape. The growing systems of national parks and national forests were joined by national monuments, a system of national wildlife refuges managed by the US Fish and Wildlife Service (USFWS), and lands managed by the Bureau of Land Management (BLM) (Skillen 2009). In the 1960s, the list expanded with the addition of wilderness areas, national wild rivers, and scenic and historic trails.

As noted above, the lion's share of US public lands lies in the western half of the continental United States and Alaska, mostly due to the fact that much of the territory lying east of the Rocky Mountains was privately owned at the time of designation. Generally speaking, national parks and forests were carved out of unclaimed land in the public domain. However, in numerous instances, these "unclaimed" lands were in fact claimed by Indigenous peoples who were forcibly removed to Indian Reservations (Spence 1999). In the eastern US, most federal lands resulted from property donated by private owners or acquired by state governments for that purpose.

Today, efforts to balance out multiple land uses—especially those that are incompatible (i.e., mining, logging, hunting)—frequently serve as points of conflict on public lands. Meanwhile, public lands *not* managed by multiple use ideals (i.e., wilderness areas, most national parks, some national monuments), struggle with their own issues of balancing recreational visitation with environmental and historical protection. The historical foundation of US public lands continues to play a significant role in these challenges. Since the beginning, managing public lands has required compromise between the competing interests of economic development and environmental protection; between viewing nature as a commodity, distinct and separate from human society, and as something with inherent or cultural value. Below, we illustrate the tangible legacy of this ongoing push and pull within the public/private divide.

Issue One—Ecosystem-based Collaborative Resource Management

As noted above, the boundaries of many federal public lands do not necessarily align with ecosystems or watersheds. National parks and forests established in the nineteenth and early twentieth century often reflected political alignments rather than natural landscape features (Wilkinson 1992). But as early as the 1880s, conservation advocates realized that in places like Yellowstone, the bison and elk that migrated out of the park in winter became vulnerable to hunters. Put simply, Yellowstone's boundaries weren't sufficient for saving the very wildlife species it was intended to preserve.

By the 1980s, a growing number of federal resource management agencies advocated ecosystem-scale management, often defined in terms of major watersheds or wildlife migration routes. But a decade later, just as managers were starting to build partnerships across jurisdictional boundaries, new challenges arose. As growing numbers of amenity migrants moved into rural western communities, they often brought with them environmental values that clashed with older residents' dependence on resource extraction (Travis 2007). The intensification of public land disputes helped usher in new ideas for resource management that included conflict resolution. Collaborative resource management represented one approach designed to overcome local conflict while addressing the need for cross-boundary ecosystem management (Wondolleck and Yaffee 2000).

The most effective forms of collaborative management have both "horizontal" and "vertical" dimensions, reflecting cooperation among agencies operating at the same level of government *and* among those functioning at different scales: from federal to state to local. Insofar as many large-scale ecosystems encompass both public and private land, including in some cases, entire towns or cities, it is paramount to include the voices of varied interests (or "stakeholders") at the table.

According to proponents of collaboration, the goal is to encourage open dialogue to identify common management priorities (Daniels and Walter 2001). When successful, collaboration reduces conflict and generates local support for the implementation of resource management plans. However, when it fails, collaboration carries the potential for local interests to override national laws and standards (Coggins 2001). Nonetheless, in recent years, most federal land agencies have incorporated collaboration into formal management protocols.

One of the most famous early examples of ecosystem-scale collaboration is in the Greater Yellowstone Ecosystem. In 1964, the Greater Yellowstone Coordinating Committee instigated a collaboration between the NPS and USFS, later expanded to include regional BLM and USFWS units. Efforts to develop an ecosystem-based planning vision in the 1990s failed in large part due to the lack of involvement by other stakeholders outside of federal agencies (i.e., the absence of vertical collaboration). But in recent decades, progress has been made in developing ecosystem-wide plans for specific issues, including wildfire, grizzly bear protection, chronic wasting disease in elk, and brucellosis in bison (Keiter 2020).

At the grass-roots level, the Quincy Library Group in California's Sierra Nevada Mountains is one of the most famous examples (Marston 2001). After decades of heavy logging, new national regulations in the early 1990s called for drastic reductions in timber production in favor of environmental protection. In response, local residents sought a compromise that might save local logging jobs. The resultant forest plan re-opened some second-growth areas to logging, employed local loggers in forest thinning projects, but continued to protect roadless areas. Called the Community Stability Plan, it impacted over 1.5 million acres across three different national forests. The effort was controversial to say the least. Although favored by many locals, the plan was criticized for weakening environmental protection by USFS officials and leaders of national environmental organizations. In the end, the Community Stability Plan was adopted after congressional action forced the Forest Service to accept it. The Quincy case thus illustrates both sides of the collaboration coin. It demonstrates the potential for collaboration as a grass-roots mechanism for compromise across an ecosystem scale, while at the same time showing the danger it poses for weakening national regulatory standards.

More recent ecosystem-scale efforts have taken root in the Chesapeake Bay Ecosystem via the Chesapeake Bay Program; in the Greater Everglades Ecosystem via the federally mandated South Florida Ecosystem Restoration Task Force; and in the Grand Canyon region through the work of the Grand Canyon Trust. While such efforts have struggled at times to incorporate the voices of all stakeholders and to achieve intended goals, collaborative approaches appear to offer a key tool for advancing and implementing ecosystem-scale management plans for public lands and resources.

Issue Two—Public Lands for Whom?

Access and visitation are another set of major challenges facing the public lands system in the United States. Public lands are often touted as natural laboratories for physical scientists and scholars. But they are also complex social-environmental systems shaped by contested histories, land ownership legacies, political geographies, and uneven visitor access and use (Dilsaver 2009).

Although early conservation movements that fueled public lands designation and protection in the United States emphasized democratic and egalitarian ideals, some observers challenge this origin story by asking who these lands serve and how they were acquired into the public domain. We will focus on three issues shaping contemporary scholarship on this issue: environmental and social justice, overcrowding, and attempts to privatize the public domain.

Environmental and social justice issues are arguably the most pressing challenges facing the social and cultural scene of federal public lands management, interpretation, and use. There is growing evidence that this system may not equally serve a diverse American public composed of different race, class, ethnic, gender identity, age, and ability backgrounds (Germic 2001; Weber and Sultana 2013; Flores et al. 2018). Geographers and other public lands scholars suggest several concerns on this theme. First, the geographic setting of many federally owned and managed public lands is often in rural locations that are far from urban population centers such as Yellowstone NP. The economic costs of entrance fees and lost wages for lower income families to take extended vacation days may discourage visitors. Public and affordable transportation inside national forests, wildlife refuges, and park sites is often minimal or lacks a connection to larger transportation networks (Youngs, White, and Wodrich 2008; Xiao et al. 2018). Even when public lands agencies maintain large recreation units in or near cities—such as the Santa Monica Mountains National Recreation Area (NPS) and San Gabriel Mountains National Monument (USFS) near Los Angeles, California—other factors, including perceptions about who the parks were intended to serve, can negatively affect visitor diversity.

There is a growing body of scholarship that calls attention to the history of underrepresented use, access, and discrimination for Native Americans, African Americans, Asian Americans, and Hispanic peoples in public lands. Some scholars trace this challenge to the history of Native American land dispossession and forced migration in the late nineteenth and early twentieth century (Spence 1999). Others draw from the environmental justice movement to explore how legacies of slavery, Jim Crow segregation, and racial violence shape African American experiences in outdoor environments (O'Brien and Njambi 2012; Finney 2014). These historical contexts may influence contemporary spatial patterns of minority peoples' public lands use. For example, scholars note that African American visitation to national forests is low, even in the American South where high concentrations of this population group live nearby USFS sites (Johnson et al. 2007).

New collaborations between scholars, communities, and public lands managers are creating programs to address minority visitation use and access challenges. Role models in outdoor recreation pursuits—as members of the public lands staff and scientific workforce or as leaders of philanthropic organizations that support protected lands—can lead to increased engagement, use, and environmental stewardship that is essential for the future of America's public lands (Mills 2014; Sene-Harper et al. 2021). Innovative programs such as Outdoor Afro connects African Americans with public lands. Latino Outdoors builds access opportunities and a sense of environmental belonging for America's fastest growing minority group. Environmental organizations and agency partnerships are also being reimagined. For example, Bears Ears National Monument in Utah was created in 2016 as a jointly administered unit of the BLM and the USFS. An Intertribal Coalition of leaders from the Hopi Tribe, Navajo Nation, Ute Mountain Ute Tribe, Pueblo of Zuni, and Utah Indian Tribe advocated for the creation of the monument and developed a land management plan for the site.

A second challenge facing many federal public lands is overcrowding and environmental impacts of visitor use. During the COVID-19 pandemic in 2020, for example, many public lands saw new levels and spatial patterns of visitation as Americans sought socially distanced outdoor recreation. Some locations—such as Cuyahoga Valley National Park—experienced distinct surges in visitation (NPS 2021). Pandemic-related health and travel restrictions resulted in distinct visitor use patterns such as a sharp decrease in group and bus tours, reduced international travelers, park closures, and limited services and facilities. The pandemic also brought increasing pressure on wilderness and backcountry areas. Adventure outdoor recreationalists such as rock climbing, backcountry skiing, and river rafting are gaining popularity, but visitors are often unprepared for the rigors of these activities in rugged outdoor situations (Youngs 2020). Search and rescue teams struggled to keep pace with increasing visitor injuries and fatalities, already stretched thin by diminished budgets and limited staff during the pandemic.

Issues of overcrowding and resource degradation are shaped by several factors including the mission of the managing agency, approaches to visitor use management, and the geographical imagination of visitors. For example, lands managed by the USFS are focused on multiple use strategies, conservation, natural resource extraction, and motorized as well as non-motorized recreation. A very different approach guides the National Park Service. The dual mission of the agency focuses only on the preservation of natural and cultural resources and recreation. Recreation in national parks is often non-motorized (e.g., hiking) or passive (e.g., scenic driving or wildlife viewing). Activities such as hunting and extractive industries are discouraged or illegal. Popular imagery is another factor that has a long history of influencing visitor use and impact. At Grand Canyon NP, for example, early twentieth-century postcard manufacturers created scenes that emphasized certain locations, colors, subjects, and activities in a select set of locations along the canyon's south rim and central trails while overlooking other sites (Youngs 2012). This process carries forward to today as landscapes emphasized in photographs and documentary films or geo-tagged in image-based social media fuel popular geographic imaginations and propel visitation to newly iconic sites on public lands.

Public lands managers across all agencies mitigate the impact of overcrowding through visitor use guidelines informed by environmental science. For example, the concept of carrying capacity—how much recreational and visitor use public lands can accommodate while also maintaining the protection or conservation mission—is key. The NPS developed a Visitor Experience and Resource Protection (VERP) framework that relies on indicators and standards that balance natural and cultural resource protection with visitor experience quality (Manning 2001). You might see this policy in the landscapes you encounter as the size and location of campgrounds, hiking trail closures to protect fragile environments, or the use of hiking permits that track visitor use.

A third and very different challenge regarding public land access and visitation is reflected in decades-long efforts to privatize public lands. Rooted in the Sagebrush Rebellions of the 1940s and 1970s, advocates call for the transfer of public lands from federal to state control in order to sell them to private individuals. Driven by reactionary political ideologies and economic concerns marked by the decline of resource extraction industries, these efforts mistakenly view regulatory restrictions on public land access as an affront to their constitutional rights. The failed effort of armed militants to take possession of the Malheur National Wildlife Refuge in Oregon in 2016 exemplifies these views (Walker 2018).

Issue Three—Impacts of Climate Change

Global climate change is another major challenge facing America's public lands. Scholars recognize three areas of concern here. First, as ecosystem regimes shift with climate change, public lands of the future will be different from the ones managed in the past. Public lands managers must adapt to these changing conditions with new approaches, tools, and strategies that embrace resistance, resilience, and transformation (Clifford et al. 2020).

Second, the parks, forests, and open spaces that make up the public lands system are vital environmental reservoirs and carbon sinks that can reduce the effects of greenhouse gases in the atmosphere. Scholars, scientists, and land managers assert that expanding the network of protected areas is a key element of conserving the diversity of flora and fauna by providing climatic refuges for species movement. Higher elevation parks and protected areas can provide a haven as lower elevation habitats diminish due to warming temperatures. Today these climatic changes and their effects are detected in episodes such as increasing size and frequency of wildfires in California, melting glaciers in Glacier and Grand Teton national parks, tree mortality, loss of bird species, and sea level rise.

Third, new levels of water scarcity—or abundance—influenced by climate change may imperil public lands in desert settings as well as coastal zones. Parks and protected areas located in the Southwest will be challenged to maintain riparian areas, streams, and other water resources as extended drought conditions persist. Conversely, coastal public lands may face rising sea levels that dramatically alter or submerge these places (Keiter 2013). The importance of ecosystem-based collaborative management schemes that work effectively across multiple agencies and land ownership boundaries carry a vital importance moving into the future as climate change shifts ecological boundaries, species distribution, and habitat availability.

Concluding Thoughts

Four main lessons emerge from this discussion of public lands in the United States. First, public lands protection is not immutable. Public land borders, both conceptually and physically, are porous and dynamic. The historical evolution of the nation's public land system illustrates how the meaning, boundaries, and levels of protection can change over time in response to economic and political conditions, land exchanges, or ecosystem scale drivers such as climate change. Second, to be most effective, public lands management increasingly requires transboundary, collaborative partnerships organized upon an ecosystem scale. Such approaches involve engagement with diverse stakeholders who may perceive, value, and use nature in different ways. Third, public lands are also social and cultural landscapes that are intended as resources for all members of society. They must be managed in a way that guards against over-use and efforts to undermine their protection for private gain, yet also provides ready access to everyone, including historically marginalized groups. Finally, climate change arguably represents the greatest challenge to the public land system. As seasonal temperature and precipitation patterns change, landscapes and species for which certain parks or refuges were established to protect may no longer exist. Nonetheless, public lands also offer reservoirs of hope as carbon sinks, venues for renewable energy production or sites of public education. In each of these ways, public lands will continue to hold a unique and significant place in the American landscape.

References

Clifford, K., L. Yung, W.R. Travel, R. Rondeau, B. Neely, I. Rangwala, N. Burkardt, and C. Wyborn. 2020. "Navigating Climate Adaptation on Public Lands: How Views on Ecosystem Change and Scale Interact with Management Approaches." *Environmental Management* 66 (1): 614–628.
Coggins, G.C. 2001. "Of Californicators, Quislings, and Crazies: Some Perils of Devolved Collaboration." *Chronicle of Community* 2: 27–33.
Daniels, S.E., and G.B. Walker. 2001. *Working through Environmental Conflict: The Collaborative Learning Approach*. Westport, CT: Praeger.

Dilsaver, L.M. 2009. "Research Perspectives on National Parks." *The Geographical Review* 99 (2): 268–278.

———. 1997. *America's National Park System: The Critical Documents*. Lanham, MD: Rowman & Littlefield Publishers.

Finney, C. 2014. *Black Faces, White Spaces: Reimagining the Relationship of African American to the Great Outdoors*. Chapel Hill: The University of North Carolina Press.

Flores, D., G. Falco, N.S. Roberts, and F.P. Valenzuela III. 2018. "Recreation Equity: Is the Forest Service Serving Its Diverse Publics?" *Journal of Forestry* 116 (3): 266–272. https://doi.org/10.1093/jofore/fvx016.

Gates, P.W. 1968. *History of Public Land Law Development*. Washington, DC: Government Publishing Office.

Germic, S.A. 2001. *American Green: Class, Crisis, and the Deployment of Nature in Central Park, Yosemite, and Yellowstone*. Lanham, MD: Lexington Books.

Haines, A.L. 1996. *The Yellowstone Story*. Vol. 1, Rev. ed. Boulder: University Press of Colorado.

Hays, S.P. 1959. *Conservation and the Gospel of Effienciency: The Progressive Conservtion Movement*. Albuquerque: University of New Mexico Press.

Johnson, C.Y., J. Bowker, G. Green, and H. Cordell. 2007. "Provide it… But Will They Come?" A Look at African American and Hispanic Visits to Federal Recreation Areas. *Journal of Forestry* 105 (5): 257–265.

Keiter, R. 2013. *To Conserve Unimpaired: The Evolution of the National Park Idea*. Washington, DC. Island Press.

———. 2020. "The Greater Yellowstone Ecosystem Revisited: Law, Science, and the Pursuit of Ecosystem Management in an Iconic Landscape." *University of Colorado Law Review* 91 (1): 1–181.

Manning, R. 2001. "How Much is Too Much? Carrying Capacity of National Parks and Protected Areas." In Proceedings of the Conference on Monitoring and Management of Visitor Flows in Recreational and Protected Areas, Bodenkultur University, Vienna, 306–313.

Marston, E. 2001. "The Quincy Library Group: A Divisive Attempt at Peace." In *Across the Great Divide: Explorations in Collaborative Conservation and the American West*, edited by P. Brick, D. Snow, and S. Van de Wetering, 79–90. Washington, DC: Island Press.

Mills, J. 2014. *The Adventure Gap: Changing the Face of the Outdoors*. Seattle, WA: Mountaineers Books.

Nash, R.F. 2014. *Wilderness and the American Mind*, 5th ed. New Haven, CT: Yale University Press.

NPS (National Park Service). 2021. Social Science – Annual Visitation Highlights. www.nps.gov/subjects/socialscience/annual-visitation-highlights.htm. Last accessed April 12, 2021.

O'Brien, W.E., and W.N. Njambi. 2012. "Marginal Voices in 'Wild' America: Race, Ethnicity, Gender and 'Nature' in *The National Parks*." *The Journal of American Culture* 35 (1): 15–25.

Sene-Harper, A., M. Floyd, and A.S. Hicks. 2021. "Black Philanthropy and National Parks: Giving Green to Give Black." *Journal of Park and Recreation Administration* 39 (1) (online first) doi.org/10.18666/JPRA-2021-10666.

Skillen, J.R. 2009. *The Nation's Largest Landlord: The Bureau of Land Management in the American West*. Lawrence: University Press of Kansas.

Spence, M.D. 1999. *Dispossessing the Wilderness: Indian Removal and the Making of the National Parks*. New York: Oxford University Press.

Steen, H., ed. 1992. *The Origins of the National Forests*. Durham, NC: Forest History Society.

Travis, W.R. 2007. *New Geographies of the American West: Land Use and the Changing Patterns of Place*. Washington, DC: Island Press.

Walker, P. 2018. *Sagebrush Collaboration: How Harney County Defeated the Takeover of the Malheur Wildlife Refuge*. Corvallis: Oregon State University Press.

Weber, J., and S. Sultana. 2013. "Why Do So Few Minority People Visit National Parks? Visitation and the Accessibility of 'America's Best Idea.'" *Annals of the Association of American Geographers* 103 (3): 437–464.

Wilkinson, C.F. 1992. *Crossing the Next Meridian: Land, Water, and the Future of the West*. New York: Island Press.

Wilson, R.K. 2020. *America's Public Lands: From Yellowstone to Smokey Bear and Beyond*, 2nd ed. Lanham, MD: Rowman & Littlefield Publishers.

Wondolleck, J.M., and S.L. Yaffee. 2000. *Making Collaboration Work: Lessons in Innovation in Natural Resource Management*. Washington, DC: Island Press.

Xiao, X., L. Aultman-Hall, R. Manning, and B. Voigt. 2018. "The Impact of Spatial Accessibility and Perceived Barriers on Visitation to the US National Park System." *Journal of Transport Geography* 68: 205–214. https://doi.org/10.1016/j.jtrangeo.2018.03.012.

Youngs, Y. 2012. "Editing Nature in Grand Canyon National Park Postcards." *Geographical Review* 102 (4): 486–509.

———. 2020. "Danger Beyond This Point: Visual Representation, Cultural Landscapes, and the Geography of Environmental Hazards in U.S. National Parks." *GeoHumanities* 6 (2): 314–346. doi:10.1080/2373566X.2020.1784017.

Youngs, Y., D. White, and J. Wodrich. 2008. "Transportation Systems as Cultural Landscapes in National Parks: The Case of Yosemite." *Society and Natural Resources* 21 (9): 797–811.

6

AMERICAN LANDSCAPES OF ENVIRONMENTAL INJUSTICE

Ryan Holifield

Environmental injustice in the US refers to inequitable or unfair burdens borne by people of color, Indigenous Nations, and low-income communities with respect to environmental harms and benefits. When we think of *landscapes* of environmental injustice in this country, any number of images might come to mind. We might, for example, picture an impoverished and racially segregated inner-city neighborhood, surrounded by hazardous waste sites, polluting factories, and truck traffic, with limited access to green space. Or we might imagine contaminated rivers and degraded ecosystems in traditional tribal lands where mining has taken place or where oil and gas pipelines are proposed. We might think about a small community of immigrant farmworkers exposed to pesticide drift. Or we might picture an informal border settlement or neglected mobile home park lacking adequate water and sewer infrastructure.

In combination, images like these raise important questions about landscapes of environmental injustice. First, how have such landscapes come to be? What forces and factors generated these visible combinations of people and environments? Second, how do these landscapes vary in different parts of the country, and with what implications? Third, what does it mean to call them "landscapes of environmental injustice"? How do people—both those who live within them and those who do not—see these landscapes, talk about them, experience them, invest them with meaning, and represent them in political activism or artistic production?

Such questions, suggested by thinking specifically in terms of *landscapes* of environmental injustice, are in many ways distinctive from those addressed by the kind of environmental justice research that arguably dominates the field in the United States: quantitative analyses of spatial patterns. Quantitative studies, which have played key roles in mobilizing activism and providing the evidence base for policy, typically focus on unequal distributions of people and environmental characteristics in space. In contrast, scholarship on landscapes of environmental injustice investigates the tangible, visual, physical environments in which disproportionately burdened communities live, asking how these specific environments took shape; how they carry meaning for people who see and experience them; and how we might engage with them in order to rectify injustices.

This chapter briefly introduces and surveys this literature, addressing the production, representation, and remediation of unjust environmental landscapes. Despite the fuzziness and ambiguity both of "regions" in the US and of the divide between "urban" and "rural," I suggest that these geographic distinctions can help us understand variation in landscapes of environmental injustice in different parts of the country, as well as the forces that produce them. The processes that produce landscapes of environmental injustice are often connected closely with processes of regional differentiation and urbanization, and the problems facing communities in cities are often quite different from those facing rural populations. Strategies for addressing the country's diverse landscapes of environmental injustice vary, but many emphasize a common theme: making them *visible*, both for "insiders" who live within them and "outsiders" who can provide resources to help transform them. I conclude by proposing a conceptual framework and research agenda that connect the dimensions of production, representation, and remediation together in a way that is sensitive to geographic differentiation.

Landscapes of Environmental Injustice in the US

Thinking of the US and its landscapes as belonging to distinctive regions—and as "urban," "suburban," "exurban," or "rural"—is commonplace but, at least in academic discourse, controversial. While I aim to avoid reifying either the region or the urban/rural distinction, I use them as shorthand for historical and biophysical processes shaping landscapes of environmental injustice in distinctive ways. For example, although regional definitions like the "Rust Belt" or "Sun Belt" are fuzzy and debatable, they point to different historical and climatic processes producing landscapes of environmental injustice. Similarly, with respect to globalization and other important political, economic, and sociocultural dynamics, the landscapes of densely populated and heavily polluted inner-city neighborhoods are positioned differently from the more lightly inhabited landscapes of tribal reservations or small towns with predominantly agricultural or extractive economies.

Landscapes of Environmental Injustice in Cities

Although the processes that have produced landscapes of environmental injustice in US cities are widespread, they often have distinctive regional dimensions. For example, both the histories and present-day conditions of institutional discrimination and racial residential segregation vary regionally. Similarly, regions have diverged in their experiences of economic dynamics of industrialization and deindustrialization. By most measures, the most racially segregated cities in the US lie primarily in the Northeast and Midwest, in the loosely defined Rust Belt that formerly constituted the heart of the country's manufacturing and heavy industry (Logan 2013). Many of these cities were destinations for southern African Americans during the Great Migration from the World War I era to the end of the 1960s. Although such cities were already highly segregated at the turn of the twentieth century, the Great Migration intensified such patterns (Logan et al. 2015).

By the end of the 1970s, a decade in which the US passed many of its most important and stringent environmental laws, it became increasingly clear that African Americans and other people of color left behind in inner cities—both within and beyond the Rust Belt—were bearing the brunt of both ongoing industrial pollution and the contaminated legacies of gradual deindustrialization. For example, in Gary, Indiana, suburbanizing white residents found themselves increasingly able to escape the most toxic air pollution from steel manufacturing, while inner-city African American communities faced not only continued exposure to contamination but also steadily deteriorating neighborhood conditions (Hurley 1995). In many industrial cities, zoning ordinances and decisions played a central role in generating landscapes of inequitable exposure to pollution (Boone and Modarres 1999; Wilson et al. 2008).

Segregated city landscapes are racialized, or imbued with racial identities and meanings, which can in turn influence patterns of investment and mobility that further entrench inequalities. The creation of homogenous white suburban spaces in Los Angeles differentiated property values in ways that consolidated white privilege and ensured that hazardous manufacturing landscapes would be concentrated in inner-city communities of color (Pulido 2000). Conversely, the discursive construction of Detroit's inner-city Black neighborhoods as empty, neglected "frontier" landscapes, ripe for homesteading and green redevelopment, has threatened their residents with dispossession, gentrification, and displacement (Safransky 2014). Meanwhile, in other settings, environmental justice activists of color seeking investment and attention to hazards have faced *denial* that their landscapes are racialized—especially from white discourses of "color-blindness" that dismiss legacies of racial exclusion (Blanton 2011).

Landscapes of urban environmental amenities, such as parks and gardens, have a complex relationship with racial segregation and racialized space. For example, although Baltimore's African American inner-city population has more parks in walking distance than do white suburbanites, this increased access masks the former's more congested landscape of greenspace and their history of struggle to desegregate recreational spaces (Boone et al. 2009; Grove et al. 2018). Meanwhile, many city parks in the US that underwent racialization in the wake of urban unrest in the 1960s became landscapes of environmental injustice as they declined in both maintenance and safety (Brownlow 2006). In contrast, current initiatives to transform the landscapes of communities of color through "green amenities," from parkways to organic food stores, have frequently sparked contestation and resistance from residents concerned about gentrification and a loss of place identity (Anguelovski 2015).

Some of the most well-known landscapes of environmental injustice lie at the intersection of environmental contamination, racialized poverty, and natural disasters. Although all US cities face vulnerability to natural hazards, many of the most vulnerable lie within the Sun Belt. The southeastern states are most vulnerable to hurricanes and severe weather; southern central states face some of the highest risks from tornadoes and flooding; and the western states face the highest risks from earthquakes and wildfires (Borden and Cutter 2008). The Lower 9th Ward of New Orleans, which Colten (2007) called a "landscape of tragedy" because of a notorious contaminated landfill, became one of the country's most prominent landscapes of environmental injustice in the wake of flooding from Hurricane Katrina in 2005.

Landscapes of Environmental Injustice Beyond the City

Beyond cities, landscapes of environmental injustice in the US are even more diverse and regionally variable. The same injustices that affect dense urban neighborhoods also afflict landscapes in many small towns and remote settlements. However, other rural or small-town landscapes of environmental injustice differ significantly from those in cities, both in the processes that generated them and in the questions of meaning and representation that they raise. Here, I briefly consider such landscapes in three broadly defined "regions": the predominantly African American "Black Belt" of the Southeast; Indigenous territories of the Midwest, West, and island territories; and the Latinx/Hispanic Southwest. Although these regions are internally diverse, each was shaped by distinctive historical processes that have generated characteristic landscapes of environmental injustice.

The Black Belt: Small-Town and Rural US South

The rural and small-town landscapes of environmental injustice in the southeastern Black Belt—which refers both to the color of the soils and the prevalent African American population—are in many ways the crucible of environmental justice activism, policy, and scholarship. This region stands apart because of its unique history of plantation-based slavery, followed by a century of Jim Crow laws enforcing racial segregation. The vast coastal plain extending from the Mississippi Delta to the Tidewater region of eastern Virginia remains home to the country's largest population of rural and small-town African Americans. Consequently, in the Southeast, landscapes of environmental injustice associated with racial segregation are common beyond cities. For example, "colorblind" policies of sea-level adaptation planning in Georgia's remote Sapelo Island overlook the island's history of erasing and displacing its Black Gullah/Geechee population and cultural heritage (Hardy et al. 2017).

Some of the country's most iconic landscapes of environmental injustice lie within this region (see, e.g., Bullard 2008). The struggle against the siting of a PCB landfill in rural Warren County, North Carolina led both to national studies of environmental inequity and to the coining of the term *environmental racism* (McGurty 2009). One well-studied area is Louisiana's "Cancer Alley," a notorious stretch of the Mississippi River famous for landscapes marked by small-town residential neighborhoods sharing fence-lines with petrochemical, nuclear, and other industrial plants (e.g., Roberts and Toffolon-Weiss 2001; Allen 2003; Kurtz 2003; Pezzullo 2003; Blodgett 2006; Davies 2018). Beyond such famous landscapes, lesser-known rural and small-town landscapes of environmental injustice, from those marked by the odor plumes of hog farms (Wing 2000) to those hosting wood pellet production facilities (Koester and Davis 2018), are common throughout the region.

Indigenous Landscapes of Environmental Injustice

Landscapes of environmental injustice carry several unique connotations in *Indian country*, a legal designation that refers to Native American reservations, Alaskan Native villages, and other federally recognized tribal lands. Although all regions of the country have tribal land, most lies in the Upper Midwest, the northern Great Plains, and the western states. In Indigenous settings, environmental justice concerns encompass not only contamination and environmental health, but also such issues as encroachments on tribal sovereignty and threats to treaty rights to practice tribal traditional lifeways (Vickery and Hunter 2016).

One distinctive environmental injustice issue facing tribes in the US is the abuse of sacred landscapes, which are inextricably connected with Indigenous cultural identity (Ornelas 2007, 2011; Dickinson 2012; Curti and Moreno 2014; Cladis 2019). In one sense, every landscape in the US represents an injustice, since virtually all have been expropriated and altered for the use of immigrant settlers and their descendants. However, even though all land is regarded sacred, many specific landforms, waterways, and other sites hold distinctive significance. Some sites, such as the "wounded landscape" of Yucca Mountain in Nevada, have been treated as "sacrifice zones" for nuclear waste or testing (Houston 2013). Compounding the injustice in many cases has been the subjection of the "sacred" to Western norms of measurement and evaluation, often in order to justify economic or infrastructural development (Wainwright and Robertson 2003; Milholland 2010). In addition, legal protections for sacred landscapes have been insufficient, fragmented, and easily undermined (Benson 2014; Middleton 2014; Dunstan 2017).

The Southwest and the Spanish-Speaking US

Although Latinx/Hispanic populations live throughout the US, the highest concentrations of rural and small-town Spanish speakers are in southwestern states that were once part of Mexico, including Texas, California, Arizona, and New Mexico. The small farmworker settlement of Kettleman City, California, the site of a famous struggle against a toxic waste incinerator, is another iconic location for the environmental justice movement (Cole 1994). The Southwest's distinctive landscapes of environmental injustice include the agricultural landscapes of Latin American migrant workers and the *colonias* along the US–Mexico border.

The landscapes of migrant agricultural workers in the Southwest, especially in California, have long been recognized as shaped by injustice. Mitchell (1996) documented how California's rural landscape—as both material environment and cultural representation—took shape through class struggle over the unjust working and living conditions of migrant workers. Occupational exposure to pesticides in the San Joaquin Valley was a central issue in the Delano grape strike organized by César Chávez and Dolores Huerta in the mid-1960s (Pulido 1996). California's vast agricultural lands have remained contested landscapes of environmental injustice, often focused on the effects of pesticide drift on undocumented Latin American immigrants (Harrison 2011).

Another landscape of environmental injustice characteristic of this region is the *colonia*. *Colonias*, which number over 2,000 and serve as home for close to half a million people, are defined as communities in the US–Mexico border region with inadequate infrastructure and housing. Among the poorest communities within the US, *colonias* attracted Mexican immigrants with cheap land and promises of basic infrastructure that in most cases have never materialized. They are often located near hazardous land uses, such as agricultural pesticides or waste dumps, or near arroyos that make residents vulnerable to seasonal flash floods (Johnson and Niemeyer 2008; Henkel 2009; Núñez-Mchiri 2009). In addition to the preponderance of dilapidated houses and trailer homes, frequently serviced by outhouses or cesspools, the landscapes of *colonias* are marked with distinctive features like water vending machines (Jepson and Brown 2014).

Engaging with Landscapes of Environmental Injustice

How can we engage with these landscapes to transform them for the better? A prominent strategy for mobilizing engagement is to make them *visible* both to their own residents and to outsiders who might have the power to make a difference. As Mels (2016) suggests, the cultural and political representation of landscapes connects closely with the pursuit of environmental justice. Here, I consider a few examples of the complexity of seeing, sensing, and representing landscapes of environmental injustice.

One way of representing landscapes of environmental injustice to reach and mobilize people who do not live within them is to use media such as documentary films and photography. However, finding the most effective and appropriate ways to use such media can be challenging. To take one example, an analysis of photos on environmental justice published on the web identified the "fence-line photo"—characterized by references to places where children play, a fence, and a visible source of pollution—as an important genre (Gabrielson 2019). Despite the power of such photos, Gabrielson (2019) argues that in isolation, they risk downplaying the systematic nature of environmental injustice, reinforcing the stigmatization of neighborhoods, and implying the passivity of poor people of color. Media representations of such landscapes are at their strongest when they amplify resident voices and contextualize local problems.

Another activist strategy is "toxic tourism," which enables outsiders to witness landscapes of environmental injustice directly (Di Chiro 2001; Pezzullo 2003, 2009; Houston 2013). Throughout the country, grassroots activists have organized such tours for decades, at scales from the neighborhood to the broader region, in order to raise awareness, build community, and mobilize action. Unlike "status quo" practices of representing hazardous landscapes, toxic tourism functions not to sanitize or aestheticize these landscapes, but to educate and spark concern (Rosenfeld et al. 2018). Toxic tours require a high level of planning, coordination, and commitment, but they provide a powerful means through which residents can make landscapes of environmental injustice visible on their own terms.

Can learning to "read" a landscape of environmental injustice be a way to address it? Davies (2018) highlights the "slow observation" of signs—dying vegetables, damage to house foundations—through which residents learn to decipher their own toxic landscapes. But others contend that transforming unjust landscapes requires looking beyond negatives. Spirn (2005, 395) argues for the concept of *landscape literacy* as a way of remediating injustices that occur when planners, designers, or even residents "focus on a neighborhood's problems and fail to recognize its resources." Learning to read a landscape of injustice means looking beyond its surface characteristics, such as vacant land or "sagging porches and crumbling foundations," to uncover its underlying stories and historical processes (Spirn 2005, 395). It also means perceiving its hidden resources, imagining its future, and taking action to transform it.

As for landscapes sacred to Indigenous populations, making them visible is more often not about transforming them, but about protecting them. For example, protecting culturally significant landscapes may require making them more "visible" in relevant environmental laws (Middleton 2013) or the calculations of environmental risk assessments (Harris and Harper 1999). Making a sacred landscape visible through media publicity and networks of scientists helped mobilize opposition to a proposal to build a new observatory on the summit of Mauna Kea in Hawai'i (Borrelle et al. 2020). Meanwhile, establishing tribal national parks enables tribes to make sacred landscapes visible to the public while protecting them using locally appropriate approaches (Carroll 2014).

Conclusion: A Conceptual Framework and Research Agenda

Although scholarship addressing landscapes of environmental injustice in the US has become both vast and rich, it remains fragmented. In order to suggest a coherent approach to selecting research questions, I conclude by sketching a conceptual framework and research agenda. The aim is not to provide a unifying theory that would erase the complexity and diversity of these landscapes, but to offer a way of thinking that brings this diversity to the foreground. Engaging productively with landscapes of environmental injustice requires understanding both what connects them and what distinguishes them from each other.

First, analyzing landscapes of environmental injustice starts with identifying *geographically specific combinations of historical factors and processes that have generated them*. With respect to cities and urbanization, landscapes in the US are positioned on a complex spectrum ranging from densely populated, industrialized (or deindustrialized) urban cores to lightly inhabited areas that may export resources to urban centers or host urban tourists (or garbage). They are also situated in historically and biophysically distinctive regions. This encompasses the country's dramatic regional differences in climate, soil, hydrology, and topography, but it also includes different forms of historical oppression and marginalization: the plantation-based slavery and Jim Crow laws that shaped southeastern landscapes; the westward displacement, confinement, and mistreatment of tribal nations through settler colonialism; the exploitation of Latin American migrant labor in and beyond the Southwest; and the relations of imperialism that characterize many island territories. A comparative historical approach to research on landscapes of environmental injustice would suggest a variety of questions. For example, how might similar historical and biophysical factors generate different kinds of landscapes of environmental injustice in the same place? Or conversely, how might similar landscapes of environmental injustice emerge in regions shaped by widely divergent factors?

Second, analyzing landscapes of environmental injustice involves investigating *their meanings and affective potential both for residents and for outsiders*. Again, this implies addressing geographic specificity, recognizing that although some experiential elements unite different landscapes of environmental injustice, meanings and experiences associated with a contaminated, racialized fence-line community differ profoundly from those attached to a degraded sacred site on tribal land. It also suggests new questions about how different media represent landscapes of environmental injustice for outsiders: for example, how do the effects of toxic tours, documentary films, or photographs on participants contrast with the effects of long-term training for landscape literacy? It might imply experimenting with methods that investigate not only effects, but also *affects*, in the sense of connections among human and nonhuman bodies that precede and resist representation (Pile 2010). Concerns about affect feature prominently in the broader landscape literature (e.g., MacPherson 2010; Waterton 2013), but they remain relatively unexplored in research on landscapes of environmental injustice.

Box 6.1 Focus on Milwaukee

Milwaukee is one of many Rust Belt cities with landscapes of environmental injustice now undergoing transformation. One such landscape is a residential stretch of the Kinnickinnic River in the predominantly working-class Latinx/Hispanic South Side (Holifield and Schuelke 2015). During the twentieth century, the stretch was straightened and filled with concrete, exacerbating flood risk in the adjacent neighborhood. Recently, the local sewerage district deconstructed houses to clear the floodplain, and it is now removing the concrete to restore the river's natural meander. Although some residents expressed concerns that the greening of the river would spark gentrification, local agencies and organizations are collaborating with residents to introduce new environmental amenities while preserving the neighborhood's affordability.

Overall, the most pressing need for research is to advance the rectification of landscapes of environmental injustice. Perhaps the most daunting challenge is identifying effective strategies for repairing and restoring toxic or degraded landscapes without creating new injustice through displacement. Although the approach of making such landscapes "just green enough" to avoid gentrification has received considerable attention (Curran and Hamilton 2017), others argue for strategies aiming to ensure that all communities have access to environments of high quality (e.g., Rigolon et al. 2020). In all likelihood, we will need a "toolbox" as diverse as the country's landscapes of environmental injustice.

References

Allen, B.L., 2003. *Uneasy Alchemy: Citizens and Experts in Louisiana's Chemical Corridor Disputes*. Cambridge: MIT Press.

Anguelovski, I. 2015. "Alternative Food Provision Conflicts in Cities: Contesting Food Privilege, Injustice, and Whiteness in Jamaica Plain, Boston." *Geoforum* 58: 184–194.

Benson, M.H. 2014. "Enforcing Traditional Cultural Property Protections." *Human Geography* 7 (2): 60–72.

Blanton, R. 2011. "Chronotopic Landscapes of Environmental Racism." *Journal of Linguistic Anthropology* 21: E76–E93.

Blodgett, A.D. 2006. "An analysis of pollution and community advocacy in 'Cancer Alley': Setting an example for the environmental justice movement in St James Parish, Louisiana." *Local Environment* 11 (6): 647–661.

Boone, C.G., G.L. Buckley, J.M. Grove, and C. Sister. 2009. "Parks and People: An Environmental Justice Inquiry in Baltimore, Maryland." *Annals of the Association of American Geographers* 99 (4): 767–787.

Boone, C.G., and A. Modarres. 1999. "Creating a Toxic Neighborhood in Los Angeles County: A Historical Examination of Environmental Inequity." *Urban Affairs Review* 35 (2): 163–187.

Borden, K.A., and S.L. Cutter. 2008. "Spatial Patterns of Natural Hazards Mortality in the United States." *International Journal of Health Geographics* 7 (1): 64.

Borrelle, S.B., J.B Koch, C.M. MacKenzie, K.E. Ingeman, B.M. McGill, M.R. Lambert, A.M. Belasen, J. Dudney, C.H. Chang, A.K. Teffer, and G.C. Wu. 2020. "What Does it Mean to Be for a Place?" *Pacific Conservation Biology* https://doi.org/10.1071/PC20015.

Brownlow, A. 2006. "An Archaeology of Fear and Environmental Change in Philadelphia." *Geoforum* 37 (2): 227–245.

Bullard, R.D. 2008. *Dumping in Dixie: Race, Class, and Environmental Quality*. Boulder, CO: Westview Press.

Carroll, C. 2014. "Native Enclosures: Tribal National Parks and the Progressive Politics of Environmental Stewardship in Indian Country." *Geoforum* 53: 31–40.

Cladis, M.S. 2019. "Sacred Sites as a Threat to Environmental Justice?: Environmental Spirituality and Justice Meet among the Diné (Navajo) and Other Indigenous Groups. *Worldviews: Global Religions, Culture, and Ecology* 23 (2): 132–153.

Cole, L.W. 1994. "The Struggle of Kettleman City: Lessons for the Movement." *Maryland Journal of Contemporary Legal Issues* 5 (1): 67–80.

Colten, C.E. 2007. "Environmental Justice in a Landscape of Tragedy." *Technology in Society* 29 (2): 173–179.

Curran, W., and T. Hamilton. 2017. *Just Green Enough: Urban Development and Environmental Gentrification*. London and New York: Routledge.

Curti, G.H., and C.M. Moreno. 2014. "Introducing Traditional Cultural Properties (In Need of Critical Geographies)." *Human Geography* 7 (2): 1–10.

Davies, T. 2018. "Toxic Space and Time: Slow Violence, Necropolitics, and Petrochemical Pollution." *Annals of the American Association of Geographers* 108 (6): 1537–1553.

Di Chiro, G. 2001. "Toxic Tourism: A New Itinerary for the Environmental Justice Movement." *Orion Afield: Working for Nature and Community* 5 (2): 34–37.

Dickinson, E. 2012. "Addressing Environmental Racism through Storytelling: Toward an Environmental Justice Narrative Framework." *Communication, Culture & Critique* 5 (1): 57–74.

Dunstan, A. 2017. "Legislative Ambiguity and Ontological Hierarchy in US Sacred Land Law." *American Indian Culture and Research Journal* 41 (4): 23–43.

Gabrielson, T. 2019. "The Visual Politics of Environmental Justice." *Environmental Humanities* 11 (1): 27–51.

Grove, M., L. Ogden, S. Pickett, C. Boone, G. Buckley, D.H. Locke, C. Lord, and B. Hall. 2018. "The Legacy Effect: Understanding How Segregation and Environmental Injustice Unfold over Time in Baltimore." *Annals of the American Association of Geographers* 108 (2): 524–537.

Hardy, R.D., R.A. Milligan, and N. Heynen. 2017. "Racial Coastal Formation: The Environmental Injustice of Colorblind Adaptation Planning for Sea-Level Rise." *Geoforum* 87: 62–72.

Harris, S., and B. Harper. 1999. "Environmental Justice in Indian Country: Using Equity Assessments to Evaluate Impacts to Trust Resources, Watersheds, and Eco-Cultural Landscapes." *Proceedings of "Environmental Justice: Strengthening the Bridge between Tribal Governments and Indigenous Communities, Economic Development and Sustainable Communities."*

Harrison, J.L. 2011. *Pesticide Drift and the Pursuit of Environmental Justice*. Cambridge: MIT Press.

Henkel, D.S. 2009. "Upholding Environmental Justice in the Colonias: A New Mexico Approach." *Journal of Borderlands Studies* 24 (1): 62–75.

Holifield, R., and N. Schuelke. 2015. "The Place and Time of the Political in Urban Political Ecology: Contested Imaginations of a River's Future." *Annals of the Association of American Geographers* 105 (2): 294–303.

Houston, D. 2013. "Environmental Justice Storytelling: Angels and Isotopes at Yucca Mountain, Nevada." *Antipode* 45 (2): 417–435.

Hurley, A. 1995. *Environmental Inequalities: Class, Race, and Industrial Pollution in Gary, Indiana, 1945–1980.* Chapel Hill: University of North Carolina Press.

Jepson, W., and H.L. Brown. 2014. "'If No Gasoline, No Water': Privatizing Drinking Water Quality in South Texas *Colonias.*" *Environment and Planning A* 46 (5): 1032–1048.

Johnson, M.A., and E.D. Niemeyer. 2008. "Ambivalent Landscapes: Environmental Justice in the US–Mexico Borderlands." *Human Ecology* 36 (3): 371–382.

Koester, S., and S. Davis. 2018. "Siting of Wood Pellet Production Facilities in Environmental Justice Communities in the Southeastern United States." *Environmental Justice* 11 (2): 64–70.

Kurtz, H.E. 2003. "Scale Frames and Counter-Scale Frames: Constructing the Problem of Environmental Injustice." *Political Geography* 22 (8): 887–916.

Logan, J.R. 2013. "The Persistence of Segregation in the 21st Century Metropolis." *City & Community* 12 (2): 160–168.

Logan, J.R., W. Zhang, R. Turner, and A. Shertzer. 2015. "Creating the Black Ghetto: Black Residential Patterns Before and During the Great Migration." *The Annals of the American Academy of Political and Social Science* 660 (1): 18–35.

Macpherson, H. 2010. "Non-representational Approaches to Body–Landscape Relations." *Geography Compass* 4 (1): 1–13.

McGurty, E. 2009. *Transforming Environmentalism: Warren County, PCBs, and the Origins of Environmental Justice.* New Brunswick, NJ: Rutgers University Press.

Mels, T. 2016. "The Trouble with Representation: Landscape and Environmental Justice." *Landscape Research* 41 (4): 417–424.

Middleton, B.R. 2013. "'Just Another Hoop to Jump Through?': Using Environmental Laws and Processes to Protect Indigenous Rights." *Environmental Management* 52 (5): 1057–1070.

———. 2014. "ChuChuYamBa/Soda Rock: Toward an Applied Critical Geographic Perspective on Traditional Cultural Properties (TCPs)." *Human Geography* 7 (2): 11–28.

Milholland, S. 2010. "In the Eyes of the Beholder: Understanding and Resolving Incompatible Ideologies and Languages in US Environmental and Cultural Laws in Relationship to Navajo Sacred Lands." *American Indian Culture and Research Journal* 34 (2): 103–124.

Mitchell, D. 1996. *The Lie of the Land: Migrant Workers and the California Landscape.* Minneapolis: University of Minnesota Press.

Núñez-Mchiri, G.G. 2009. "The Political Ecology of the *Colonias* on the US-Mexico Border: Human-Environmental Challenges and Community Responses in Southern New Mexico." *Journal of Rural Social Sciences* 24 (1): 67–91.

Ornelas, R.T. 2007. "Understanding Sacred Lands." *Great Plains Research* 17: 165–171.

———. 2011. "Managing the Sacred Lands of Native America." *International Indigenous Policy Journal* 2 (4): Article 6.

Pezzullo, P.C. 2003. Touring "Cancer Alley," Louisiana: Performances of community and memory for environmental justice. *Text and Performance Quarterly* 23 (3): 226–252.

———. 2009. *Toxic Tourism: Rhetorics of Pollution, Travel, and Environmental Justice.* Tuscaloosa: University of Alabama Press.

Pile, S. 2010. "Emotions and Affect in Recent Human Geography." *Transactions of the Institute of British Geographers* 35 (1): 5–20.

Pulido, L. 1996. *Environmentalism and Economic Justice: Two Chicano Struggles in the Southwest.* Tucson: University of Arizona Press.

———. 2000. "Rethinking Environmental Racism: White Privilege and Urban Development in Southern California." *Annals of the Association of American Geographers* 90 (1): 12–40.

Rigolon, A., S.J. Keith, B. Harris, L.E. Mullenbach, L.R. Larson, and J. Rushing. 2020. "More than 'Just Green Enough': Helping Park Professionals Achieve Equitable Greening and Limit Environmental Gentrification." *Journal of Park and Recreation Administration* 38 (3): 29–54.

Roberts, J.T., and M.M. Toffolon-Weiss. 2001. *Chronicles from the Environmental Justice Frontline.* Cambridge: Cambridge University Press.

Rosenfeld, H., S. Moore, E. Nost, R.E. Roth, and K. Vincent. 2018. "Hazardous Aesthetics: A 'Merely Interesting' Toxic Tour of Waste Management Data." *GeoHumanities* 4 (1): 262–281.

Safransky, S. 2014. "Greening the Urban Frontier: Race, Property, and Resettlement in Detroit." *Geoforum* 56: 237–248.

Spirn, A.W. 2005. "Restoring Mill Creek: Landscape Literacy, Environmental Justice and City Planning and Design." *Landscape Research* 30 (3): 395–413.

Vickery, J., and L.M. Hunter. 2016. "Native Americans: Where in Environmental Justice Research?" *Society and Natural Resources* 29 (1): 36–52.

Wainwright, J., and M. Robertson. 2003. "Territorialization, Science and the Colonial State: The Case of Highway 55 in Minnesota." *Cultural Geographies* 10 (2): 196–217.

Waterton, E. 2013. "Landscape and Non-representational Theories." In *The Routledge Companion to Landscape Studies*, edited by P. Howard, I. Thompson, and E. Waterton, 66–75. London: Routledge.

Wilson, S., M. Hutson, and M. Mujahid. 2008. "How Planning and Zoning Contribute to Inequitable Development, Neighborhood Health, and Environmental Injustice." *Environmental Justice* 1 (4): 211–216.

PART II

Social, Cultural, and Popular Identities in the American Landscape

Alyson L. Greiner

Introduction

With the American landscape, as with any landscape, what we see is *not* what we get. Rather, much remains hidden, compelling us to probe beneath the surface in order to excavate and analyze the processes that create, perpetuate, re-make, or un-make landscapes. A multitude of entanglements that have cultural, political, economic, social, and symbolic dimensions shape the American landscape. Recently, hyperglobalization, the ostensible omnipresence of the internet and corporate power, as well as related changes in our societal habits including the ways we communicate, date, shop, and seek entertainment have increased the complexity of these entanglements.

The chapters in this section consider social, cultural, and popular identities in the American landscape through a focus on race, gender and sexuality, ethnicity and transnationalism, religion, film, music, sports, and science fiction. Since we could conceivably devote an entire volume to any one of these topics, our approach is necessarily selective and partial. These chapters provide brief, eye-opening synopses intended to whet your curiosity and mediate a process of reflection and prospection: looking back over the past 25 years while pressing forward with new and multivocal insights from diverse scholars of the American scene.

These chapters share several themes related to identity, the production of space, and imaginative geographies. The essays in this section articulate diverse ways of understanding identity: as fluid, contingent, embodied, and performative. Identity and power intersect through the process of othering, as well as through classist, racist, and gendered practices that are reflected both in lived realities and science fiction. This collection of chapters also traces the contours of identity at different scales, from bodily to regional and transnational extents, underscoring the point that the entanglements of the American landscape result in part from the interconnections between local and global processes.

Our contributors disrupt the idea that places and landscapes are fixed and bounded, choosing instead to emphasize placemaking, place packaging, and the production of space. Placemaking may yield tangible, inscribed landscapes, but it also involves the creation of relational, cinematic,

DOI: 10.4324/9781003121800-9

virtual, and racialized spaces. We build networks of connections—themselves inflected by power—through our work, the social and familial relationships we create and maintain, as well as processes of sacralization. Placemaking provides a lens enabling us to reveal and critically examine the entanglements of space, time, and movement, all of which constitute durable themes in the shaping and reshaping of the American landscape. Explicitly or implicitly, these chapters illustrate the role of capitalism as a driver of processes affecting the American landscape, its spatial forms, and networks of interaction. These landscapes of capitalism are Janus-faced, with a flip side that includes the ongoing changes wrought by so-called "creative destruction."

We cannot fully understand the processes shaping the American landscape unless we also consider the role of imagination, particularly how we envision different places and layer them with our collective values and expectations. Imaginative geographies have long been associated with the American landscape, from the idealized and romanticized New England village to the Land of Zion for the Latter-day Saints. Much scholarly work in this vein has focused on the American West and examined the impacts of myth and idealized perceptions on the region, its places, and inhabitants. Imaginative geographies not only express and reflect identities, they also shed light on belonging and processes of world-building, both fictional and real. As the chapters in this section show, the American landscape continues to reflect a tension between fiction and reality.

In the first chapter that follows (Chapter 7), Aretina R. Hamilton uses the 2021 siege on the US Capitol to show that placemaking in the American context cannot be separated from the everyday violence of institutionalized and systemic racism. Introducing the concept of the white unseen, she demonstrates that our status quo landscapes privilege whiteness and that this both disguises and naturalizes the violence that these landscapes perpetuate against Black, Indigenous, and People of Color. Christina E. Dando's chapter on the oil landscapes of western North Dakota (Chapter 8) also associates landscapes with violence, particularly against women. Dando not only examines the gendering of the landscape and the performance of hypermasculinity as the fracking boom led to the rapid establishment of man camps, she also illustrates how imaginative geographies that are anchored to the feminization of the American landscape frame our social and environmental interactions. An accompanying box within the chapter provides an overview of ecofeminism.

The next two chapters address American pluralism within ethnic and religious contexts. Through the study of Indian migrants in Phoenix, Emily Skop and Stephen Suh (Chapter 9) show the importance of transitory and transnational spaces to the negotiation of immigrant identity and belonging, and remind us that migrant landscapes are not necessarily visibly inscribed landscapes. They also show how gender, multiple allegiances, and imagined communities can work to affirm American identities among Korean Americans who return to Korea. Samuel Avery-Quinn (Chapter 10) traces the increasing religious diversity in the US, an important hallmark of the country. His chapter demonstrates the salience of lived religion, including embodied and everyday religious practices, for exploring religious pluralism, its concomitant tensions, and implications for coexistence.

Turning to the American cinematic landscape, Chris Lukinbeal (Chapter 11) documents a shifting urban and regional geography of film production away from Los Angeles. His use of "landscape (@)work" calls attention to processes of film production that hide the labor that produces the American cinematic landscape. Specifically, he details how the American cinematic landscape generates a "fifth wall" that serves to completely separate actors and audiences from the processes of production. If the hegemony of Los Angeles has been challenged in ways by the film production industry, the city has (according to some data) gained in the national rankings for pop music. In Chapter 12, David J. Keeling and Thomas L. Bell explore regional and national trends through a consideration of changes in the production, distribution, and consumption of music. They reflect on the ubiquity and territoriality of music while addressing the creative destruction that has shuttered record stores as digital streaming has expanded. Music and sports overlap in their use of similar venues such as parks and stadiums, and this brings us to John Lauermann's essay (Chapter 13). In many respects our landscapes of sport are microcosms of broader society, reflecting class, gender, and racial disparities as well as political struggles over standing or taking a knee during the national anthem. As Lauermann also attests, professional sports run on commodification that enables the harvesting of revenues not just from the stadium or venue, but from vast broadcasting hinterlands.

With landscapes of the North American future in mind, Fiona Davidson (Chapter 14) identifies and analyzes different science fiction subgenres as well as post-climate change science fiction. These alternative visions of the future variously depict America as an idyllic Arcadia, hypercapitalist cyberpunk realm, or dystopia. Maria Lane's accompanying box on landscapes of Mars reveals that similar technocratic imaginings about landscape improvement were projected on both the American and Martian landscapes. These spatial imaginaries justified Manifest Destiny and settler colonialism in the name of progress, echoing key points made in the first chapter in this section. Taken together, these chapters offer an instructive and critical appraisal of the evolving American landscape that, we hope, prompts additional studies.

7

"THIS IS NOT THE AMERICA I KNOW"

Reading the White Unseen Within the American Landscape

Aretina R. Hamilton

As the rioters forcibly entered the Capitol and hunted down the elected officials whom they blamed for stealing the election from Donald Trump, commentators repeated: "This is not the America I know." As Richard Barnett posed for pictures in the office of House Speaker Nancy Pelosi, the mantra "This is not the America I know" continued to resonate through the media. The smoke wafting in front of the Capitol with the American flag waving in the background paralyzed many Americans with cognitive dissonance and dissociation. It seemed that America was crumbling before their very eyes.

However, for many African Americans, the Capitol siege was part of a larger continuum of white racial violence that has plagued our nation since its founding. It could not be separated from the murders of George Floyd, Breonna Taylor, Ahmaud Arbery, attacks on voter rights, or the rapid gentrification that continues to destroy Black communities and displace residents in the name of "progress." As Missouri Congresswoman and activist Cori Bush commented, "Many have said that what transpired on Wednesday was not America. They are wrong. This is the America that Black people know" (Bush 2021).

How did it come to be that white and Black Americans had such different reactions to the Capitol siege? How is white supremacy rendered invisible in landscapes? Why are some landscapes marked as the property of whiteness, while others—e.g., Black neighborhoods—are labeled "problems" and targeted for surveillance? What lenses are needed to read race in the landscape? In this chapter, I introduce the *white unseen*—which I define as an intentional thought pattern and epistemological process in which acts of white violence and the everyday terrors, trauma, and tensions faced by Black, Indigenous, and People of Color (BIPOC) are rendered invisible—as a conceptual lens for reading race in the landscape. The *white unseen* brings into focus the ways in which white supremacy works as a cartographic tool and technology that undermines democracy and promotes place-based racial inequalities.

DOI: 10.4324/9781003121800-10

Reading Media Representations: Making the White Unseen Visible

How is Whiteness Rendered Invisible in Landscapes?

Art is an important way of speaking to and about race in American landscapes. From the Hudson Valley to the Schuylkill River to the hills of Appalachia, painters have curated images of the United States as a quaint, vibrant, and prosperous *white* country. Portraits of early American landscapes absent of people have been prized by wealthy benefactors who, seduced by the charm of open terrain, erased the Indigenous people who occupied the land. As the Western Frontier opened, artists moved from portraying nature as a utopian blank slate to featuring economic pursuits and colonial conquests. An example of this can be seen in *American Progress*, John Gast's painting of the United States as an angelic white woman who, bathed in light, guides the settlers toward their manifest destiny of reclaiming the Western frontier from the American Indians (Figure 7.1). This and many other famous early American paintings strongly shaped perceptions of race, landscape, and American national identity. The ideology of American civilization and progress is based on a white gaze that defines physical landscapes, and our national identity as a whole, as the domain of whiteness, while ignoring the violence of colonialism. Images like these are still romanticized in textbooks and reify the whiteness of the American landscape.

Boston's Freedom Trail is a tourist attraction consisting of 16 historical sites that, according to the Freedom Trail Foundation (2022), "tell the story of the American Revolution." One of the sites along the Freedom Trail—Copp's Hill Burying Ground—is one of Boston's oldest

Figure 7.1 John Gast painted *American Progress* in 1872 and it was reproduced in guides about the West. Image courtesy Library of Congress, Prints & Photographs Division, LC-USZC4-668.

cemeteries. While many of Boston's most notable merchants, artisans, and craftspeople from the colonial era to the 1850s are buried in the main grounds, across the street is a less curated burial ground with unmarked graves, in which another set of Boston's founders were laid to rest: free African Americans. Like the potter's field they are buried in, these founders are not highlighted in Freedom Trail marketing—nor are their contributions to Boston or to America. The Black burial space at Copp's Hill and cemeteries like it stands as a stark reminder of how social codes, policies, and norms erase Black lives and reify the white unseen. The white unseen does not just shape the American narrative, it is embedded in the soil and the landscape itself. It shapes Black contributions during their lives, and in the afterlife as well.

Another stop on the Freedom Trail is Boston's Faneuil Hall. Although the Freedom Trail dubs Faneuil Hall as the "Cradle of Liberty" and the "Home of Free Speech," just a few steps away from the sites of the first Town Meetings sit markets in which merchants sold the enslaved Africans whom they had trafficked across the Middle Passage from West Africa. Peter Faneuil financed at least two slaving voyages and several suspected others, and he enslaved men and women in his home. In addition, Faneuil built his financial empire on a complex trading system that relied on the institution of slavery. So, while Boston is valorized as the birthplace of freedom, democracy, progress, and American civilization, through another lens it can be seen as a space of violence, economic exploitation, and captivity. By uncovering the multiple narratives embedded in these spaces, and asking whose stories are privileged and whose stories are lost, geographers can expose how the white unseen continues to shape the American narrative. In addition, the reminders and remnants of the violence of colonial Boston endure in the process of gentrification that is currently shrinking Black populations, as well as the alarming racial wealth inequality reflected in studies showing that Black families possess only $8 in median net worth, compared to the $247,500 median net worth of a white family (Meschede, Hamilton, Muñoz, Jackson, and Darity 2016).

More recently, we have started to see representations of US history that disrupt the white unseen. HBO's *The Gilded Age* begins to tackle this question through the character of Peggy Scott, an African American writer who works as a personal secretary for a wealthy white Manhattan family in the late 1880s. The presence of Peggy Scott in the show is significant because it is rare for the experiences of educated middle-class African Americans to be centered, particularly in period dramas. While certainly there were significant populations of educated, middle-class Black Americans in the late 1800s, Peggy Scott's presence disrupts the assumption of the elite white space. For viewers who have been conditioned to implicitly associate wealthy New York landscapes with whiteness, Peggy's presence represents a disjuncture in the carto-graphic imagination. Simply through Peggy Scott's presence, this episode begins to make a crack in the white unseen.

But it is not just Peggy's presence that disrupts the white unseen. It is also how she moves through and interacts with other characters in the space. While it would have been easy for the writers to frame Peggy's character through a trope of Black female servitude, *The Gilded Age* reminds the audience that everything is not what it seems. During a visit to her parents' posh home in Brooklyn—which was a wealthy, predominantly Black community in the late 1800s— Peggy's interaction with Marian Brook, a young white woman, further exposes the white unseen. As Marian Brook pays a visit to Peggy's home in Brooklyn, she steps off her carriage, with a dust bag in tow. She is met with the puzzled gazes of African American passersby. She looks up at the massive brownstone with an air of surprise and confusion before ringing the doorbell. Upon being greeted by the maid and waiting for a formal introduction, she gawks at the majestic home. The art. The family photos. The decor. The fabrics. It is evident by her nervousness that she feels out of place.

A few minutes later, Peggy and her parents walk into the parlor where Peggy asks, "What are you doing here?" Peggy's father echoes, "Why are you here uninvited?" Despite Peggy's mother's efforts to smooth over the tension, a feeling of unease lingers. While Marian's unannounced visit flouted social norms at the time, this was not her biggest faux pas. Toward the end of the conversation, Marian reveals that she had brought some used boots in her bag, an act (or so she thought) of charitable benevolence. Peggy confronts Marian outside the house, saying:

> What are you doing here? And the shoes. What was that? Because we're colored… we must be poor?! I loaned *you* the train fare! You don't know anything about me. About my life. About my situation. I live in a different country from the one you know.
>
> *(Gilded Age 2022, Episode 4)*

It is clear that Marian and Peggy are living in two separate realities, shaped by their backgrounds but also by the cultural landscapes that they are permitted to occupy. For Peggy, no amount of wealth can open up or alter geographies that will not allow her entry. For Marian, whose father recently died leaving her penniless, her social capital as a white woman still expands, opens up, and transforms the spaces she seeks to enter. Perhaps she should have noticed the social cues suggesting that Peggy was from an elite background. But in the land of the white unseen, Black agency—indeed, Black life—is frequently rendered invisible. In this interaction, Peggy gives the lie to the viewer's assumption of white space. This representation is exceptional because the white unseen is too often shored up within television, cinema, and everyday life.

While cultural geographers have begun to examine race as a social construct, a tool of power and subordination, and a scene of congestion, white supremacy remains elusive. However, it is not "natural"—nor is it by accident—that race is muted within landscapes. To rectify this whitewashing, cultural geographers need to pay attention not only to the material and physical, but the symbolic elements of a landscape. Part of the magic of the white unseen is that it primes the viewer to see white experiences, white accomplishments, and the white lens as the only—or at least the most important—factors defining American spaces and American identity.

Critical Historical Geographical Inquiry: The Past is Always Present

Let's return to the January 6th siege on the US Capitol. Why was it that many white Americans were shocked by the white mob violence—and the Capitol Police's weak response to it—while many Black Americans were not surprised at all? The answer is complicated, but suffice it to say that the white unseen is at the heart of it.

It is important to note that the siege on the US Capitol was not the first, nor the only, act of white supremacist political violence in US history. US history is riddled by white racial violence. From its origins as a colonial frontier for the British crown, the earliest settlers used power, force, and coercion to massacre, subdue, squelch, tame, and "civilize" the natives. The genocide of Indigenous peoples and the theft of their land set the stage for a system of colonial order based on racial and ethnic violence and inequality. The enslavement of Africans further solidified these hierarchies, perpetuating an unjust and unequal system that came to define race relations in the United States.

In early America, BIPOC people were viewed as obstacles to the colonists' ability to take possession of and occupy the "great woods." The idea that Indigenous peoples and their land must be tamed, along with the colonial compulsion to seek out and dominate other cultures, set the foundation for this ideology. The epistemologies that supported the Turner thesis, manifest destiny, and the opening of the western frontier are built on notions of white entitlement and ownership. And yet, these massive movements frequently ignore African Americans, at least in the eras following the abolition of chattel slavery. The white unseen as a conceptual framework cannot be confined to one space or historical era because it is a constant and continuous feature of the American landscape. This narrative holds that white people made America what it is. A corollary assumption is that whites own the landscape. In turn, the history of white racial and spatial violence is often whitewashed from the dominant narrative promoted by our schools, media, and political institutions.

White supremacist placemaking and spatial violence are central to the American landscape. Seldom are they disrupted, in part because they are seen as necessary to defend and advance democracy. Within famous portrayals of the American landscape such as John Gast's *American Progress*, emptying out and eliminating the disorder and chaos of the western frontier is synonymous with the elimination of Black, Indigenous, and People of Color. Thus, any violence that is enacted upon BIPOC bodies and BIPOC communities is not seen as violence at all. Because the white unseen is concerned with maintaining the white space and the white supremacist ideologies that underlie it, white racial violence literally gets written out of the landscape. White racial violence is perceived not as a massacre, but as the price of progress.

White Supremacist Placemaking: The Tulsa, OK, and Wilmington, NC Race Massacres

Through the prism of what scholar George Lipsitz calls the "Black spatial imaginary," the desire for "pure and homogeneous spaces" against the impurity of the racialized other created cultural sites marked by spatial violence (2011, 13, 29). For Lipsitz, the quintessential sites of racial violence were the plantation, hypersegregated Black communities, and the prison. In writing about these spaces, Lipsitz and other scholars have noted the ways in which Blackness and Black people have been relegated to certain spaces. The institution of chattel slavery made an imprint on the cultural and physical landscapes of the United States not only in the South, but throughout the United States as a whole.

The Tulsa race massacre is an example of this phenomenon. Tulsa had a thriving and prominent business district where African Americans enjoyed unprecedented economic and political independence and success. And yet, their community was destroyed by mob violence enabled by the US government. In the attempt to escape the specter of white violence and expulsion, many African Americans have created their own spaces of belonging—areas of economic, political, social self-sufficiency where they can claim their full humanity. Yet, as we saw in Tulsa, African Americans rarely escape the panopticon of whiteness. White racial violence is constant, continuous, and frequently impossible to escape.

White racial violence in the United States has not only persisted over time, but across space as well. From 1883 to 1941, white mob violence was directed towards African Americans, ethnic whites, and new immigrants, as well as Indigenous people and extended to every region of the United States (Seguin and Rigby 2019). In an analysis of 2,850 lynching records, the researchers found that lynchings spanned 33 states during this period; only Maine, Massachusetts, New Hampshire, Rhode Island, and Vermont reported no cases (Seguin and Rigby 2019). Go to https://davidrigbysociology.com/lynchings-in-the-us-1883-1941 to see these maps and their data.) Why is violence so invisible when it is perpetuated by white people? It seems as though white violence is everywhere and nowhere. When people say, "this is not the America I know," how can we see what has been lost from the dominant American narrative?

One of the most infamous and largely forgotten political coups and race massacres in US history occurred in Wilmington, North Carolina in 1898. The Wilmington Race Riot of 1898 was not so much a riot, but a massacre that stemmed from the Democratic Party's white supremacist plan in North Carolina. During Reconstruction, the city of Wilmington had achieved a moderate degree of racial progress and integration. There existed a vibrant and successful African American middle class that dominated various trades and business ventures, a growing contingent of Black elected officials, and an interracial political coalition. Yet, before long, these successes by African Americans were met with white violence and backlash.

To subdue the growth of Black political and economic power, local newspapers frequently wrote about the threat of miscegenation, or the threat of Black men to white women's sexual purity. In August of 1898, Alexander Manly, the editor of Wilmington's *Daily Record* responded to a speech given by a white Georgian woman, Rebecca Latimer Felton, who said: "If it takes lynching to protect women's dearest possession from drunken, ravenous human beasts, then I say lynch a thousand a week." Many publicly disputed Felton's ideas, pointing out that many white women are secretly attracted to Black men. The media used this to stoke white fears of Blackness and purported Black male rapists. At the same time, they leveraged the outrage against the Fusion party, the interracial political coalition that was dominant in local politics, to rally white citizens to take up arms and chase Black residents out of town. Nine Black people were shot and killed in the massacre and many others were chased out with nowhere to relocate.

White supremacist politicians and leaders called for the murders of Black people and the destruction of their communities and businesses in Wilmington. Representative Alfred Waddell, a North Carolina Democrat, said bluntly:

> If we have not the votes to carry the election, we must carry it by force. If you find the Negro out voting, tell him to leave the polls. If he refuses, kill him, shoot him down in his tracks. We shall win tomorrow if we have to do it with guns.
>
> *(Tyson 2006)*

Newspapers in North Carolina and across the country reported on the white violence matter-of-factly and often vilified Black people as the "problem." For example, an article in the *Atlanta Constitution* blamed the Carolina riots on a "long series of provocations and an epidemic of crime, the victims of which were respectable white citizens, and the perpetrators were blacks and low whites" (*Atlanta Constitution* 1898).

A news clipping published on November 11, 1898 in the *New York Herald* titled "Whites Kill Negroes and Seize City of Wilmington" tacitly endorsed the massacre and white insurrection, framing it as a return to law and order. Black political representation was seen as a threat to democracy and social order. This view was not only in the South but was held widely as evidenced by the thousands of white citizen councils, vigilante groups, and laws that were passed to control the Black population after the end of slavery. The narrative of Black people as a foreign element who would overturn white power and annihilate whites and all of their property gained traction. At the same time, whites crafted a "White Declaration of Independence" calling for the removal of voting rights for Blacks and the overthrow of the newly elected interracial government.

The white fear of Black politicians and Black economic power reflected the belief that Black people had no right to full citizenship. The animalistic imagery found in political cartoons at the time evokes the common Jim Crow ideology of Blacks as subhuman and menacing. As sociologists remind us, the most effective way to justify violence against a group is to define them as less than human. In addition, the *New York Times* article and other media narratives of the time perpetuated the idea of a large-scale conspiracy where Black Americans were determined to kill white people and establish Black rule. In the political influence and representation of African Americans, white citizens saw their country descending into a place they could no longer own or claim as theirs. These white citizens justified rising up against the Black residents and members of the Fusion party as politically necessary for their own self-defense and the defense of American democracy. Through this ideological sleight of hand, white Americans successfully flipped a script to cast themselves and the nation as a whole as victims of Black violence.

The Enduring Effects of White Supremacist Placemaking

In his highly touted article "The First White President," Ta-Nehesi Coates' (2017) identified Trump's ideology as white supremacy, in all its truculent and sanctimonious power. Trump branded his Presidential campaign by casting himself as the defender of white maidenhood against Mexican "rapists," only to later be accused by multiple women, and by his own proud words, as a sexual predator himself. White supremacy is often oversimplified as grotesque actions perpetrated by a few extremist individuals. However, as Trump and his loyal devotees demonstrated throughout his term, white supremacy cannot be so easily demarcated. It is, in fact, a system that delineates and infects the environment that it is attached to. In this way, white supremacy transforms from a random act of violence to a process called "white supremacist placemaking practices" which describes the active process of spatial violence that promotes the preservation of white spaces and communities. It is both direct violence and structural through policies and practices. The tools of white supremacist placemaking have shaped urban planning, the development of racial covenants, gerrymandering, gentrification, urban revitalization, but also in the development of a new nationalized white identity. Hence, by the time that insurrectionists stormed the Capitol on January 6th, Trump appropriated the ideology of white supremacy to build a new country that wielded the weapons of white supremacy.

In the face of his pattern of contradicting action, many people held onto the hope that he would act in a presidential way and exit office gracefully. Even at the start of the siege, as rioters and looters stormed the Capitol calling for revolution, the media waited with baited breath for Trump to quell the madness. To stop the violence. He finally released a recorded statement telling protesters to "go home." He later added:

> I know you're in pain. I know you're hurt. We had an election that was stolen from us. It was a landslide election, and everyone knows it, especially the other side. But you have to go home now. We have to have peace. We have to have law and order. We have to respect our great people in law and order. We don't want anybody hurt.
> *(WBUR 2021)*

This denialism of the facts was a trademark of Trump's Presidency, much like it is the foundation of the white unseen. And yet, it would be naive to blame the rise of white supremacy and the storming of the Capitol on Trump alone. In fact, the reemergence of virulent white criminality was born from the idea that American spaces were the natural birthright of white people. A few weeks before the siege, Representative Marjorie Taylor Greene, the Republican US Congresswoman from Georgia, urged Americans to "mobilize" and take the Capitol, because it is "yours."

In this way, domestic racial terrorism acts as a placemaking practice, a policy, and an ideology that is justified and encouraged. For the insurrectionists, the siege was a culmination of white angst, anger, and entitlement. In interviews with the media, white insurrectionists exclaimed, "this is our place," "this is our country," and "we are taking it back." This language echoed the sentiments of the white insurrectionists in Wilmington over a century ago. Embedded in the mantras of "Make America Great," "Stop the Steal," and the general sense of fear, panic, and anger surrounding the inauguration of Joe Biden as president and Kamala Harris as vice president, were the age-old assumptions that Black Americans, and those associated with them in interracial coalition, must be stopped in defense of the nation. As a cross was raised and a noose erected on the West lawn of the US Capitol, these iconographies of a racist past and present were reminders that in this America, white rule dominates.

Lastly, in understanding how the white unseen can change and shape narratives, an examination of how law enforcement officers interacted with the white insurrectionists is necessary. In numerous accounts and recorded interactions, Black Lives Matter activists have been forcefully suppressed, arrested, and injured by the police. The January 6th insurrectionists, on the other hand, were treated quite differently. They were presumed innocent. Despite flagrantly violating local and federal laws, these individuals were not seen as threats, criminals, or terrorists, nor were they treated as such. In video after video, the insurrectionists fought the police, sprayed them with pepper spray, took selfies with the police, destroyed and stole property, threatened the lives of elected officials, killed a police officer, threatened to topple the government, and threatened to kill Vice President Mike Pence. Notably, all of these acts were met with little to no retribution from the police. Even after the siege ended, some politicians defended their acts of violence and stated that we should sympathize with the frustrated and disenchanted. As Joshua Inwood has noted:

> This is the heart of neoliberal racism. One of the things that makes neoliberal racism so difficult to confront is that it takes the overt white supremacy of previous generations and repackages it in a language and context that offer 'plausible deniability' to those who profit from it.
>
> *(2015, 45)*

And yet, it is clear that the insurrectionists were not just a few bad apples or disenfranchised citizens. These events, like the Wilmington and Tulsa massacres before them, spoke volumes about how white entitlement is static and still goes unchallenged. Consequently, the white unseen remains etched into the landscape because it was built that way.

Conclusion

The "white unseen" exposes the ways in which physical and cultural landscapes render whiteness and white supremacy invisible. Let's return to the White House for a final example of how it came to be that white and Black Americans had such different reactions to the January 6th siege on the US Capitol.

In 2016, President Barack Obama completed his second term as the first African American President of the United States. For Americans who had grown cynical regarding politics, the economy, and the ongoing wars in Iraq and Afghanistan, Obama initially provided a renewed sense of hope and change. His idyllic family which was enhanced by his wife Michelle, a statuesque lawyer who was an alumna of Princeton and Harvard, provided some white Americans with a counterexample to the stereotypical images that they were barraged with constantly. Many white Americans saw themselves in the Obamas without seeing race, thus reinforcing the myth that America had become a post-racial society. For other Americans, eight years of a Black president had built up feelings of angst, rage, and disenfranchisement. For these Americans, Trump's election was a welcome interruption to Black progress and to the inter-racial coalition that threatened what they viewed as their "natural" place in the American social and political order.

When Donald Trump became the 45th President of the United States, it quickly became clear that Trump and the Republicans were obsessed with "making America great again," which effectively meant erasing all noticeable markers of a Black presidency and reinstating white supremacy. This was apparent in the handing over of the White House, which is both a cere-monial and practical act. Incoming first lady Melania Trump refused to move into the White House for six months, reportedly until Michelle Obama's bathroom shower and toilet had been ripped out (Lippman 2020). Melania described herself as "passionate about the historic pres-ervation of the White House and its grounds," yet she eviscerated the historic rose garden and removed all the Obama's décor, which she called "old" and "shabby" (Lippman 2020). Melania and Donald Trump's transition of power thus reified a narrative that erased Blackness and positioned whites as restoring order, purifying a racialized landscape, and returning America to an untainted Garden of Eden.

Thus, the January 6th siege, like the 1898 coup d'état in Wilmington before it, is symbolic of how whiteness erodes and mutes anything that is not white. Trump's election and transi-tion into the White House was about turning back the progress of racial representation and reinstating white supremacy. Yet, for many white Americans, this largely went unseen. For them, the mantra "Making America Great Again" evoked sentimental images that harkened back to simpler, safer times. But for whom? In considering these events, one should ask: What is the America I know, and how is that America shaped by my racial lens?

References

Atlanta Constitution. 1898. "Cause of Carolina Riots: Crime, Insult, and Corruption." *Atlanta Constitution,* November 11, p. 3. Accessed April 25, 2022. www.newspapers.com/image/26913483/.

Bush, C. 2021. "Opinion: This is the America that Black People Know." *Washington Post,* January 9. Accessed March 14, 2022. www.washingtonpost.com/opinions/2021/01/09/cori-bush-capitol-mob-white-supremacy-government/.

Coates, T-N. 2017. "The First White President." *The Atlantic.* Accessed March 14, 2022.www.theatlantic.com/magazine/archive/2017/10/the-first-white-president-ta-nehisi-coates/537909/.

Freedom Trail Foundation. 2022. *The Freedom Trail.* Accessed April 25, 2022. www.thefreedomtrail.org.

Gilded Age. 2022. "A Long Ladder." Season 1, Episode 4. Directed by M. Engler and S. Richardson. Aired February 14, 2022 on HBO.

Inwood, J. 2015. "Neoliberal Racism: the 'Southern Strategy' and the Expanding Geographies of White Supremacy." *Social and Cultural Geography* 16 (4): 407–423.

Lippman, D. 2020. "8 Juicy Details from the New Melania Trump Tell-All Book." *Politico.* August 29 Accessed March 14, 2022. www.politico.com/news/2020/08/29/new-melania-trump-tell-all-book-404926

Lipsitz, G. 2011. *How Racism Takes Place.* Philadelphia, PA: Temple University Press.

Meschede, T., D. Hamilton, A. P. Muñoz, R. Jackson, and W. Darity Jr. 2016. "Wealth Inequalities in Greater Boston: Do Race and Ethnicity Matter?" *Federal Reserve Bank of Boston, Community Development Discussion Papers.* No. 2016–02. www.bostonfed.org/commdev.

New York Herald. 1898. "Whites Kill Negroes and Seize the City of Wilmington." November 11, 1898.

Seguin, C. and D. Rigby. 2019. "National Crimes: A New National Data Set of Lynchings in the United States, 1883 to 1941." *Socius.* doi: 10.1177/2378023119841780.

Shaw, W. S., R. D. K. Herman, and G. R. Dobbs. 2006. "Encountering Indigeneity: Re-Imagining and Decolonizing Geography." *Geografiska Annaler. Series B, Human Geography* 88(3): 267–276. www.jstor.org/stable/3878372.

Tyson, T. 2006. "The Ghosts of 1898: Wilmington's Race Riot and the Rise of White Supremacy." *Raleigh News & Observer*, November 17, p. 1H. media2.newsobserver.com/content/media/2010/5/3/ghostsof1898.pdf.

WBUR Newsroom. 2021. "Go Home: Trump Tells Supporters Who Mobbed Capitol to Leave, Again Falsely Claiming Election Victory." January 6. Accessed April 27, 2022. www.wbur.org/news/2021/01/06/go-home-trump-supporters-us-capitol-transcript.

8

GENDER, SEXUALITY, AND LANDSCAPE

Christina E. Dando

Alexandre Hogue's powerful 1936 painting *Erosion No. 2—Mother Earth Laid Bare* is dominated by the contours of a woman's body emerging in the eroded soil (Figure 8.1). When we think of "landscape," we could be speaking of any number of things, such as a location taken in by the human eye, a location shaped by human actions, or a painting or photograph communicating a human's experience of a location. The human eye and the brain behind it project onto landscapes their thoughts, hopes, dreams, and cultural perceptions, as well as their imagination. Landscapes have long been infused with gender and sexuality by their viewers. With Hogue's painting, the viewer is positioned as voyeur, their gaze directed onto the woman's form, assaulted by the plow, left exposed to wind and water. *Erosion No. 2* captures Hogue's impression of the Dust Bowl era and cultural traditions linking women to landscape ("Mother Earth") (Kalil 2011).

This chapter considers the ways the American landscape has been gendered and sexualized, the ways gender and sexuality impact how it is experienced, and reflects on gender-specific spaces in it. From representations through diverse gendered spaces, gender and sexuality are written and performed on the American landscape.

Gender, at times defined as the social construction of characteristics for boys and girls, men and women, can also be described as a regulated performance that is not a strict binary but is often "contested and troubled" (Browne et al. 2010, 573). Sexuality is not only who an individual is attracted to but also the practices and spaces associated with these desires. Today, we acknowledge the great spectrum and fluidity of gender and sexuality. Like gender, sexuality is a performance "inscribed on the body and the landscape," that takes place as well as makes place (Bell and Valentine 1995, 8; Oswin 2013, 106). Even heterosexuality is a performance situated in place, takes place, makes place, and varies from place to place. Normative heterosexuality regulates bodies and insists individuals be either male or female, with one constructed in opposition to the other, with one attracted only to the other. These regulated practices extend beyond the bodies to the ways individuals move through their landscapes.

Erosion No. 2 reflects old cultural traditions personifying the earth, landscapes, and nations as feminine. European explorers and settlers described the American landscape in feminine terms (Kolodny 1975). Cartographers included decorations of nude women for the pleasure of their European male audience, such as on the title page of Abraham Ortelius' *Theatrum Orbis Terrarum*

DOI: 10.4324/9781003121800-11

Figure 8.1 Alexandre Hogue (American, 1891–1994). *Erosion No. 2–Mother Earth Laid Bare*, 1936. Oil on canvas, 101.6 × 139.7 cm. Philbrook Museum of Art, Tulsa, Oklahoma. Museum Purchase, 1946.4. © Olivia Marino.

(1573 edition) where the continents are depicted as women, only Europe is clothed and North America is an Indigenous woman (Harley 1988, 299). The United States has been depicted as Columbia and as Liberty, Greek goddess-like figures (Johnson 2016; Groseclose 2001, 350). Less common are masculine personifications: the US was first personified as Uncle Sam in the early nineteenth century (Figure 8.2).

In considering gender, sexuality, and landscape, this chapter draws on examples from the American Great Plains. An 1872 promotional newspaper letter describes Nebraska as "her":

> Although [Nebraska's] development has just begun, her natural advantages are such that the casual observer cannot fail to see the great and glowing future that awaits her when all her rich lands shall be under cultivation, and when within her vast borders are gathered together the brain and talent, the muscle and sinew, of the hardy industrious emigrant from the too thickly populated regions of our country and Europe.
>
> *(Quoted in Davis 1975, 183)*

On an 1890 poster, South Dakota is an attractive young woman being presented by Uncle Sam to a crowd of men (refer again to Figure 8.2). But the Plains are not always the maiden waiting to be transformed by the masculine touch. In Ole Rolvaag's *Giants of the Earth*, the giant is actually a giantess: "Monsterlike the Plain lay there... She would know, when the time came, how to guard herself and her own against him!" (Rølvaag 1929, 249). A "virgin" land to be broken by the yeoman farmer and his plow, a Mother Earth stripped and abused, a monstrous

Figure 8.2 "Free homes, government lands, and cheap deeded lands in South Dakota," circa 1890. In this South Dakota's Immigration Commission poster, South Dakota is being presented at the World's Fair (background) by Uncle Sam (white top hat) to European "suitors." The text at the bottom extolls "her" virtues, "Her Wheat, Corn, Oats . . . are of First Quality. Her hills and mountains yield GOLD, SILVER, COPPER . . ." Beinecke Rare Book and Manuscript Library, Yale University.

giantess plotting her revenge: personifying the Plains as feminine is problematic for both landscape and actual women. This chapter will survey work in gender, sexualities, and landscape before considering a case study on man camps where all three elements come together.

A Historiography of Gender, Sexuality, and Landscape

Early landscape work was written by men, for men, about the ways in which men had seen and constructed the world around them. Western society's patriarchal structure had created a world and an epistemology that placed men at the center. With the 1960s came increased interest in women's experiences and perspectives. Academic research began to consider the linkage of earth, landscape, and femininity. Annette Kolodny's *The Lay of the Land: Metaphor as Experience and History in American Life and Letters* examined the construction of the American landscape as feminine in early explorer and settler accounts (1975). This construction was "expressed through an entire range of images, each of which details one of the many elements of that experience, including eroticism, penetration, raping, embracing, enclosure, and nurture, to cite only a few" (Kolodny 1975, 150). These images continue to be utilized today.

Geographers, such as Janice Monk, engaged in scholarship considering the linkages of the domination of women to the domination of nature, of women and nature, and of women and space (Monk 1984, 24–25). Monk's work included calls for more diversity in women's landscape responses, studies of different contexts, and deeper conceptual work tied to ethical discourses on humans and their environment (Monk 1984, 29). She contributed to this growing literature with her edited collection with Vera Norwood: *The Desert Is No Lady: Southwestern Landscapes in Women's Writing and Art* (1987).

Gillian Rose (1993) critiqued geographic scholarship as historically masculinist, with men gatekeeping what could be considered "geographical knowledge" and who could produce it (Rose 1993, 2). With landscape, Rose focused on the importance of gaze in geography and on landscape's gendering:

> The female figure represents landscape, and landscape a female torso, visually in part through their pose: paintings of Woman and Nature often share the same topography of passivity and stillness. The comparison is also made through the association of both land and Woman with reproduction, fertility and sexuality, free from the constraints of Culture. Incorporating all of these associations, both Woman and Nature are vulnerable to the desires of men… the sensual topography of land and skin is mapped by a gaze which is eroticized as masculine and heterosexual.
>
> *(Rose 1993, 96–97)*

Rose suggests it is the "intersections of voyeurism and narcissism, then, [that] structure geography's gaze at landscape": to gaze as a geographer is then to gaze from the white bourgeois heterosexual masculine vantage point (Rose 1993, 108). Rose concludes her consideration of landscape reflecting on feminist work and suggesting an alternative feminine position of seeing landscape and the possibilities of more relational alternative gazing (Rose 1993, 111–112). Since its publication, other geographers, such as Karen Morin (2008), have considered such alternative gazing, exploring how different women engage with landscape and communicate their experiences, contributing to their audiences' geographical imaginations. While gender is often used as a shorthand to address women and women's issues, scholars have also considered the construction of masculinities and their construction in relation to femininity (Pratt and Erengezgin 2013, 77).

The American landscape has historically been associated with explorers, adventurers, and cowboys, a mythology of heroic masculine figures roaming this landscape. The collection *Across the Great Divide: Cultures of Manhood in the American West* explores the relationship between the Western landscape and masculinity, and considers their histories as well as their representations in popular culture (films, novels, music) (Basso et al. 2001). Portions of the landscape once described as "frontier" continue to be viewed as frontier and associated with the performance of masculinities. Dando considered the cinematic construction of masculinities in an essay on the film *Boys Don't Cry* (2005), arguing the Plains were still constructed as a frontier in the film where the construction of masculinity is fluid (Dando

2005, 17). Maureen Hogan and Timothy Pursell explored the ways the contemporary Alaskan landscape is intrinsically tied to masculinity and the control and conquest of nature, marginalizing or repudiating "that which is female, urban, and Native" (Hogan and Pursell 2008, 81). Through these performances "real Alaskans" are constructed as masculine, both for themselves as well as for the nation (Hogan and Pursell 2008, 82).

Work on sexualities and landscape developed after feminist geographies, with a dynamic body of research emerging after 1995 considering the performance of sexual identities and their inscription on the body and the landscape (Oswin 2013). Bell and Valentine's *Mapping Desire: Geographies of Sexualities* was one of the first collections to consider the "spaces of sex and the sexes of place" and includes essays addressing landscapes and how they can be either "liberatory or oppressive sites for the performance of our sexed selves" (Bell and Valentine 1995, 1, 98).

Queer theories, emerging during the same period, challenge heteronormativity and its construction of homosexuality as its binary opposite, acknowledging the importance of sexualities to identity and the incredible diversity of experiences and identities, including issues of race and ethnicity (Brown, Browne, and Lim 2007). Scholars such as Larry Knopp and Jon Binnie actively engaged with queer theories to critique the narrow heteronormative view of traditional geographical knowledge, working towards a "queering of the geographical imagination" (Binnie 1997; Knopp 2007, 27).

Surveying the body of research in geographies of sexualities, urban and rural landscapes have received the greatest amount of attention. But recent work has found the stereotypical urban "gay" neighborhood is not the urban experience for many lesbians and queers (transgender and gender non-conforming people). Jen Jack Gieseking's (2020) research draws on extensive interviews to find they produce spaces Gieseking describes as "constellations"—scattered and fleeting locations connected by lines (subways, bus routes, walks)—that are an "alternative geographic imagination" of urban space (2020, 942).

One of the most dynamic areas in geography to develop recently has been the area of Black Geographies, with the discipline finally beginning to consider diversity in landscape responses considering race as well as gender and sexuality. (See Chapter 7 as well as the Special Essay: Bridging Social and Political Landscapes.) In 2014, the journal *Gender, Place and Culture* featured a two-part themed section on "Gender and Sexual Geographies of Blackness," largely addressing the American landscape (Bailey and Shabazz 2014, 319). LaToya Eaves' work addresses sexualities and landscape, with one paper exploring a young Black queer woman's experiences moving to a rural Southern community to consider the "overlapping dialectics that link institutions, power, and knowledge" and to deconstruct "queer sexual identities, spatial productions, and sociospatial interactions" (Eaves 2017, 87). Similarly, work on Latinx geographies has been increasing and includes work on sexualities.

Scholarship has widened to encompass a rainbow of sexualities, genders, and identities and its spatial construction and negotiation, such as the recent collection *Identities and Place: Changing Labels and Intersectional Communities of LGBTQ and Two-Spirit People in the United States* (Crawford-Lackey and Springate 2020). In particular, recent work has considered the experiences of transgender persons who face unique landscape challenges, from the small scale (bathrooms and locker rooms) to the large scale (communities, states, nations) (Browne 2004; Robbins and Helfenbein 2018). Persons identifying as transgender live lives departing from conventional gender as assigned at birth, that "cross over (trans-)" gender categories (Stryker 2020, 89). Some historical periods were more tolerant, some cultures and/or some landscapes today are more tolerant of transgenders than others, resulting in an uneven geography across time and space (Stryker 2020). Historically, frontiers were locations where individuals could reinvent themselves, crossing gender boundaries to suit their own purposes (Boag 1998; Stryker

2020). In the twenty-first century, transgender individuals are more visible in the landscape, a result of transgender people "finally becoming visible." As Stryker notes, "gender itself is changing radically in ways we can now scarcely comprehend" (2020, 110).

Case Study: Man Camps on the Plains

Over the last twenty years, an oil boom in western North Dakota, centered on Williston, led thousands of men to migrate there for work, quickly outgrowing available housing, and resulting in "man camps." Man camps are a gendered landscape—landscapes dominated by one gender—often associated with extractive industries or the military (Holdsworth 1995). Housing can be temporary or permanent. Most man camps include small populations of women. As western North Dakota boomed, the man camps received considerable press, with the *New York Times* declaring "man camps" one of the most important words of 2012 (Caraher et al. 2017).

Man camps are not new to the Plains. Historically, military forts were predominantly male. The cowboy profession was almost entirely male as were the cattle towns serving the industry: Dodge City, Kansas, had an estimated seven times as many men than women in 1870 (Courtwright 1996). Homesteading is stereotyped as a masculine endeavor of breaking the land, with bachelors in claim shanties or a father with his family. Scholar Peter Boag suggests the transformation of the frontier into a garden was

> … a heterosexual myth… With Indians vanquished, women subordinated to husbands, no cities in sight, and idealized family farms everywhere, this myth effectively excluded non-heterosexuals from the garden…
>
> *(Boag 1998, 329–330)*

Man camps result in pronounced masculine, heterosexual spaces on a landscape long personified as feminine.

The presence of man camps in Williston, a "testosterone-soaked town," leads to complicated gender relations on and with this landscape, including masculine violence against women and against the feminine landscape (Ernst 2014; Martin 2019). According to Williston city officials, at times there were as many as 20 men to every woman (Hagen 2019). In some man camps it is estimated 99% of residents were male (*The Economist* 2013). Long, hard work, isolation from families, available cash, and available alcohol and drugs create an environment ripe for abuse. Prostitution and strip clubs cater to the hypermasculine audience with money to burn; CNN reported strippers could make up to $2,000 a night in Williston (Ellis 2011). In this extremely skewed gender landscape, "reports of assaults and rape… increased 400 percent from 2004, and would skyrocket to a comparative 600 percent by 2017" (Hagen 2019). The few women working in the oil industry often experienced harassment: "It's cutthroat out there… It's a big dick-swinging contest" (Hagen 2019).

Scholars have suggested the issue is "rigger culture," a particular performance of masculinity linked to the oil industry, a term associated with sexism, hypermasculinity, and a disconnect from the local community (Goldenberg et al. 2008; Pippert and Schneider 2018). Female residents of Williston and neighboring communities described experiences of harassment as they went about their daily lives (Pippert and Schneider 2018).

Rigger culture has created even greater problems for Native American communities, such as neighboring Fort Berthold Indian Reservation. The oil boom brought employment opportunities but also greater risks: Native American women are reportedly victimized at 23.2 per 1,000, with more than half experiencing sexual violence in their lifetime; and they are reportedly ten times more likely to be murdered (Hagen 2019; Hylton 2020). There is growing evidence linking the disappearance and murder of Indigenous women to the fossil fuel industry. Native American activist Pamela CɘlalákɘmBond observed: "If you put a pin marker in every place where [I]ndigenous women are missing they're in very close proximity to man camps and pipelines. So that black snake doesn't just pollute the land and the water it's on—it's consuming [I]ndigenous people" (quoted in Clabots 2019).

Organizations such as the Women's Earth Alliance draw attention to the parallels between violence against women and violence against the earth. Konsmo and Pacheco's handbook for action, *Violence on the Land, Violence on Our Bodies* (2016) documents "the ways that North American Indigenous women and young people's safety and health are impacted by extractive industries" to expose and curtail their impacts on their communities and lands (Konsmo and Pacheco 2016, 2–3). Throughout the handbook are expressed the "intimate connection" between women and land, such as "The land is our Mother, so when we lose value for the land... people lose value for the women" or "each succeeding generation inherits a body burden of toxic contaminants from their mothers. In this way, we, as women, are the landfill" (Konsmo and Pacheco 2016, 4, 20).

This linkage of women and land was first discussed and critiqued by ecofeminist scholars and activists (see Box 8.1). Rigger culture explicitly engages with these connections, using double entendres connecting sexual intercourse to extraction, such as to "drill" or "lay pipe," linking the female body to oil extraction (Parson and Ray 2020, 249). In Williston, both the landscape and women are treated as commodities to be exploited, utilized until no longer economically valuable:

> They [extractive industries] treat Mother Earth like they treat women," [states] Lisa Brunner, a member of the White Earth Ojibwe Nation... "They think they can own us, buy us, sell us, trade us, rent us, poison us, rape us, destroy us, use us as entertainment, and kill us.
>
> *(Walsh 2018, 40)*

Native American and environmental justice activists also link landscape and women but focus on a positive, relational connection to landscape. This relational connection can also be found in the non-Indigenous residents who call this landscape home. In an editorial, a former oil worker described the "irreparable harm... done to the cultural landscape of a place" by extractive industries, observing "transient workers tend not to value a place in the way that those who call that place 'home' do" (Zawojski 2015). Those who call this landscape home struggled with the man camps and the boom, welcoming the growth and development but despairing at the dramatic changes the skewed gender ratio brought to their community (Pippert and Schneider 2018).

Box 8.1 Ecofeminism

Christina E. Dando

Ecofeminism developed out of the environmental and feminist movements of the 1960s/early 1970s as an approach to women and landscape. Many different cultures around the world have long linked women to nature and/or landscape, revering female personifications on the earth—the Mother-Goddess, Earth Mother, Mother-Nature—symbolizing life and fertility (Giancola

2018). With the rise of the environmental and feminist movements came an interest in critically considering the linkage of women and nature by societies. Ecofeminism views the hegemony of patriarchy and capitalism at the root of the exploitation and degradation of the environment as well as women (Buckingham-Hatfield 2000). Likewise, ecofeminists believe women's liberation cannot be achieved without also the liberation of nature (Gaard 1997).

Some ecofeminists embraced the association of women with nature, suggesting that because of their biology, women have a "particular relationship with nature … this proximity to nature qualified them to speak more eloquently on nature's behalf" (Buckingham 2004, 147). Other scholars reject this, arguing that reducing anyone to their "essence" would be problematic, basically restricting those who are "Othered," be it based on gender, ethnicity, sexuality, ability or other categories, to an essence that is unchanging and unchangeable (Buckingham 2004; Rose 1993). Ecofeminism today is both an analysis of society–nature relations and a practice working to transform these relationships and is more a matter of ecofeminisms: a diverse set of theoretical approaches linking capitalism, colonialism, ecological domination, and heteronormativity, recognizing that "many systems of oppression are mutually reinforcing" (Parsons and Ray 2020; Buckingham-Hatfield 2000; Gaard 1997, 114).

The fields of ecofeminism and environmental justice overlap, with environmental burdens being unequally distributed and poor communities being at greater risk (see Chapter 6). Women are especially active in the environmental justice movement. In the early twentieth century, American women became involved in the nascent environmental movement, extending the traditional role of women beyond that as caretakers of home and family to the larger environment. Women and children are disproportionately affected by polluted environments (home and work), particularly those unable to leave harmful locations due to social or economic restrictions, with injustice experienced at scales ranging from the body to the community (Bell 2016; Kirk 1997). Today, groups in the United States work on addressing both environmental and gender issues, such as WE ACT for Environmental Justice, Intersectional Environmentalist, Women's Earth and Climate Action Network (WE CAN), and the Indigenous Environmental Network.

In particular, the looming climate crisis has led to the action of women across the United States (and the world), in what some have termed "climate feminism." Geographer and activist Katharine Wilkinson has said: "The climate crisis is not gender-equal or gender-neutral" and "If you're going to be a feminist on a hot planet, you have to be a climate feminist" (Jackson 2021). Climate feminism illuminates the intersection of our global climate crisis and gender inequality, seeking to mobilize women in the work of climate mitigation and adaptation, as extreme weather events are exacerbating gender issues such as early marriage, sex trafficking, and domestic violence.

Ultimately, man camps themselves are transient, as many such gendered landscapes are, only lasting as long as the boom. Williston's man camps slowly closed as oil prices dropped and production slowed, the workers moving on to other landscapes (Healy 2016).

Williston's landscape became gendered with the arrival of thousands of male oil and construction workers and sexualized as heterosexual as female prostitutes and strippers followed them, catering to the captive, eager market. At the same time, the Plains landscape was and still is associated with femininity, with early white explorers and settlers as well as later oil workers sexualizing this landscape and what they do with it. Extractive industries are an extreme form of landscape consumption, "literally devouring whole sections of land and water" (Domosh

2013, 204). The gaze of the riggers on this landscape is only what they want from it: oil and income.

Gender and sexuality impact how landscape is experienced. In this landscape, what it means to be a man versus what it means to be a woman results in very different experiences of landscape. The large number of men combined with rigger culture's hypermasculinity result in a landscape where heterosexual men move easily through the space, dominating it on every level. The performance of rigger masculinity became associated with the violent penetration of the earth in pursuit of petroleum while objectifying and commodifying women. Women move uneasily through this landscape, subject to male gaze: while some may endure the attention to make money, others do not want it but are unable to avoid it (Parson and Ray 2020). Rigger culture is superficially fixated on women for heterosexual encounters but not all are: sex trafficking in North Dakota also includes males and children (Hagen 2019). The men work hard physically; they seem to desire to play equally hard.

Women and landscape/resources are viewed as having been put on this earth to be utilized by men, just as early Europeans and Americans saw the landscape two hundred years ago and as the South Dakota poster illustrates "Her hills and mountains yield GOLD" (Figure 8.2) (Kolodny 1975). As American society becomes more mindful of the problems associated with such mindsets, of the unequal power relations associated with such thinking, there are glimmers of hope we can move beyond such gendered behaviors, such constructions of gender (both of men and of women), and such ways of thinking about our earth, its landscapes, and its resources.

Native American women link the mistreatment of women and mistreatment of the earth, embracing the link between women and earth, and calling on caring for both. This requires walking a fine line between essentializing and having a thoughtful, reciprocal relationship with our landscapes: "environment" shifts from having humans outside of nature, to repositioning humans *within* nature, so "environmental concerns are simultaneously about human well-being too" (Gaard 2017, 82).

Besides the gender in this landscape and the gendering of this landscape, there is also the portrayal of this landscape and man camps in the media. The textual construction of this landscape is yet another element of landscape in the Williston story. Through media coverage, the gaze of those off the Plains (nation, international) is drawn to this hypersexualized landscape, to the "Wild West" being reconstructed in modern America.

The man camps of Williston represent gender-specific spaces in a feminized landscape, the ways gender and sexuality impact how it is experienced, and the ways the landscape has been gendered and sexualized. Despite 200 years of exploration and settlement on the Plains and 200 years of cultural evolution, there are still gendered spaces in this landscape. The similarities between past man camps and the latest in Williston is not lost on Plains residents: Williston resident Chuck Neff said "'It's like if you turned the clocks back 200 years and you got tents up on the prairie'" (Nicas 2012, 4).

Final Thoughts

Of course, not all men support masculine violence against landscape. Montana cowboy poet Wally McRae's poem "The Land" critiques masculine mistreatment of the land, writing:

> You'd ravish her with mindless lust,
> Then curse her for a whore,
> … … … …
> Tear her vitals out,

    ... .

haphazardly putting her back together,

    ... ...

leave her prostrate there.

(McRae 1986, 16)

McRae could be writing about the Williston basin or describing Hogue's painting (Figure 8.1). Yet, despite his critique of what is being done to "the land," he is still constructing the landscape as feminine.

In some ways, we have come a great distance in advancing work on gender, sexualities, and landscape. This chapter only sketches out broadly a diverse, expansive literature on gender, sexualities, and literature: much work has been done, yet much work remains. We are more attuned today to gender and sexualities and their construction, but there are gaps: gaps between landscape research and research on gender, sexualities, and race; gaps between academic discussions and American society when it comes to gender, sexualities, and especially race. Despite advances, landscapes are still being gendered, with "feminine" constructed as passive and immobile, still associated with reproduction, fertility, sexuality, open to the heterosexual male gaze, eroticized. Landscape, our earth, and how we view it and construct it, does not have a gender or sexuality. Gender and sexuality are social constructs and are performed. But landscape is also a lived experience and gender and sexualities impact how we experience it, how we view it, and how we construct it.

Hogue's *Erosion No. 2* is a dramatic scene of violence against the earth, yet it also contains hope if you take your eyes off the female form. Under the plow handles, alongside erosion forming Mother Earth's hip, along the water at the bottom of the canvas, new green grass is emerging. A bird flies over the abandoned farmstead. The landscape endures.

References

Bailey, M., and R. Shabazz. 2014. "Gender and Sexual Geographies of Blackness: Anti-Black Heterotopias (Part 1)." *Gender, Place and Culture* 21 (3): 316–321.

Basso, M., L. McCall, and D. Garceau, eds. 2001. *Across the Great Divide: Cultures of Manhood in the American West*. New York: Routledge.

Bell, K. 2016. "Bread and Roses: A Gender Perspective on Environmental Justice and Public Health." *International Journal of Environmental Research and Public Health* 13, 1005: 1–18.

Bell, D. and G. Valentine. 1995. "Introduction: Orientations." In *Mapping Desire: Geographies of Sexualities*, edited by D. Bell and G. Valentine, 1–27. London: Routledge.

Binnie, J. 1997. "Coming Out of Geography: Towards a Queer Epistemology." *Environment and Planning D: Society & Space* 15 (2): 223–237.

Boag, P. 1998. "Sexuality, Gender, and Identity in Great Plains History and Myth." *Great Plains Quarterly* 18 (Fall): 327–340.

Brown, G., K. Browne, and J. Lim. 2007. "Introduction, or Why Have a Book on Geographies of Sexualities?" In *Geographies of Sexualities: Theory, Practices and Politics*, edited by J. Lim et al., 1–18. London: Routledge.

Browne, K. 2004. "Genderism and the Bathroom Problem: (Re)materializing Sexed Sites, (Re)creating Sexed Bodies." *Gender, Place, and Culture* 11 (3): 331–346.

Browne, K., C. Nash, and S. Hine. 2010. "Introduction: Towards Trans Geographies." *Gender, Place and Culture* 17 (5): 573–577.

Buckingham, S. 2004. "Ecofeminism in the Twenty-First Century." *The Geographical Journal* 170 (2): 146–154.

Buckingham-Hatfield, S. 2000. *Gender and Environment*. New York: Routledge.

Caraher, W., B. Weber, K. Kourelis, and R. Rothaus. 2017. "The North Dakota Man Camp Project: The Archaeology of Home in the Bakken Oil Fields." *Historical Archaeology* 51: 267–287.

Clabots, B. 2019. "The Darkest Side of Fossil-Fuel Extraction." *Scientific American*. October 14. Accessed September 23, 2021. https://blogs.scientificamerican.com/voices/the-darkest-side-of-fossil-fuel-ext raction/.

Courtwright, D. 1996. *Violent Land: Single Men and Social Disorder from the Frontier to the Inner City*. Cambridge: Harvard University Press.

Crawford-Lackey, K., and M. Springate, eds. 2020. *Changing Labels and Intersectional Communities of LGBTQ and Two-Spirit People in the United States*. New York: Berghahn Books.

Dando, C. 2005. "Range Wars: Cinematic Portrayals of the Contemporary Great Plains." *Journal of Cultural Geography* 23 (1): 91–113.

Davis, J. 1975. "Constructing the British View of the Great Plains." In *Images of the Plains: The Role of Human Nature in Settlement*, edited by B. Blouet and M. Lawson, 181–185. Lincoln: University of Nebraska Press.

Domosh, M. 2013. "Consumption and Landscape." In *The Wiley-Blackwell Companion to Cultural Geography*, edited by N. Johnson, R. Schein, and J. Winders, 198–208. London: John Wiley.

Eaves, L. 2017. "Black Geographic Possibilities: On a Queer Black South." *Southeastern Geographer* 57 (1): 80–95.

The Economist. 2013. "There's Gold in Them There Wells." December 21.

Ellis, B. 2011. "Earn $2,000 a Night as a Boomtown Stripper." *CNN Money*. October 25. Accessed September 23, 2021. https://money.cnn.com/2011/10/25/pf/America_boomtown_strippers/index.htm.

Ernst, A. 2014. "The Dark Side of the Oil Boom: Human Trafficking in the Heartland." *Al Jazeera*. April 28. Accessed September 23, 2021. http://america.aljazeera.com/watch/shows/america-tonight/artic les/2014/4/28/the-dark-side-oftheoilboomhumantraffickingintheheartland.html.

Gaard, G. 1997. "Toward a Queer Ecofeminism." *Hypatia* 12 (1): 114–137.

———. 2017. "Feminism and Environmental Justice." In *The Routledge Handbook of Environmental Justice*, edited by R. Holifield, J. Chakraborty, and G. Walker, 74–88 London: Routledge.

Giancola, D. 2018. "Women, Land, and Eco-Justice." In *The Palgrave Handbook of Philosophy and Public Policy*, edited by D. Boonin, 737–747. London: Palgrave Macmillan.

Gieseking, J. J. 2020. "Mapping Lesbian and Queer Lines of Desire: Constellations of Queer Urban Space." *Environment and Planning D: Society and Space* 38 (5): 941–960.

Goldenberg, S., J. Shoveller, A. Ostry, and M. Koehoorn. 2008. "Sexually Transmitted Infection (STI) Testing Among Young Oil and Gas Workers: The Need for Innovative, Place-Based Approaches to STI Control." *Canadian Journal of Public Health* 99 (4): 350–354.

Groseclose, B. 2001. "A History/Historiography of Representations of America." In *A Companion to 19th-Century America*, edited by W. Barney, 345–358. Oxford: Blackwell Publishers. Ltd.

Hagen, C. 2019. "Women in Bad Lands." *High Plains Reader* (Fargo, ND). June 12.

Harley, J. B. 1988. "Maps, Knowledge, and Power." In *The Iconography of Landscape: Essays on the Symbolic Representation, Design, and Use of Past Environments*, edited by D. Cosgrove and S. Daniels, 277–312. Cambridge: Cambridge University Press.

Healy, J. 2016. "Built Up by Oil Boom, North Dakota Now Has an Emptier Feeling." *New York Times*. February 7.

Hogan, M. and T. Pursell. 2008. "The 'Real Alaskan': Nostalgia and Rural Masculinity in the 'Last Frontier.'" *Men and Masculinities* 11 (1): 63–85.

Holdsworth, D. 1995. "'I'm a Lumberjack and I'm OK': The Built Environment and Varied Masculinities in the Industrial Age." In *Gender, Class, and Shelter: Perspectives in Vernacular Architecture V*, edited by E. Cromley and C. Hudgins, 11–25. Knoxville: The University of Tennessee Press.

Hylton, S. 2020. "A Well of Grief: The Relatives of Murdered Native Women Speak Out." *The Guardian*. January 13.

Jackson, L. 2021. "The Climate Crisis is Worse for Women. Here's Why." *The New York Times*. August 24.

Johnson, A. 2016. "Columbia and Her Sisters: Personifying the Civil War." *American Studies* 55 (1): 31–57.

Kalil, S. 2011. *Alexandre Hogue: An American Visionary, Paintings and Works on Paper*. College Station, TX: Texas A&M University Press.

Kirk, G. 1997. "Ecofeminism and Environmental Justice: Bridges Across Gender, Race, and Class." *Frontiers* 18 (2): 2–10.

Knopp, L. 2007. "From Lesbian and Gay to Queer Geographies: Pasts, Prospects and Possibilities." In *Geographies of Sexualities: Theory, Practices, and Politics*, edited by J. Lim et al., 21–28. London: Routledge.

Kolodny, A. 1975. *The Lay of the Land: Metaphor as Experience and History in American Life and Letters.* Chapel Hill: The University of North Carolina Press.

Konsmo, E. M. and A.M. Kahealani Pacheco. 2016. *Violence on the Land, Violence on Our Bodies: Building an Indigenous Response to Environmental Violence.* Women's Earth Alliance & Native Youth Sexual Health Network. Accessed September 23, 2021. http://landbodydefense.org/uploads/files/VLVBReport Toolkit2016.pdf.

Martin, N. 2019. "The Connection Between Pipelines and Sexual Violence." *The New Republic.* October 15.

McRae, W. 1986. *It's Just Grass & Water.* Spokane, WA: Oxalis.

Monk, J. 1984. "Approaches to the Study of Women and Landscape." *Environmental Review* 8 (1): 23–33.

Morin, K. 2008. *Frontiers of Femininity: A New Historical Geography of the Nineteenth-Century American West.* Syracuse: Syracuse University Press.

Nicas, J. 2012. "Oil Fuels Population Boom in North Dakota City." *Wall Street Journal.* April 7, A 4.

Norwood, V. and J. Monk. 1987. *The Desert Is No Lady: Southwestern Landscapes in Women's Writing and Art.* New Haven: Yale University Press.

Oswin, N. 2013. "Geographies of Sexualities: The Cultural Turn and After." In *The Wiley-Blackwell Companion to Cultural Geography,* edited by N. Johnson et al, 105–117. London: John Wiley.

Parson, S. and E. Ray. 2020. "Drill Baby Drill: Labor, Accumulation, and the Sexualization of Resource Extraction." *Theory & Event* 23 (1): 248–270.

Pippert, T. and R. Zimmer Schneider. 2018. "'Have You Been to Walmart?' Gender and Perceptions of Safety in North Dakota Boomtowns." *The Sociological Quarterly* 59 (2): 234–249.

Pratt, G. and B. Çavlan Erengezgin. 2013. "Gender." In *The Wiley-Blackwell Companion to Cultural Geography,* edited by N. Johnson, R. Schein, and J. Winders, 73–89. London: John Wiley.

Robbins, K. and R. Helfenbein. 2018. "Gendered Bathrooms, Critical Geography, and the Lived Experience of Schools." *Journal of Curriculum and Pedagogy* 15 (3): 263–277.

Rølvaag, O. 1929. *Giants in the Earth.* New York: Harper & Brothers Publishers.

Rose, G. 1993. *Feminism & Geography: The Limits of Geographical Knowledge.* Minneapolis, MN: University of Minnesota Press.

Stryker, S. 2020. "Transgender History in the United States and the Places That Matter." In *Changing Labels and Intersectional Communities of LGBTQ and Two-Spirited People in the United States,* edited by K. Crawford-Lackey and M. Springate, 89–129. New York: Berghahn Books.

Walsh, K. 2018. "Moving More Than Oil: The Intimate Link Between Human Trafficking and Dirty Energy in Minnesota." *Earth Island Journal* Autumn: 39–42.

Zawojski, C. 2015. "Fracking, and the Workers It Draws, Change a Region's Cultural Landscape." *Baltimore Sun.* February 19. Accessed September 23, 2021. www.baltimoresun.com/opinion/op-ed/bs-ed-fracking-0220-20150219-story.html.

9

ETHNICITY AND TRANSNATIONALISM

Emily Skop and Stephen Cho Suh

When thinking about the American landscape, as so many chapters in this volume do, specific imagery and the materiality of form comes to mind. We immediately think of fences, sidewalks, farms, and commercial landscapes. In the case of migrant landscapes, too, we conjure those places that embody the identities of newcomers, through housescapes, restaurants, places of worship, and neighborhoods. We cannot as easily evoke the ephemeral landscapes of transnationalism, however, because of the hyper-charged back and forth nature of this phenomenon across space and time. With transnationalism, placemaking amongst migrants takes on new dimensions and remake landscapes in quite transitory and unexpected ways.

In 1992, Glick Schiller, Basch, and Blanc-Szanton wrote their highly influential piece titled "Towards a transnational perspective" (Glick Schiller, Basch, and Black-Szanton 1992). They proposed the idea of transnationalism to describe the process by which migrants forge and sustain multiple social "fields" that cross geographic, cultural, and political borders. They went further to connect this cross-national movement and connection of people, goods, information, and financial capital, describing how migrants across space foster many relationships: social, organizational, religious, and political.

Importantly, as the scholarship on transnationalism has developed, scholars have reworked traditional notions of landscape because of the ways in which migration operates across multiple scales and temporalities. Thus, in this chapter, through short summaries from the urban sociology and geography literatures and more extensive examples from the co-authors' own case studies, we untangle the American landscape through the spatial lens of transnationalism. First, we encourage a move away from seeing, sending, and receiving communities as discrete places towards the conceptualization of abstract, multidimensional spaces with unbounded, often discontinuous boundaries. Second, we suggest a shift from seeing ethnic identities as primordial and fixed in place toward a conceptualization of imagined communities and diasporic return where migrants orient their lives around more than one site in more than one place. In our conclusion, we complicate notions of group and community membership centered on loyalty and subservience to the nation-state (e.g., citizenship), exploring instead "flexible" alternatives that allow ambiguous, fluid, and multiple allegiances. Our conclusion also opens the discussion to the consequences of transnationalism for understanding evolving conceptualizations of the American landscape.

DOI: 10.4324/9781003121800-12

Unbounded Landscapes: A Literature Review

Placemaking describes the processes whereby spaces are transformed by migrants into social and communal places that they can call their own (Smith and Furuseth 2006; Teixeira et al. 2012; Frazier, Tettey-Fio, and Henry 2016). Placemaking has also been defined as the process that occurs when places meet migrants' social and cultural needs, which otherwise might go unnoticed or ignored (Rios and Vazquez 2012; Kaplan 2018). For migrants, their relationship to place—social and emotional—is integral to their own sense of self (Jones 2008; Price and Benton-Short 2008; Singer, Hardwick, and Brettell 2008; Airriess 2015). For receiving societies, placemaking is typically seen as just one step in an eventual march towards assimilation, and while romanticized in the initial phases of settlement, is expected to diminish in importance over time as migrants (and their descendants) become integrated into the dominant culture.

Placemaking has typically been affixed to absolute locations, whereby quintessential migrant communities take form in quite recognizable and visible ways. In the American landscape, scholars have spent decades identifying the character of these neighborhoods, often referred to as ethnic enclaves, by focusing on the style and exterior decoration of housing, businesses, and community centers, and identifying the skin color, speech, and surnames of residents (see overviews by Skop and Li 2003; 2011). Scholars frame these ethnic enclaves in the context of an assimilationist paradigm, and connections are made between integration, identity (re) formation, and placemaking. The idea is that while these ethnic enclaves serve a temporary sheltering function, eventually these communities should lose their utility as residents move away and become "Americanized." This is a process known as spatial assimilation and refers to the suburbanization process that defined the mid-to-late twentieth century flight of (mostly) white ethnic European minorities away from the city center, and with that flight, a shift away from their migrant identities. The idea goes further to argue that all visible markers of ethnic difference disappear in "all-American" suburbs through generational shifts and individual integration (see Murdie and Skop (2012) and Skop and Li (2017) for critical overviews of the spatial assimilation paradigm, and in particular the role of racialization in migrant incorporation).

Early twenty-first century transnationalism is rewriting the ways we think about "American" landscapes and processes of ethnic identity formation. With transnationalism, while the landscapes may still be identifiable as ethnic enclaves, the migrant communities themselves become more abstract, multidimensional spaces with unbounded, discontinuous boundaries. Peggy Levitt describes in *Transnational Villagers* (2001) how Dominican migrants build a culturally (and visibly) familiar community in New York City while simultaneously maintaining bonds that extend the homeland. Meanwhile, Li (2009) defines "ethnoburbs" as the spaces where migrant identities are negotiated and transformed, and not without controversy, as they continue to thrive and transform Southern California's suburban communities. Yu (2018, 2020), too, explores the continuously created lived spaces and experiences of the Chinese diasporic community in Flushing, New York. More recently, Kim (2018) asserts that Manhattan's Koreatown offers a unique case in which the enclave's transnationality is contingent as much upon the "nation branding" goals of the sending country (South Korea) as it is the participation of local transmigrant actors. Kim (2018) dubs this new formation a *transclave*:

> *Transclave* refers to a consumption ethnic space, lacking residencies, where people (entrepreneurs and consumers regardless of their race, ethnicity and gender), transnational corporations, and pop and consumer culture move back and forth between two nations and become embedded in a small geographic section of a global city.
>
> *(p. 291)*

Figure 9.1 Korea Sah Buddhist Temple in Koreatown, Los Angeles. Photo courtesy of Jen Mapes.

In all of these examples, migrant enclaves are visible signatures on the landscape (see Figure 9.1), but they are also fluid, transnational spaces that recognize a shift away from notions of permanency and assimilation.

Unbounded Landscapes: A Case Study

Indian migrants in Phoenix (and elsewhere) provide another way to consider the transformation of the American landscape. In this case, drawn extensively from Emily Skop's broader ethnographic project on the Indian diaspora in the US (Skop 2012, 2014, 2015, 2017), newcomers utilize a variety of alternative placemaking strategies in their attempts to recreate and maintain a sense of cohesion and solidarity despite lacking visibility in the urban landscape. These "transitory spaces" are the places and spaces of Indian community formation that employ different members of the community in the project of cultivating alternative notions of "Indian-ness" an identity neither based on here or there, nor on one monolithic conceptualization of being "Indian" but rather based on transnational and translocal notions of self and community (Skop 2012, 2015). They also include engaging on the Internet as a "thirdspace" that has become imbued with symbolic meaning as it becomes the digital terrain upon which many migrants build and reproduce important community ties and relationships on a regular basis (Skop 2016), along with engaging in their local surroundings in receiving communities through placemaking. These strategies are in reaction to a racialization process occurring in an anti-immigrant environment whereby Indian migrants and their children are negotiating their ethnic identities as a "model minority" (Skop 2017). Having "model minority" status assumes a positive connotation, but is inherently a racialized stereotype with potentially deleterious consequences: something that Indian migrants and their children increasingly recognize as they become more permanently established in their new communities.

These "transitory" spaces include movie theaters and area schools. As an example, nearly every weekend a local cinema is transformed as migrants bring their children to watch movies imported from their home states. Usually the theater hosts customers wanting to pay less to watch older-release movies. Once a week, however, migrants living in metropolitan Phoenix descend upon the theater to catch the latest releases from India. A couple began the movie events as a hobby to stay involved with the Gujarati Association. They found distributors in Chicago, New York, and San Francisco and started selecting and purchasing the most recent and popular overseas Gujarati films. Ironically, some of the latest hits reach the United States before they are even released in India (Helweg and Helweg 1990, 129). Given that India produces over 1,000 movies a year (the Indian film industry is known as "Bollywood"), the process of purchasing movies turned out to be a complicated, and time-devouring affair for the Gujarati couple. Yet, they have persisted in the venture because of their commitment to the migrant community. While the movie purchases rarely prove monetarily profitable, the couple recognizes the social rewards of bringing together members of the Gujarati Association in a more informal setting.

Meanwhile, the Maharashtra Mandal gathering, Ganeshotsav (Ganesh festival), takes place annually in late summer. Traced back to India, the festival is focused entirely on the Hindu God, Ganesh. Ganesh is the elephant-headed deity, son of Shiva and Parvati and renowned for his benevolence in granting wishes and offering protection (Knott 2000, 102). He is a popular god especially in the central and southern states of India. Traditionally, the event required community members to pay tribute to Ganesh with visits to the local temple, followed by a period of meditation and contemplation. Ganeshotsav also provided the occasion to cook special foods, and to revel in dancing and music.

In its Phoenix manifestation, Ganeshotsav is an all-day celebration at a local high school and includes an afternoon banquet lunch followed by religious services (Skop 2012). The central event of the day is a roll call of all members to present gifts to Ganesh. From the eldest participant to the youngest, everyone recommits themselves to the deity with the presentation of a special food offering. During the local rendering, Ganeshotsav serves not so much as a religious event but as a time of homecoming and remembrance for Indian immigrants from the state of Maharashtra, in central India. The celebration is tightly focused on the members of the local Maharashtra Mandal rather than on the entire Indian community living in metropolitan Phoenix. It is a relatively small affair, with just over 300 participants, most of whom claim Maharashtra as their regional homeland. Interestingly, the event has a far better turn out than the larger India Festivals sponsored by the local pan-Indian umbrella organizations, with more than half of the migrants who claim to be from Maharashtra participating in Ganeshotsav. With such active membership, Ganeshotsav serves to intensify the already tightly knit community and continuously reconnect members with their traditions. The festival's commingling of the sacred and secular also provides migrant parents with the opportunity to introduce their US-born children to traditions found specifically in their home region of Maharashtra. The festival regenerates a sense of inwardly focused "Maharashtra-ness" that differs dramatically from the collective notions of "Indian" ethnicity promoted by larger associations.

These local Indian associations are loosely organized around linguistic and regional divides (Skop 2012). The popular get-togethers form an integral part in transnational identity formation, and reinforce a process of parochialization in which subgroups such as Gujaratis, Punjabis, Tamils, Sikhs, and Keralites strengthen their particular "back home" identities rather than succumbing to broader, more nationalistic notions of "Indian-ness."

Then there is the India Association's Republic Day celebration. This is an event that utilizes a local school auditorium as a "transitory" site to commemorate its version of "Indian" ethnic heritage. Costumes, music, and folk dance capturing India's "past" commingle with modern performances to create all the trappings of a national celebration in India. The celebration also includes important "American" traditions, suggesting that Republic Day has been re-invented to accommodate to its new local setting. More than 100 families (around 500 attendees) from all areas of metropolitan Phoenix gather at a local high school for the Republic Day celebrations and is primarily a family affair. One festival coordinator shares that many of the participants play a part year in and year out. The conspicuous presence of second-generation adolescents participating in the festivities also suggests that many of the association's members have been living in the US for some time.

In many ways, the festival appears to be commanded by parents passing on some form of teachings to the younger generation for fear that traditions may be lost. The recital is mostly a costumed procession of young, second-generation, female Indian dancers moving to both classic and modern Indian music. The dances are adaptations of traditional classics, like the Maharashtra fishermen dance and the farmers of Punjab routine, along with the Gujarati "stick and round" dance. Other dances depict Bharat Natyam of South India, along with reenactments of modern movie performances, like the Bhumro, a contemporary dance from the recently released film "Mission Kashmir." Participants draw from a repertoire of music and dance familiar to the overseas Indian community and their children to showcasing their "Indian" ethnic heritage. In this event, members draw upon a more sweeping "Indian-American" identity and explicitly lay aside more specific parochial identities. This identity perhaps fits in better with the second-generation (as well as the surrounding community) ideas of what "Indians" should look and act like. In another way, the festival offers organizers the opportunity to boast the migrant group's contributions to the cultural and economic advancement of Phoenix and the nation as a whole, as special guests, including key political figures, sit in the audience. Whether this is an intentional or unwitting tactic to negotiate broader racialization dynamics is uncertain, but it appears to reinforce the idea of this group as a "model minority" (Skop 2017).

On any given weekend, Indian immigrants find numerous ways to build social relationships across multiple sites, both at home and abroad. The evidence outlined in the above case study suggests a reworking of conventional descriptions of placemaking, one that reveals the significance of unfixing geographical space and geographical boundaries in determining social identities. In these observations, the quintessential migrant landscape fails to materialize in a necessarily recognizable manner as Indian émigrés (and the organizations that represent them) construct migrant spaces that are largely ephemeral in nature (like most transitory spaces of interaction). In part, this is a reaction to Phoenix's anti-immigrant context, and placemaking is a strategy that is simultaneously empowering and isolating (Skop 2017). Indeed, what the settling of Indian immigrants in Phoenix's suburbs suggests is a newly developing transnational rewriting of American landscapes, one that does not necessarily rely upon the conspicuous transformation of place.

Shifts in Identity: A Literature Review

At their core, studies on immigrant populations in the United States, regardless of their epistemological stances, are interested in exploring and explaining the dialectical relationship that immigrants share with their new social environments. Key to this work has been comprehending how immigrants and their descendants negotiate conceptions of self and community, typically centered on constructs such as ethnicity, race, and/or nationality, within and against those of the host society. With traditional assimilationist theories, US-based urban sociologists such as Robert E. Park (1928) and Milton Gordon (1964) contended that ethnic distinctions held by new immigrant populations in the US would dissipate over time and across future generations. In this scenario, the cultural differences rooted in an immigrant population's ethnic or national origins would eventually be replaced by the dominant norms and practices of the host society until they had become assimilated or incorporated into the sociocultural and demographic landscapes of the mainstream.

Transnationalism, while acknowledging that migrants' identities change across space, pushes forth a more multidimensional and less totalistic understanding of *how* this change occurs. By contending that migrants, and to various degrees their children and grandchildren, can concurrently occupy multiple social fields even while physically only residing in one locale, transnationalism argues that the identities of migrants and migrant communities are not zero-sum nor strictly tethered to any one landscape or culture. Moreover, transnationalism asserts that adopting new sociocultural identities and norms does not necessitate the vacating of previous ones. Not only can new and existing identities be co-constituted, they can also be co-generative, engendering new ways of knowing, being, and doing that are not inherently tied to sending or receiving nations.

Countless scholars have specified ways in which the maintenance of economic, political, and/or cultural ties with a natal or ancestral homeland by migrants and migrant communities have a direct impact on their identities as well as to the American landscape in which they reside. In the seminal *Georges Woke Up Laughing*, Nina Glick Schiller and Georges Fouron (2001) provide an autobiographical and autoethnographic exploration of their "long-distance nationalism." The authors dedicate much of the book to documenting Fouron's life as a Haitian immigrant and university professor living in upstate New York. Even as visits to Haiti are infrequent, Fouron is a transmigrant in every sense of the word—he remains deeply invested in Haitian politics, he regularly sends remittances to family members in Haiti, and he maintains strong ties with friends and family in Haiti. Serving as inspiration for the book's title, Fouron often "wakes up laughing" to memories of his youth in Haiti—memories that consistently fill him with longing and nostalgia for his embattled homeland. Importantly, Fouron's "long-distance nationalism" significantly impacts his identity in the United States, with his Haitian ethnicity more centrally factoring into his sense of self than his racialization as Black in the US. "The blood remains Haitian," Fouron declares, even as his home is in the United States (Glick Schiller and Fouron 2001, 92).

In writing about the interplay between transnationalism and migrant identities, Min (2017) explores the ways in which the New York–New Jersey area Korean immigrant community's participation in cultural transnationalism aids in the transgenerational maintenance of Korean ethnic identity. Min argues that whereas political and economic forms of transnationalism are comparatively niche, cultural transnationalism can be effectively performed by all members of the migrant community given advances in trade, travel, and telecommunications. Specifically, Min elaborates on the Korean immigrant community's participation in organized transnational cultural events such as traditional artistic performances, exhibitions, and lectures, as well state-sponsored cultural programs held in Korea. Min contends that because these transnational cultural events have few barriers to entry, they more directly impact the ethnic identities of Korean immigrants and their children than other types of transnational practices.

Relatedly, Wilcox and Busse (2017) explore how two distinct immigrant communities, Chinese in Minnesota and Peruvians in New Jersey, use traditional forms of dance for national and ethnic identity construction purposes. Participation in dance not only allows these groups to maintain cultural ties to their homelands, it also accelerates their racialized incorporation into a "multicultural" American society in which their dancing bodies represent an authentic, yet easily consumable foreign culture. It is the migrants' transnationalism, then, that triggers meaningful interactions with both places of arrival and departure, thereby engendering both shifts in identities and varying inscriptions on the physical landscapes around them.

Shifts in Identity: A Case Study

In conceptualizing different manifestations of transnationalism across migrant communities, scholars have ventured to examine the ethnic return migration of diasporic descendants, or the relocation to and settlement within an ancestral homeland by individuals who have resided most if not all their lives elsewhere (Suh 2020b). For diasporic descendants with weak or fleeting ties to the ancestral homeland, ethnic return migration would seemingly denote the ultimate rendition of transnationality—the developing of meaningful ties with the ancestral homeland spawned from the actual relocation and settlement within it. Ethnic return migration also complicates the spatially bound conception of an American community or landscape, especially when beginning to consider the formation of "American" enclaves abroad. Thus, the literature on the ethnic return migration of various American ethnic populations is empirically robust, with case studies showcasing the "return" of diasporic populations across the globe.

Stephen Cho Suh has written extensively on the topic of ethnic return migration from the perspective of 1.5- and 2nd-generation Korean Americans. Drawing on in-depth qualitative interviews and ethnographic fieldwork in Seoul, Suh's research attends to the growing number of Korean Americans residing in South Korea since 1999, the year when South Korea began providing preferential immigration status to former émigrés and their descendants. With over 45,000 Korean Americans residing in South Korea as of 2019, Suh attempts to address the basis for this large-scale Korean American diasporic return, specifically seeking to answer why Korean Americans who are born and/or raised in the United States and possess weak ties with South Korea decide to relocate to the country as adults (Suh 2020a). In addition to being motivated by economic factors, a rationale well-documented in the general return migration literature, Suh finds that Korean Americans are also spurred to migrate for social identity-based reasons. Furthermore, "return" has profound impacts on the social identities of migrants.

Race and ethnicity, for instance, serve to both organize numerous aspects of their social lives and to centrally anchor their understandings of identity and belonging within the US and South Korea. In terms of race, Suh (2020a) finds that the racialization of Korean Americans as Asian American "model minorities" and "perpetual foreigners" contributes to their difficulty in claiming social and cultural citizenship within the United States. Regarding ethnicity, messaging from family members, specifically parents and grandparents, as well as from the South Korean state, reifies essentialized notions of ethnicity and Korean-ness. Race and ethnicity, then, work in tandem to connect 1.5- and 2nd-generation Korean Americans to the broader diaspora, in many instances jumpstarting their identification with Korean ethnicity as well as feeding into their transnational interests. In some cases, interest in Korean ethnicity and culture can lead to the adopting of a Korean American identity and subjectivity as well as the claiming of specific spaces within the broader American landscape. For example, Suh (2016) argues that the reimagining of Koreatown, Los Angeles, as a distinctly second-generation Korean American space has allowed for the children of Korean immigrants to also stake claims to social and cultural citizenship within a US context. In other cases, interest in Korean ethnicity and culture can jumpstart transnational yearning, thereby potentially contributing to "return" and settlement in South Korea (Suh 2020a).

Suh's (2017) research also examines the intersectional way that gender factors into the transnational interests and motives of 1.5- and 2nd-generation Korean Americans. Drawing on scholarship that expounds upon the gendered racialization of Asian American men as effeminate and asexual and Asian American women as exotic and sexually desirable, Suh contends that the "return" of Korean Americans to the ancestral homeland is intricately shaped by their conceptions of and relationship to gender. For Korean American "returnee" men, for example, their moves to South Korea are often seen in a redemptive light, with many believing themselves to be more masculine or manly than their South Korean counterparts (Suh 2017). Conversely, Suh finds that Korean American "returnee" women tend to feel anxious and less enthusiastic about their "homeland" sojourns because of long-held presumptions that Asian societies are more intently patriarchal and less gender egalitarian than Western countries. Many, then, are critical of South Korean women for not being "feminist enough," citing their propensity for cosmetic surgery and hyperfeminine behavior. These actions, they contend, showcase the ways in which South Korean women are complicit in their own subjugation.

Curiously, because Korean American "returnees" share ethnic characteristics with the local population but are not legally South Korean citizens, Suh (2019) discovers that they tend to occupy a liminal, double-edged position within the country where their foreignness prevents their full participation in key sociopolitical institutions. Yet they are not quite foreign enough to unabashedly ignore social norms as can certain other American "expat" migrants. Almost paradoxically, then, "returnees" profess more comfortably identifying as American or Korean American *upon* settling in South Korea, some even for the first time. In effect, Suh (2020b) argues that a move intended to bring migrants culturally closer to their Korean heritage often leads to the transnational affirmation of their American identities.

Taken together, Suh's work argues that transnationalism plays a central role in the ethnic identities of 1.5- and 2nd-generation Korean Americans, thus showcasing the complex ways in which migrants and migrant communities make sense of community and belonging across and between geographic spaces. Instead of being sedentary or fixed, the identities of migrants and migrant communities take inspiration from multiple locales and are constantly being negotiated and contested in ways that best accommodate their needs.

Conclusion

This chapter addresses the implications of transnationalism for ethnic and social identity (re) formation within an evolving United States landscape. The above discussion indicates that transnationalism most definitely complicates notions of group and individual memberships that are centered in loyalty and subservience to a particular nation-state and instead requires understanding "flexible" alternatives that allow ambiguous, fluid, and multiple allegiances. The above discussion also reveals the ways in which migrants fashion alternative social spaces located between many poles of "here" and "there" often framed within political dynamics that require reworkings of social identities. Indeed, the multiplicity of the migrant existence reveals a tension between lived-in, transitory, and distant spaces that are infused, at different moments in often divergent contexts, with both complex and intersectional identities. As Skop (2019) argues, current mappings of migration have become entirely destabilized in the past few decades. The binaries, categories, and borders whereby migrants' identities are continually drawn and redrawn, whether as Indian migrants settled in Phoenix, Arizona or US-raised Korean-American "returnees," nullifies previous conceptions of settlement, permanency, and eventual assimilation. Indeed, given the multistranded nature of these contemporary sojourns, Paul and Yeoh (2020) argue that using a transnational framework that limits analysis to a sending and receiving nation ignores the reality that much migration today is "multinational" and temporary in nature and that this fact has a profound impact on the identities and economic opportunities of migrants. We agree with their observations, and in the context of this chapter, offer examples of the profound (and often invisible) impacts not only on the ethnic and racial identities of migrants and their descendants, but also on the American landscapes that signal both their arrival and accommodation, as well as their otherness and outsider status.

References

Airriess, C. A., ed. 2015. *Contemporary Ethnic Geographies in America*. 2nd ed. Lanham, MD: Rowman & Littlefield.

Frazier, J. W., E. Tettey-Fio, and N. F. Henry, eds. 2006. *Race, Ethnicity, and Place in a Changing America*. Albany, NY: SUNY Press.

Glick Schiller, N., L. Basch, and C. Blanc-Szanton. 1992. "Transnationalism: A New Analytic Framework for Understanding Migration." *Annals of the New York Academy of Sciences* 645 (1): 1–24. doi: 10.1111/j.1749-6632.1992.tb33484.x.

Glick Schiller, N. and G. E. Fouron. 2001. *Georges Woke Up Laughing: Long-distance Nationalism and the Search for Home*. Durham: Duke University Press.

Gordon, M. 1964. *Assimilation in American Life: The Role of Race, Religion and National Origin*. New York: Oxford University Press.

Helweg, A. W. and U. M. Helweg. 1990. *An Immigrant Success Story: East Indians in America*. Philadelphia, PA: University of Pennsylvania Press.

Jones, R., ed. 2008. *Immigrants Outside Megalopolis: Ethnic Transformation in the Heartland*. Lanham, MD: Lexington Books.

Kaplan, D. H. 2018. *Navigating Ethnicity: Segregation, Placemaking, and Difference*. Lanham, MD: Rowman & Littlefield.

Kim, J. 2018. "Manhattan's Koreatown as a Transclave: The Emergence of a New Ethnic Enclave in a Global City." *City & Community* 17 (1): 276–295. doi: 10.1111/cico.12276.

Knott, K. 2000. "Hinduism in Britain." In *The South Asian Religious Diaspora in Britain, Canada, and the United States*, edited by H. Coward, J. R. Hinnells and R. Williams, 89–108. Albany, NY: State University of New York Press.

Levitt, P. 2001. *The Transnational Villagers*. Berkeley: University of California Press.

Li, W. 2009. *Ethnoburb: The New Ethnic Community in Urban America*. Honolulu: University of Hawaii Press.

Min, P. G. 2017. "Transnational Cultural Events among Korean Immigrants in the New York-New Jersey Area." *Sociological Perspectives* 60 (6): 1136–1159.

Murdie, R. and E. Skop. 2012. "Immigration and Urban-Suburban Settlements." In *Immigrant Geographies of North American Cities*, edited by C. Teixeira, W. Li, and A. Kobayashi, 48–68. Don Mills, ON: Oxford University Press Canada.

Park, R. E. 1928. "Human Migration and the Marginal Man." *American Journal of Sociology* 33 (6): 881–893.

Paul, A. M. and B. S. A. Yeoh. 2020. "Studying Multinational Migrations, Speaking Back to Migration Theory." *Global Networks* 21 (1): 3–17.

Price, M. and L. Benton-Short, eds. 2008. *Migrants to the Metropolis: The Rise of Immigrant Gateway Cities*. New York: Syracuse University Press.

Rios, M. and L. Vazquez, eds. 2012. *Diálogos: Placemaking in Latino Communities*. New York: Routledge.

Singer, A., S. W. Hardwick, and C. B. Brettell, eds. 2008. *Twenty-First Century Gateways: Immigrant Incorporation in Suburban America*. Washington, DC: The Brookings Institute.

Skop, E. 2012. *The Immigration and Settlement of Asian Indians in Phoenix, Arizona 1965–2011: Ethnic Pride vs. Racial Discrimination in the Suburbs*. New York: Edwin Mellen Press.

———. 2015. "Asian Indians and the Construction of Community and Identity." In *Contemporary Ethnic Geographies in America*, 3rd ed., edited by C. A. Airriess, 325–345. Lanham, MD: Roman and Littlefield.

———. 2016. "ThirdSpace as Transnational Space." In *Indian Transnationalism Online: New Perspectives on Diaspora*, edited by H. de Kruijf and A. Sahoo, 81–102. Farnham: Ashgate Publishing.

———. 2017. The Model Minority Stereotype in Arizona's Anti-Immigrant Climate: SB 1070 and Discordant Reactions from Asian Indian Migrant Organizations. *GeoJournal* 82 (3), 555–566. doi: 10.1007/s10708-016-9704-4.

———. 2019. "Geography and Migration." In *Sage Handbook on International Migration*, edited by C. Inglis, B. Khadria, and W. Li, 108–123. New York: Sage.

Skop, E. and W. Li. 2003. "From the Ghetto to the Invisiburb: Shifting Patterns of Immigrant Settlement." In *Multicultural Geographies: The Changing Racial/Ethnic Patterns of the United States*, edited by J. W. Frazier and F. Margai, 113–124. New York: Academic Publishing.

———. 2011. "Urban Patterns and Ethnic Diversity." In *21st Century Geography, A Reference Handbook*, edited by J. P. Stoltman, 314–324. Thousand Oaks, CA: Sage Publications.

———. 2017. "Ethnicity." *The Wiley-AAG International Encyclopedia of Geography: People, the Earth, Environment, and Technology*, edited by D. Richardson, N. Castree, M. F. Goodchild, A. Kobayashi,

W. Liu, and R. A. Marston. New York: Wiley International. www.wiley.com/en-us/Internatio nal+Encyclopedia+of+Geography%3A+People%2C+the+Earth%2C+Environment+and+Technol ogy%2C+15+Volume+Set-p-9780470659632.

Smith, H. and O. Furuseth, eds. 2006. *Latinos in the New South: Transformations of Place*. Burlington, VA: Ashgate.

Suh, S. C. 2016. "Introducing K-Town: Consumption, Authenticity, and Citizenship in Koreatown's Popular Reimagining." *Journal of Asian American Studies* 19 (3): 397–422.

———. 2017. "Negotiating Masculinity across Borders: A Transnational Examination of Korean American Masculinities." *Men and Masculinities* 20 (3): 317–344.

———. 2019. "Living as 'Overseas Koreans' in South Korea: Examining the 'Differential Inclusion' of Korean American 'Returnees.'" In *Transnational Mobility and Identity in and out of Korea*, edited by Y. Ahn, 99–114. New York: Lexington Books.

———. 2020a. "Racing 'Return': The Diasporic Return of US-raised Korean Americans in Racial and Ethnic Perspective." *Ethnic and Racial Studies* 43 (6): 1072–1090.

———. 2020b. "Nostalgic for the Unfamiliar: Korean Americans' 'imagined affective connection' with the ancestral homeland." *Ethnicities*. doi: 10.1177/1468796820981667.

Teixeira, C., W. Li, and A. Kobayashi, eds. 2012. *Migrant Geographies of North American Cities*. Toronto: Oxford University Press Canada.

Wilcox, H. N. and E. Busse. 2017. "'Authentic' Dance and Racialized Ethnic Identities in Multicultural America: The Chinese in Minnesota and Peruvians in New Jersey." *Sociology of Race and Ethnicity* 3 (3): 355–369.

Yu, S. 2018. "That is Real America!: Imaginative Geography among the Chinese Immigrants in Flushing, New York City. *Geographical Review* 108 (2): 225–249. doi: 10.1111/gere.12248.

———. 2020. "Becoming in/out of Place: Doing Research in Chinatown as a Chinese Female Geographer in the Era of Transnationalism." *The Professional Geographer* 72 (2): 272–282. doi: 10.1080/00330124.2019.1633368.

10

INTERPRETING THE LANDSCAPES OF A NEW RELIGIOUS AMERICA

Samuel Avery-Quinn

In 1959, Eric Isaac described the geography of religion as the study of religion's motivational role in human modifications of the landscape. However, within a decade, the American religious landscape and how geographers study religion began to change radically. Following the passage of immigration reform in 1965, millions of immigrants reconfigured America's religious diversity. Mainline Protestant denominations declined, growing numbers of Americans embraced alternative spiritualities or began identifying as agnostics, atheists, or of no religion (Wuthnow 1998). Amid this new religious landscape, the geographic study of religion drifted away from Isaac's emphasis on landscape. Early scholarship treated religious landscape as religious territory consisting of sacred spaces and religiously-shaped culture regions (Kong 2010). By the 1990s, an emphasis on the representational qualities of space and the spatialities of class, gender, and race in sacred spaces emerged. Although the geopolitical crisis that followed the 9/11 terrorist attacks and the development of non-representational theory renewed interest in the geography of religion, the new literature did not articulate landscape as well as other spatial categories (Olson 2013).

This chapter argues for renewed attention to landscape in studying what Eck (2002) has described as a new religious America. Landscape can be a concept for bridging the spatialities of religion in a diverse and pluralistic society. Landscape is one of Geography's most deployed concepts for understanding intersections of cultural expression, embodied practice, and the built and natural environments (Wylie 2007). However, religion is a contested term (McCutcheon 1997), an epistemological category covering a broad constellation of meaningful expressions and practices (Ivakhiv 2006). Religious landscapes comprise those spaces people declare as religious, spiritual, or manifesting otherworldly dimensions. Deploying the concept of landscape as a framework for understanding the religious spatialities of the new religious America, this chapter begins with a reflection on the author's work with an organization that promoted pluralism in the aftermath of 9/11, traces religious diversity and pluralism in America emphasizing everyday interactions that constitute a lived religious landscape, and concludes by exploring common scales for reading America's religious landscape.

DOI: 10.4324/9781003121800-13

Pluralism and the Post-9/11 Landscape of Colorado

Ten days after the terrorist attacks of September 11, 2001, 3,000 people gathered at the Colorado Muslim Society. The vigil, organized by the Interfaith Alliance of Colorado (TIACO), transformed the center's parking lot into an emplacement of the Denver metropolitan area's religious diversity. Attendees surrounded the center carrying signs declaring "LOVE" or "UNITY." Religious leaders condemned discrimination and expressed solidarity with the Muslim community. The vigil attendees, they claimed, were America's better angels demonstrating unity in diversity and the power of pluralism. As the gathering ended, Mohammed Jodeh, representing the center, was overwhelmed. "This crowd," he said, "*this* is the real America!" (*Denver Post* 2001).

In America, promoting interfaith understanding is as old as the 1893 World Parliament of Religions (Fahy and Bock 2020). The crisis following 9/11 was, however, a watershed moment for the interfaith movement. In 1980, there were fewer than 50 interfaith organizations in America, and, by 2006, there were over 600. For organizations like the Interfaith Alliance of Colorado (TIACO), post-9/11 violence against Muslims and Sikhs and new legislation that threatened to erode civil liberties posed acute challenges. Liberal religious activists founded the Interfaith Alliance (TIA) in 1994 to counter the Christian Coalition (Knutson 2005). TIA tracked legislation, urged candidates to sign a Code of Civility against negative campaigning, and worked to increase media awareness that conservative Christians were not the only religious voices in America. By the late 1990s, TIA had chapters across the country mirroring, at the state and local level, the national organization's advocacy of campaign reform and legislative monitoring. Most chapters were small volunteer groups but by 2000, the Interfaith Alliance of Colorado had over 800 members, a volunteer executive director, and two paid staff.

In the aftermath of 9/11, TIACO struggled to expand its interfaith efforts. A few days after 9/11, Executive Director Chuck Mowry called Mohammed Jodeh and began planning the vigil. Within weeks, TIACO met with partner organizations, media outlets, and the director of community relations in the FBI's Denver Field Office. However, the limits of the organization's interfaith network were soon evident. TIACO expanded its interfaith task force, organized an interfaith concert, and promoted a weekend of events with Buddhist monk Thích Nhất Hạnh. However, after a year of pursuing political advocacy and interfaith dialogue, TIACO's interfaith network had not expanded far beyond predominantly white and politically moderate or progressive Jewish, Protestant, and Unitarian Universalist congregations. TIACO maintained ties to the Colorado Muslim Society, and Sheik Ibrahim Kazerooni of the Ahl-ul Bayt Shi'ite mosque participated in several TIACO events. However, outreach to other religious communities, such as the Denver Buddhist Temple or the Hindu Temple of the Rockies, did not progress past initial conversations.

Although liberal political advocacy and a predominantly white network limited TIACO's interfaith work, the crisis of 9/11 and the subsequent growth of interfaith organizations across the country represented a transformation in the interfaith movement in America (McCarthy 2007). The efforts of TIACO and other interfaith organizations are part of a long history of negotiations over the parameters of coexistence in a religiously diverse society. For many of these organizations, the embrace of pluralism has provided a means for negotiating religious differences and promoting the common good. However, religious differences in America are more than a dimension of civil society or something manifest in the space of formal interfaith dialogues. Religious diversity is a defining feature of contemporary American life, materialized in space, negotiated in everyday encounters, and navigated across a religious landscape.

Religious Diversity and Pluralism

Religious diversity, or the coexistence of religious groups in a secular state, has long been a demographic fact in America, but pluralism is a recent ethic in a contentious history of responses to that diversity (Hutchinson 2008; Cohen and Numbers 2013). The country's first normative response emerged in a colonial context of the delegitimization of Indigenous traditions amid genocidal dispossession of native lands, forced conversions of enslaved Africans, a tenuous tolerance of Catholics and Jews, and a notion of religious liberty, enshrined in the Constitution, shaped by enlightenment ideals and the interests of white Protestant sects. Under the weight of nineteenth-century immigration, an ethic emerged stressing tolerance and cultural assimilation, but every surge of immigrants also spurred nativist movements, exclusionary immigration laws, and violence (Lippy 2015). By the 1950s, America embraced myths that American values were predicated on a "Judeo-Christian" tradition, thus nominally admitting white Catholics and Jews into the majoritarian fold and that America was a melting pot in which immigrants absorbed majoritarian values. By the 1960s, a counterculture that questioned majoritarian norms unsettled the consensus that national myths promulgated. At the same time, movements promoting Black, Chicano, and Native American identity, civil rights, women's liberation, and gay rights challenged discriminatory societal structures.

In the unsettled social context of the 1960s, Congress passed the 1965 Hart-Celler Immigration and Naturalization Act. The law ushered in a wave of immigration and a radical expansion of religious diversity. Between 1965 and 2015, over 58 million documented and undocumented immigrants arrived in the United States (Gjelten 2016). Previously sustained by older waves of immigration and natural birth rates, religious communities experienced dramatic changes (Jones and Cox 2017). Bolstered by immigrants from Eastern Europe and Russia, America's Jewish population increased to 5.7 million (m) adherents. Immigrants swelled the ranks of Buddhism to between 2.5m and 4m, including native-born converts, expanded America's Muslim population to about 3.3m, and increased the Hindu population to over 2.7m. The new immigration also brought a stream of Bahai's, Daoists, Jains, Sikhs, and Zoroastrians. However, over two-thirds of new immigrants were Christians (Warner 2006).

The new religious diversity necessitated a new normative response. The cultural politics fomented by the 1960s, paired with the interests of a neoliberal state, have fostered pluralism as a normative response to religious difference in which "individuals and communities recognize each other as parallel forms of the phenomenon called religion" (Klassen and Bender 2010, 1). As a means of defining the parameters of coexistence, pluralism is a process of negotiation and a system of spatial interaction in specific emplacements of religion, such as congregations, and in everyday encounters. Although the number of congregations in America fluctuates, in 2010, there were at least 344,894 congregations emplacing religion on the American landscape (Figure 10.1) (ARDA 2014). This included 191,122 Conservative or Evangelical Protestant, 77,760 Mainline Protestant, 20,589 Catholic, 17,754 Black Protestant, 14,393 Mormon, 3,464 Jewish, and 2,551 Orthodox Christian congregations. Non-western traditions included at least 2,854 Buddhist centers and temples, 2,106 mosques, 1,558 Hindu congregations, and 246 Sikh gurdwaras. Although houses of worship anchor religions on the landscape, interactions among the faithful occur more often in everyday encounters.

American cities and towns have become heterogeneous spaces of encounter. Religious groups maintain ties across vast distances via communications and travel, while media depictions of the religious other saturate even the most homogenous town. In the new religious America, the local is global, and neighborhoods, schools, stores, and workplaces are spaces of encounter. In these encounters, Giles (2019) finds "emergent" or "lived religious pluralisms." Emergent

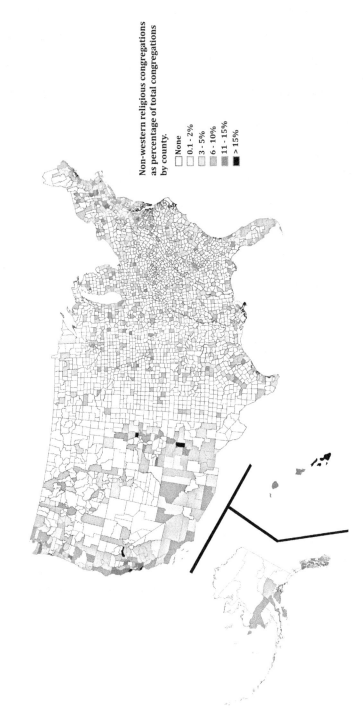

Figure 10.1 Diversity of religious congregations—Baha'i, Buddhist, Daoist, Hindu, Jain, Muslim, Shinto, Sikh, and Zoroastrian congregations as a percentage of total congregations by county. Data source: ARDA (2014). Map by author.

religious pluralisms operate at a micro-social level, fostered by friendships and hospitality between neighbors. In a society shaped by individualism and declining congregational membership, attention to the elements shaping religion in everyday life, including emergent religious pluralisms, is crucial for understanding the American religious landscape.

Lived Religious Landscapes

A concern for everyday life bears a strong affinity for Jackson's (1984) emphasis on vernacular landscapes and follows a growing number of geographers (Dwyer 2016) who have drawn on scholarship by historians, religious studies scholars, and sociologists on "lived religion" (Orsi 1985; Hall 1997; Ammerman 2006; McGuire 2008). As Ammerman claims, to study lived religion is to attend to the "domains of life where sacred things are being produced, encountered, and shared" (2016, 89). This approach focuses on the expressions and practices of religion. As Orsi writes, what matters is what people do with religion, as there is "no religion that people have not taken up in their hands" (2003, 172). If landscape is to be a useful concept for understanding the emplacements of religion and spaces of interfaith interaction in the new religious America, a landscape approach must attend to the understandings of individuals and communities as they "take up" religion in everyday life at multiple scales. As Olson notes, "it is only through a grounding in space and scale that lived religion transforms into lived landscape" (2013, 78).

Studying lived religious landscapes begins by attending to the way people, through embodied practices, sacralize everyday spaces and create physical, social, and mythic connections between those spaces. These practices and the stories people tell about these spaces are embodied acts of landscaping (Lorimer 2005). How such actions are materialized on the landscape varies spatially and temporally. In religious buildings, lived religious practices are means of contestation and enchantment (Chidester and Linenthal 1995). Lived religious practices are not restricted to buildings designated as religious but encompass a multitude of locations. Everyday practices and interactions not only shape these spaces but connect them as "articulated moments in networks that stretch across space," time, and cosmology (Schein 1997, 662). However, a landscape approach to understanding the geography of the new religious America must do more than privilege places and trace connections between them. Lived religious spaces are not so orderly. Rather, the study of lived religious landscapes should ground religion in diverse spaces and at relevant scales. In the new religious America, these scales include the body, household, congregation, and metropolitan area.

Bodies

The ways the human body articulates with religion are foundational for religious landscapes (Tweed 2006). Bodily engagements generate the spatial metaphors through which interactions, performances, places, and landscapes are understood (Knott 2008). Visual, olfactory, tactile, and proprioceptive stimuli guide the body's engagement with religious space. Embodied sensorial performances, mobility, and the corporeal inscriptions of discursive formations constitute the micro-geography of lived religious landscapes. For embodied actors, rhythms of scriptural recitation, song, and symbolic gestures enliven a sacred space and color its affective registers (Bell 1997). Ritual can attune participants to the sacrality of a space, shape the tenor of their interactions, and recognize the presence of gods and spirits.

Religions are mobile, carried through space by people bearing the inscriptions of multiple signifiers through clothing, costumes, masks, tattoos, and their ascribed gender, race, and sexual orientation. In interfaith interactions, emergent religious pluralisms are not negotiated along lines of religious difference alone but are embodied encounters shaped by intersecting identities. Embodied expressions of faith reinforce and sometimes subvert gendered and racialized religious identities. Among American Catholics, for example, practices of devotion to saints via street festivals (Orsi 1985) or novenas at Marian shrines (Tweed 1997) provide gendered repertoires for identity construction (Maldonado-Estrada 2020). Religious communities may also establish boundaries based on assertions of heteronormative sexuality, undergirded by a bricolage of scriptural arguments filtered by political ideology. For LGBTQIA Americans, such boundaries make religious spaces locations for confronting homophobia, pursuing inclusion, and crafting affirming expressions of faith, if not as points of disaffiliation (Thumma and Gray 2005).

Embodied practices establish the dimensions of lived religious landscapes, enlivening spaces and articulating connections between people, places, objects, and divine agents present, absent, and present in their absence. Embodied practices, whether liturgy, ritual, or acts of sacralization deploy objects to bring the material dimensions of sacred space into being (Morgan 2010). Things comprising sacred spaces can be fixed or portable. For instance, the athame and other Wiccan ceremonial objects are mobile, deployed wherever a practitioner conducts magical rites (Hume 1998). Rituals and their assemblages of embodied actors and symbolic materials are spatially extensible, forming landscapes marked by ritual locales. The sacrality of such locales can endure through the efficacy of fixed objects and repeated use or may manifest only so long as participants are present. In and between ritual locales, people sacralize everyday moments through such fleeting acts as prayer or a spontaneously uttered "inshallah" (McGuire 2008). Such actions can infuse ordinary spaces with a sense of the sacred and reinforce the connective dimensions of religious landscapes.

Households

While embodied actors carry religions across the landscape, households can be intimate emplacements of religion (Seamon 2018), loci for lived religion. Households have central roles in family life, religious socialization, and devotions. Household members can structure the ways children learn religious beliefs, behaviors, and values. Households can inculcate an individual's sense of membership in a religious community, and how to interact with others across lines of difference. For immigrants, households can be nodes in transnational networks, sending and receiving units bifurcated by vast distances while connected by communications and travel, and spaces for constructing religious identities in a new land (Trinka 2019).

In America, religious practices occur more often in households than in houses of worship. For example, less than 25% of American Muslims regularly attend mosque for Friday prayers (Kahera 2002). Of those in attendance, typically few are women, thus making the Muslim household a gendered space for worship. Across traditions, households provide space for devotional practices, reading scripture, prayer, observance of holy days, and ritualized meals. The material culture of domestic religion buttresses these practices. Religious iconography can make gods, saints, and spirits symbolically present. Religious art, such as depictions of sacred sites can locate a household in a religious geography and foster nostalgia for a spiritual homeland. Concentrating these material assemblages in a home altar is a common practice. In many Hindu households in America, statues of revered deities and other iconography mark a space set aside for the devotional practice of puja (Figure 10.2) (Mazumdar and Mazumdar 1993). Likewise, in Santería, a boveda bears candles, flowers, and vessels of water for ancestral spirits (Murphy 2010). In the American religious landscape, households locate embodied actors in intimate

Figure 10.2 Sri Mahalakshmi puja altar. Ervins Strauhmanis, www.flickr.com/photos/76523360@ N03/17118589092. Creative Commons license CC-BY-2.0, https://creativecommons.org/licenses/ by/2.0/.

units, provide a repertoire of religious expressions and practices, and integrate people into religious communities maintained, in part, through the congregational life of houses of worship.

Houses of Worship

Houses of worship are common spatial aspects of congregational life in religious communities. However, they are not a universal feature of that life as congregations also worship in bars, movie theaters, and school gymnasiums. For the congregations that rely on them, houses of worship are material manifestations of a congregation's spatial practices, a religious tradition's design elements, and architectural trends. More than mere expressions of theology, houses of worship are community spaces rife with economic forces, political ideologies, gender and racial formations, and contests over authority.

Houses of worship are tangible signs of religious diversity and ongoing reconfigurations of congregational life. As visual elements of the landscape, the architecture of a house of worship declares a building's identity and hints at the spaces within. Inside, ritual locations work with other design elements to shape spatial experience. Boundaries, pathways, and thresholds choreograph movement through space, while dimensions, lighting, visual qualities of materials, and iconography foster architectural moods (Barrie 2010). These elements are reconfigurable with changing congregational needs, architectural trends, and shifting cultural and religious contexts. For example, following the reforms of the Second Vatican Council (1962–1965), many American Catholic congregations embraced modernist architecture and amphitheater-style sanctuaries (Stroik 2012). Over the same time, Muslim women, asserting their rights to equitable mosque access have spurred a range of architectural responses. Apart from establishing women's mosques, a growing number of mosques seeking to maintain gender-separate worship have installed partitions in their prayer halls (Bagby 2009).

As a space for the construction of community, a house of worship, whether in a store-front or a gleaming new temple, provides a venue for shaping congregants' identities through shared experience (Figure 10.3). Congregations are connected communities (Ammerman and Farnsley 1997), drawing members from a limited geographic area and thus reflecting local concerns, demographics, and economic conditions. Congregations provide moral guidance and spiritual nurture, and offer space for establishing personal connections and building social capital. Congregational activities also foster religious socialization and community-building processes through celebrations and family-centered rites such as baptisms, bar/bat mitzvahs, funerals, weddings, ritual occasions, worship, and children's religious education.

Post-1965 immigration has reconfigured congregational life. For established congregations, responses to changed local conditions may mean closure, relocation, or accommodating a multi-cultural and multi-ethnic congregation through such measures as offering services in Korean, Filipino, or Spanish (Ebaugh and Chafetz 2000). Ethnic and linguistic differences are not the only internal diversity congregations face. Many Hindu temples in America, for instance, serve religiously diverse communities and must accommodate diasporic gods and goddesses from different regional Indian traditions (Waghorne 2004). Further, some congregations, out of an ethic of pluralism, if not financial need, have opened their doors to interfaith space sharing

Figure 10.3 A repurposed building serving as the Annoor Mosque, Knoxville, TN (front); billboard overlooking the mosque advertising the Muslim community's civic engagement (left) and a separate advertisement for a private Christian school (right). Photo by author.

(Numrich 2016). Across religious traditions, such diversifying contours of congregational life most often occur in metropolitan landscapes.

Metropolitan Areas

Congregations operate on an American landscape transformed by over a century of the reconfiguration of metropolitan areas. From the Great Migration (Weisenfeld 2018) to the mid-twentieth-century "white flight" of Mainline Protestant congregations out of cities (Kruse 2005), to the establishment of immigrant communities in the suburbs (Lung-Amam 2017), metropolitan geographies have reshaped the contours of the American religious landscape. To the degree that class, ethnicity, and race have sorted metropolitan areas, such sorting has guided the emplacements of religion, neighborhood by neighborhood.

In metropolitan areas, corner bodegas, freeways, shopping malls, and subdivisions are the spaces in which religious communities build social networks and negotiate moral orders (Prickett 2021; Thomas 2021). Protestant megachurches, for example, are a common feature of these landscapes. These predominantly white, middle-class, suburban emplacements of Evangelicalism create campus-like environments with arena-sized sanctuaries and spaces that mirror secular venues such as coffee shops, conference centers, gyms, and libraries. Wilford attributes the growth of megachurches to Evangelicalism's capacity to reframe metropolitan areas so that "the freeway, the office cubicle, the soccer field … are made holy by incorporating them into Evangelical narratives of grace, salvation, and holiness" (2012, 4). Discourses that sacralize mundane spaces are not the only means by which religious communities navigate the metropolitan landscape. Across America, Orthodox Jewish communities have established eruvs or symbolic lines imprinting the spiritual borders of their communities (Rapoport 2011). In other communities, religious dimensions course through neighborhood gatherings. In the predominantly Latino city of Santa Ana, California, for instance, many neighborhood parties are extended celebrations of Catholic sacraments, such as an infant's baptism or a girl's quinceanera (Calvillo 2020). Metropolitan areas have been the most common spaces for lived religious practice and everyday interfaith encounters, making them significant spaces in the American religious landscape.

Conclusion

The American religious landscape has been buffeted by the forces of late modernity, religious diversity, and negotiations over the parameters of religious coexistence. Approaching the religious dimensions of America through landscape foregrounds the material and spatial dimensions of religion and opens an epistemological space for understanding the ways individuals and communities have navigated and shaped the contours of the new religious America. In this landscape, religion is part of everyday life at multiple spatial scales. Although the body, household, congregation, and metropolitan area are the most common scales, religions are emplaced beyond the metropolitan frontier in spaces that are outside this chapter's scope. In rural communities, seasonal tourism workers, migrant farm labor, and immigrants and refugees in meat or poultry plants have made rural communities sites of negotiation (Leitner 2012). In Native American communities, lived religious practices are shaped by the maintenance of traditional ceremonies and worldviews, kinship networks that extend beyond reservation boundaries, conflicts over sacred lands and waters, and the economic, political, and social challenges faced by Indigenous communities (Martin 2001; Dunbar-Ortiz 2014). Elsewhere, Hispanic Catholic traditions shape everyday religion in communities along the United States' border with Mexico, a border entangled in religious mappings of the region (Machado, Turner, and

Wyller 2018). Beyond these spaces, and sometimes crossing them, are pilgrimage routes (Swatos and Tomasi 2002) and religious tourist destinations (Bielo 2018; Messenger 2000).

As Tweed notes, religions "move across time and space. … They leave traces. They leave trails" (2006, 62). The traces of religions in America are part of the fabric of everyday life and leave trails at multiple scales. In everyday spaces, lived religious practices shape how individuals, households, congregations, and religious communities orient themselves within broad religious geographies and vast cosmologies. In a religiously diverse society, lived religious spaces are, as the members of the Interfaith Alliance of Colorado acutely experienced after 9/11, spaces of encounter across lines of difference and spaces from which new religious pluralisms may emerge. As manifest in recent American politics, they may also be spaces of conflict, flashpoints for expressions of white fragility and rage, assertions of America as a Christian Nation, and concomitant acts of antisemitism, Islamophobia, racism, and xenophobia (Whitehead and Perry 2020). These spaces of pluralism and anti-pluralism are defining features of contemporary American society. Attending to landscape provides a way to explore these spaces, identify religion's traces, and follow religion's trails.

References

Ammerman, N. and A. Farnsley. 1997. *Congregation & Community*. New Brunswick: Rutgers University Press.

Ammerman, N. 2016. "Lived Religion as an Emergent Field: An Assessment of Its Contours and Frontiers." *Nordic Journal of Religion and Society* 29 (2): 83–99.

———. 2006. *Everyday Religion*. New York: Oxford University Press.

Association of Religion Data Archives. 2014. "US Religion Census: Religious Congregations and Membership Study, 2010 (State File)." Accessed March 2021. www.thearda.com/Archive/ChState.asp.

Bagby, I. 2009. "The American Mosque in Transition: Assimilation, Acculturation, and Isolation." *Journal of Ethnic and Migration Studies* 35 (3): 473–490.

Barrie, T. 2010. *The Sacred In-Between*. New York: Routledge.

Bell, C. 1997. *Ritual: Perspectives and Dimensions*. New York: Oxford University Press.

Bielo, J. 2018. *Ark Encounter*. New York: NYU Press.

Calvillo, J. 2020. *The Saints of Santa Anna*. New York: Oxford University Press.

Chidester, D. and E. Linenthal, eds. 1995. *American Sacred Space*. Bloomington: Indiana University Press.

Cohen, C. and R. Numbers, eds. 2013. *Gods in America*. New York: Oxford University Press.

Denver Post. 2001. "Vigil Crowd Overwhelms Muslim Site 3,000 Visit Mosque in Support." September 22, A7.

Dunbar-Ortiz, R. 2014. *An Indigenous Peoples' History of the United States*. Boston: Beacon Press.

Dwyer, C. 2016. "Why Does Religion Matter for Cultural Geographers?" *Social & Cultural Geography* 17 (6): 758–762.

Ebaugh, H. and J. Chafetz. 2000. "Dilemmas of Language in Immigrant Congregations: The Tie That Binds or the Tower of Babel?" *Review of Religious Research* 41 (4): 432–452.

Eck, D. 2002. *A New Religious America*. San Francisco: HarperCollins.

Fahy, J. and J. Bock, eds. 2020. *The Interfaith Movement*. New York: Routledge.

Giles, D. 2019. "Uncovering Neglected Emerging Lived Religious Pluralisms." In *Emergent Religious Pluralisms*, edited by J. Bock, J. Fahy, and S. Everett, 145–166. New York: Palgrave Macmillan.

Gjelten, T. 2016. *A Nation of Nations*. New York: Simon & Schuster.

Hall, D., ed. 1997. *Lived Religion in America*. Princeton University Press.

Hume, L. 1998. "Creating Sacred Space: Outer Expressions of Inner Worlds in Modern Wicca." *Journal of Contemporary Religion* 13 (3): 309–319.

Hutchinson, W. 2008. *Religious Pluralism in America*. New Haven: Yale University Press.

Ivakhiv, A. 2006. "Toward a Geography of 'Religion': Mapping the Distribution of an Unstable Signifier." *Annals of the Association of American Geographers* 96 (1): 169–175.

Jackson, J. B. 1984. *Discovering the Vernacular Landscape*. New Haven: Yale University Press.

Jones, R. and D. Cox. 2017. "America's Changing Religious Identity." PRRI. Accessed April 2021. www.prri.org/research/american-religious-landscape-christian-religiously-unaffiliated/.

Kahera, A. I. 2002. *Deconstructing the American Mosque*. Austin: University of Texas Press.

Klassen, P. and C. Bender. 2010. "Introduction." In *After Pluralism*, edited by C. Bender and P. Klassen, 1–30. New York: Columbia University Press.

Knott, K. 2008. "Spatial Theory and the Study of Religion." *Religion Compass* 2 (6): 1102–1116.

Knutson, K. 2005. "Interfaith Advocacy Groups in American Politics." In *The Changing World Religion Map*, Vol. V, edited by S. Brunn, 3585–3600. New York: Springer.

Kong, L. 2010. "Global Shifts, Theoretical Shifts: Changing Geographies of Religion." *Progress in Human Geography* 34 (6): 755–776.

Kruse, K. 2005. *White Flight*. Princeton University Press.

Leitner, H. 2012. "Spaces of Encounters: Immigration, Race, Class, and the Politics of Belonging in Small-Town America." *Annals of the Association of American Geographers* 102 (4): 828–846.

Lippy, C. 2015. *Pluralism Comes of Age*. New York: Routledge.

Lorimer, H. 2005. "Cultural Geography: The Busyness of Being 'More-Than-Representational.'" *Progress in Human Geography* 29 (1): 83–94.

Lung-Amam, W. 2017. *Trespassers?* Oakland: University of California Press.

Machado, D., B. Turner, and T. E. Wyller, eds. 2018. *Borderland Religion*. New York: Routledge.

Maldonado-Estrada, A. 2020. *Lifeblood of the Parish*. Brooklyn. NYU Press.

Martin, J. 2001. *The Land Looks After Us*. New York: Oxford University Press.

Mazumdar, S. and S. Mazumdar. 1993. "Sacred Space and Place Attachment." *Journal of Environmental Psychology*, 13 (3): 231–242.

McCarthy, K. 2007. *Interfaith Encounters in America*. New Brunswick: Rutgers University Press.

McCutcheon, R. 1997. *Manufacturing Religion*. New York: Oxford University Press.

McGuire, M. 2008. *Lived Religion*. New York: Oxford University Press.

Messenger, T. 2000. *Holy Leisure*. Philadelphia: Temple University Press.

Morgan, D, ed. 2010. *Religion and Material Culture*. New York: Routledge.

Murphy, J. 2010. "Objects That Speak Creole: Juxtapositions of Shrine Devotions at Botanicas in Washington, DC." *Material Religion* 6 (1): 86–108.

Numrich, P. 2016. "Space-Sharing by Religious Groups." *Practical Matters Journal* 9: 10–30.

Olson, E. 2013. "Myth, *Miramiento*, and the Making of Religious Landscapes." In *Religion and Place*, edited by P. Hopkins, L. Kong, and E. Olson, 75–93. Dordrecht: Springer.

Orsi, R. 1985. *The Madonna of 115th Street*. New Haven: Yale University Press.

———. 2003. "Is the Study of Lived Religion Irrelevant to the World We Live in?" *Journal for the Scientific Study of Religion* 42 (2): 169–174.

Prickett, P. 2021. *Believing in South Central*. University of Chicago Press.

Rapoport, M. 2011. "Creating Place, Creating Community: The Intangible Boundaries of the Jewish 'Eruv.'" *Environment and Planning D: Society and Space* 29 (5): 891–904.

Schein, R. 1997. "The Place of Landscape: A Conceptual Framework for Interpreting an American Scene." *Annals of the Association of American Geographers* 87 (4): 660–680.

Seamon, D. 2018. *Life Takes Place*. New York: Routledge.

Stroik, D. 2012. *The Church Building as a Sacred Place*. Chicago: Hillenbrand Books.

Swatos, W., and L. Tomasi, eds. 2002. *From Medieval Pilgrimage to Religious Tourism*. Westport: Praeger.

Thomas, T. 2021. *Kincraft*. Durham: Duke University Press.

Thumma, S. and E. Gray, eds. 2005. *Gay Religion*. Lanham: Altamira Press.

Trinka, E. 2019. "Migration and Internal Religious Pluralism: A Review of Present Findings," *The Journal of Interreligious Studies* 28: 3–22.

Tweed, T. 1997. *Our Lady of the Exile*. New York: Oxford University Press.

——— 2006. *Crossing and Dwelling*. Cambridge: Harvard University Press.

Waghorne, J. 2004. *Diaspora of the Gods*. New York: Oxford University Press.

Warner, R. S. 2006. "The De-Europeanization of American Christianity." In *A Nation of Religions*, edited by S. Prothero, 233–255. Chapel Hill: University of North Carolina Press.

Weisenfeld, J. 2018. *New World A-Coming*. New York: NYU Press.

Whitehead, A. and S. Perry. 2020. *Taking America Back for God*. New York: Oxford University Press.

Wilford, J. 2012. *Sacred Subdivisions*. New York: NYU Press.

Wuthnow, R. 1998. *After Heaven*. Berkeley: University of California Press.

11

THE AMERICAN CINEMATIC LANDSCAPE INSCRIBED @WORK

Chris Lukinbeal

When we look out upon the American cinematic landscape, we do not see a continuously unfolding view as if staring off into the distance watching someone enter from the horizon. Nor can we stand on the hillside and gaze out over a quaint cinematic town or spectacular wilderness. Rather, to see the American cinematic landscape is to see a patchwork quilt, a bricolage or montage, an ever-changing and continuously unfolding assemblage of scenes. The American cinematic landscape (ACL) unfolds as sequences of moving images stitched together for a specific narrative event. Landscapes ground narrative action and allow events to *make* place. In cinema, image events take space and transform them into places wherein narrative events unfold. We are guided through the cinematic landscape like a projected mental map cast outward by the director and film production team. But rather than a mental map we can navigate by, this map is preordained and takes us on a topographical journey through a series of unfolding scenes. Oddly enough, when we venture to, and seek out, the ACL we are often disappointed that the unfolding and continuous view of what we see does not reflect the stitched together patchwork quilt of the narrativized place.

This essay focuses on two current trends involving the making of the ACL: the textual metaphor and a cinematic landscape @work. The two different approaches reflect the trend in interpreting landscape as an inscription or a practice. With an inscriptive view we follow the linguistic tradition of "reading" the landscape where the focus is either approaching the landscape through a particular theoretical perspective or through a regional descriptive approach of narrating place. In contrast, a landscape @work follows Don Mitchell's reading of landscape as part of a process of labor. The use of "@" in this essay prefaces "work" as being a location where landscape is staged, framed, and digitally commodified for its imagery. Here, landscape is an ongoing production that stages what can be seen/scene by visitors.

Reading the American Cinematic Landscape

The diversity of approaches in research on the cinematic landscape suggests that there is no one theoretical framework in which to situate geographic knowledge. Much of the research to date can be situated under the rubric of the textual metaphor. Yet the multiplicity of theories interrelating text, landscape, and cinema are enormous (Sharp and Lukinbeal 2015). Humanistic readings of cinematic landscape expound upon various genres where landscape

DOI: 10.4324/9781003121800-14

is conceived as pre-existent and fully formed (Harper and Rayner 2010). Critical readings of cinematic landscape posit that both space and texts are social products which mediate identities, power relations, and actively produce and reproduce landscapes and our understanding of them (Craine and Aitken 2004).

A few problems exist with the conceptualization of the cinematic landscape as text, mainly: (1) the "epistemological trap" of the reel/real binary; and (2) a hermeneutics of suspicion in relation to the dominance of the use of poststructural linguistic theories and their application to representational analysis. The reel/real binary is where we situate cinematic landscape within a bifurcated ontology where the reel world is a re-presentation of the real, offscreen world (Sharp and Lukinbeal 2015). In contrast, Anderson (2019, 1121) notes that the hermeneutics of suspicion is where the "presumption that people's access to the world was primarily an interpretive one always already mediated by 'signifying systems.'" Anderson uses this idea to explain the shift in human geography from representational studies in the 1990s to non-representational studies in the 2000s. Anderson's criticism follows an emerging trend of emphasizing more-than-representational or representation-in-relation-to approaches. Lukinbeal (2019, 98) argues that "there has been a shift from what representations stand for to what they do."

Reading the ACL relies on the textual metaphor which structures a way of seeing that "encapsulates the notion of fixity—of a text already written" (Cresswell 2003, 270). Such a conceptualization favors the sense of landscape as a representation, a view, a way of viewing. This has also been the dominant mode of inquiry into the ACL, where researchers examine how specific cultural and social themes are represented and relate to place or region. Whereas Alpers (1983, 136) suggests that landscape is "both what the surveyor was to measure and the artist to render," Casey (2002) argues that landscape is caught between the science of mapping and the art of painting, positioning it as placescape and sitescape. Whereas placescape is "an arena of open spectacle, a mise-en-scène of the world as it spreads out before and around us, giving us its layout," sitescape is linked to scopic and implies the prospect of the view (Casey 2002, 272). Scopic implies both scope and vision, and derives from the Latin *scopium* meaning to look at or inspect.

The French film critic Christian Metz coined the term *scopic regime* to contest the notion that vision is universal. According to Fleckenstein (2006, 70), "[a] scopic regime consists of the visual conventions that determine what we see and how we see… Like a discursive formation, scopic regimes exercise a formative power over what is knowable, sayable and doable." Scopic regimes reflect dominant institutionalized and naturalized ways of seeing but are always competing with other regimes "for organizational power within any culture" (Fleckenstein 2006, 70). Further, scopic regimes are historically and spatially contingent and change over time due to technological innovations and the cultural politics embedded in production and reception.

Inscriptive and textual logic focuses on sight, the voyeur, the immobile spectator and a "pre-existing pattern, template or programme, whether genetic or cultural … 'realized' in a substantive medium" (Ingold 1993, 157). Representation frames an inscriptive mode of logic, where *form* is pre-given and prioritized over the essential formative *process*. According to Shiel (2001, 3), "the larger part of Film Studies over the years has concerned itself primarily with the language of cinema… as a powerful signifying system… focused on the individual, the subject, identity, representation—for the most part, with the reflection of society in films." Here, film is conceived as a *text* to be examined hermeneutically and its form is already realized in a substantive medium. The metaphor of text cast onto representation acts to naturalize a binary logic of real/imaginary or material/non-material. From this perspective, the focus on the ACL is about narration: explaining the cultural traits or contesting naturalized meanings. Conceptualized as such, the aim of studying the relationship between film and landscape is to interpret the cultural systems represented. Inscription provides a path towards understanding film form, but it closes off the immanent openness through which they are made meaningful.

The popularity of inscriptive logic continues; however, most of this work is not being done in geography. Current research on the cinematic landscape focuses on regions, nations, or cities and "reads" the landscape through a particular perspective. Themes vary in recent readings including a focus on pastoral landscapes (Mazey 2020), Indigenous landscapes (Bertrand 2019), the Western genre (Broughton 2020), and anti-heimat cinematic landscapes (Ashkenazi 2020). Heimat films were popular in post-WWII Germany for their portrayal of traditional cultural themes and values, folklore, and rural idealism (Zimmermann 2008). Recent readings of American cinematic cities include San Francisco (Gleich 2018); San Diego (Lukinbeal 2012); New Orleans (Parmett 2019); and Los Angeles (Sheil 2012) to name a few.

The American Cinematic Landscape @*Work*

Labor practices continually shape the American cinematic landscape. W. J. T. Mitchell (2002, 1) suggests that we need to "change 'landscape' from a noun to a verb" to distinguish it "not as an object to be seen or a text to be read, but as a process by which social and subjective identities are formed." In this sense the ACL is never an end product, but rather, a process of continual spatial production and identity negotiation. The *work* of the ACL involves the people in the film industry and a stage for the continuation of that work. The cinematic landscape does *work* and part of that involves removing all traces of production from visitors. Cinema is produced from soundstages and sites with crews based out of regional production centers. Streets, homes, and businesses serve as the sites from which image facts are harvested while larger districts and neighborhoods with unique architectural and landscape aesthetics become backlots for day-to-day production.

Don Mitchell (2000) argues that if landscape is a product of human labor, then it is also a commodity in the sense that it actively hides the labor that goes into its making. In cinema, realism can refer to a representational practice that attempts to hide its mode of production, its labor and technology, while linking the diegesis to a location in an effort to suspend viewers' disbelief and accept diegetic geography as mimetic (Lukinbeal 2005b). Mitchell (2003, 8) refers to this process as an aesthetic issue where landscape functions as a stage which frames characteristics the author wants the viewer to see and hides other aspects off stage:

> The transition, one could say, is from a city of public spaces, a city of people at home in the city, not wealthy, but not abject either, to a city as *landscape*, a place devoid of residents, a landscape in which people are only visitors and welcome mostly as consumers (especially of imagery). The city-as-landscape does not encourage the formation of community or of urbanism as a way of life; rather it encourages the maintenance of surfaces, the promotion of order at the expense of lived social relations, and the ability to look past distress, destruction, and marginalization to see only the good life (for some) and to turn a blind eye towards what that life is constructed out of.

Landscape stages a spectacle in which unwanted aspects of daily life, such as homelessness, poverty, and blight, are screened out. In the case of homelessness, Mitchell (2003, 186 emphasis added) writes:

> "Indeed, homeless people's constant intrusion onto the stages of the city seems to threaten the carefully constructed *suspension of disbelief* on the part of the "audience"

that all theatrical performances demand, thereby seemingly turning that audience away and toward other entertainments: the suburban mall or the theme park."

The same suspension of disbelief occurs with the cinematic landscape: it requires that labor is hidden from the inscribed product.

From Primary Production Centers to Film Ranches

Landscape studies have a strong tradition of looking to origins to explain the relationship between shape/form and seen/scene. But rather than looking at sixteenth-century paintings or cartography, with cinema we turn to the (post)modern mobile gaze. Narrative cinema is a constant process of placemaking: the conversion of location imagery harvested from production centers to ground diegetic spaces. The ACL is always becoming and will continue as long as there are crews out filming. The landscapes chosen to ground narrative events that allow placemaking to occur reflect a complex process of economic and aesthetic decisions and the legacy of the production processes that underlie each image fact in a film. Regional production centers stitch together the ACL as part of a decentralized industry that extracts locational imagery and converts it into digital products. The ACL cannot be divorced from the larger political economy of film production and consumption. Equally important is the cultural economy of the film industry and the labor of its workers. Research on the economic geography of film production shows how location decisions are tied to outsourcing as well as networks, guilds, unions, and workers. The Fordist-based production model in the Golden Age of filmmaking centralized production in Los Angeles (LA) with New York playing a central role in financing and television show production. Other cities and states produced shows for local consumption or were used as occasional sites for the travelling Hollywood-based film production crews.

The origin of the ACL lies in New York City (NYC) where the film industry developed. NYC today is home to all the major television networks with abundant soundstage infrastructure, as well as the US center of finance, fashion, marketing, and advertising. NYC and LA are the oldest developed film production centers in North America. These centers reflect a significant level of infrastructural investment, as well as economic and cinematic agglomerations. If LA is the fabled backlot of Hollywood, then NYC is its fabled front lot open for touring (Katz 2004). As of 2019 NYC had approximately 1.8 million square feet of soundstage placing it fourth in North America with another one million currently under development. At the same time NYC had the highest number of streaming original productions on Netflix and Amazon reflecting 23% of the city's total production. This explains the filming of shows like *Punisher* in Brooklyn where Netflix has set up a production hub (FilmLA 2019).

LA, where the American film production industry migrated after leaving NYC, is a bricolage of regional icons to announce its presence in diegetic space, overlaying film neighborhoods and districts, and local cinematic gems rented out for film production. LA's cinematic landscape has evolved since Buster Keaton filmed *Sherlock Jr* (1924) in Hollywood (Figure 11.1). Sheil (2012) provides an integration of the cultural history and geography of the film industry in the LA basin whereas Scott's (2005) *On Hollywood* delves into the geography of economic activity and labor practices. Today, LA's 30-mile film zone (TMZ) dictates wages and per diem for union workers and represents the most densely filmed location anywhere on the planet (Lukinbeal 2004). Roland (2016, 1) refers to the TMZ as the film industry's "cartographic self-portrait that helps it organize its armies of workers." The film industry has institutionalized the labor practices of the TMZ for use in other regional centers. The TMZ reflects a standard labor relationship between unions so there is an expectation that when union members travel there will be similar types of policies

Figure 11.1   Sherlock Jr. filmed in Hollywood on trolley line (film in public domain since 2020).

in place. It also shows how runaway production reflects outsourced aspects of location production where local talent is used for below-the-line work (for example, makeup artists or camera operators) and above-the-line workers (for example, screenwriters or producers) are brought in from LA or NYC. This makes location production workers more vulnerable to the variable flows of production. So, with labor precarity and out-of-town producers, an institutionalized practice based on the centroid of the production office if outside LA is often used as a local TMZ. Within these TMZs a specific cinematic land use based on architectural styles and zoning can be found that covers the needs of a narrative being produced.

Some LA locations have representational legacies from being locational stars in blockbusters like the Bradbury Building, the Stahl or Ennis House, the Biltmore Hotel, Union Station, Hollywood Forever Cemetery, City Hall, or the LA Times Building, just to name a few. Others are equally as cinematic, however, the camera lens shifts from the aesthetics of exteriors to interiors. Many LA buildings maintain an ongoing relationship with the locations industry, like the Quality Café which has been filmed thousands of times. These buildings specialize in niche representational markets. In LA there is a plethora of micro studios specializing in a fake hospital, office, or police station, and open warehouse spaces like that of Willow Studios (Figure 11.2). The LA Warehouse District and neighboring Pico Gardens is ground zero for micro studios. Other famous warehouse spaces include the Nate Starkman Building, constructed in 1908 and whose current purpose is film production (Figure 11.3). Also known as the Pan Pacific Warehouse, this building has been depicted for decades in films such as *Repo Man*, *The Fast and Furious* series, and *The Fighter*. Also in LA are large to small independent and studio-owned film ranches loosely transecting across northern LA County up through Four Aces Movie Ranch in Palmdale over the Transverse Mountains in the high desert.

Film ranches grew out of the need for western genre backlots during the Golden Era of filming and can be found throughout the Western US but mainly in California, Nevada, Utah, and Arizona. Where Lone Pine and Big Bear were favored for their driving proximity to LA, the largest western backlot facility outside of LA was Old Tucson Studios (Figure 11.4). Old Tucson Studios (OTS) rose from the Sonoran Desert in the late 1930s just west of Tucson for the film *Arizona* (1939). It would later be taken over by Bob Shelton, who from the 1950s through the early 1980s, would host all the major film stars and directors associated with the Western genre. Interpersonal network connections play a strong role in the film production industry and part of the early success of OTS belongs to Jane Loew who was once married to Shelton during the height of OTS. Jane is Hollywood royalty, being the granddaughter of both Marcus Loew,

Figure 11.2   Willow Studios advertisement from LA's Warehouse District. Photo by author.

Figure 11.3   Nate Starkman Building. Photo by author.

Figure 11.4 Old Tucson Studio main street featuring the Hotel Del Toro. Photo by author.

who created MGM, and Adolph Zukor, who founded Paramount Pictures and conceived of the "star system." OTS was unique in that it was a short plane trip from LA, had three different Western backlots including a town with main streets and a mission east of Tucson, a remote high desert small town with vast vistas by Mescal west of Tucson (Figure 11.5), and a remote outpost on what was once the largest cattle ranch in southern Arizona, the Empire Ranch and OTS's "Slash Y" set. OTS would suffer a similar fate as other burgeoning film locations: film tax incentives were cut by the state, the Western film genre became less popular, and ultimately a fire would burn most of the main backlot in 1995. While OTS did continue as a tourist attraction and theme park with some filming taking place, the COVID-19 pandemic finally shut it down permanently. Fires were the death of many western movie ranches including, most recently, the Paramount Movie Ranch, that burned in 2019 in the Woolsey Fire.

Chasing the Dream of Hollywood

By the 1990s regional film production centers began competing for Hollywood based productions that could easily be lured away because of the development of small- to medium-sized independent production companies with their own backlots like Stu Segal Productions in San Diego or Dino DeLaurentis in Wilmington, North Carolina. These independent production companies found steady work through a formula of television series and television movies for the newly forming cable broadcasting networks like USA and WB. Segal had many early hit series like *Renegade* and *Silk Stalkings*, with the former playing a fictional "Bay City, CA" and the latter playing Palm Springs, Florida. Where *Renegade* highlighted the diversity of urban, rural, and open space afforded by San Diego's environs, *Silk Stalkings* was one of many productions to use San Diego as a double for Florida with the most famous example being *Some Like It Hot* filmed at the Del Coronado Hotel. In Wilmington, perhaps the most known of the television shows coming out of DeLaurentis studios was *Dawson's Creek* which took place in a fictional town of

Figure 11.5   Old Tucson Studio's Mescal Set with film historian Marty Freese showing the famous shoot-out location with Union Soldiers in *Outlaw Josie Wales*. Photo by author.

Capeside, Massachusetts. Where DeLaurentis Studio led to calls of a "Hollywood East," and San Diego hoped for a "South," both would soon be passed over as better motion picture incentives (MPI) were offered elsewhere, most notable for the early 2000s was New Mexico.

The first to offer MPIs to lure productions away from LA was Vancouver and Canada more generally. The first types of productions to move were ones with generic cityscapes, ones where it did not matter which city you were filming in but rather that it had a central business district (CBD), skid row, waterfront or other landscape spectacle, an "Anywhere USA," and other basic urban land use elements. "Anywhere USA" is a phrase used in the locations industry to reflect residential neighborhoods that could substitute for the East Coast. Any city wishing to have a regional production center had to have this and a generic CBD. In LA, Anywhere USA would be Altadena, in San Diego it's Mission Hills, in New Orleans it's neighborhoods near Lake Pontchartrain.

By the mid-1990s if you had to guess where a "B" movie or any television show playing the ACL was being produced and you guessed Vancouver you were probably right more than 75% of the time. The rise of Vancouver as part of the ACL is well documented by Mike Gasher in his book *Hollywood North*, wherein he states that while business was booming and continues to do so to this day, Vancouver's landscapes are harvested and used by Hollywood and rarely does the city get a chance to play itself. Following Canada's lead, US states began passing MPIs with the hopes that they too could garner a bit of Hollywood and develop their own local film production industry.

Nearly all states in the US have since tried an MPI with most finding that the return-on-investment was not viable. However, the glitz of Hollywood and seeing other states trying their luck at being a film production center drove many legislatures to pass bills in favor of the film production industry. More recently, it has become fairly easy to chart the "race to the bottom" as MacDonald (2011, 89) calls it: the race to see which state can offer the best tax incentive package, gain production, subsidize an industry that will leave the moment the subsidies stop, and send the bill to the local tax-paying citizens who are disproportionately affected depending on whether they reside in the film production center's primate city or in the hinterland where producers rarely venture. In the early 2000s, New Mexico—primarily Albuquerque—rode the wave of their MPI with *Breaking Bad*, a flurry of Westerns and other productions. Then, Louisiana—primarily New Orleans—stepped up and dropped billions of dollars in film subsidies for television shows (*NCIS New Orleans, Treme, American Horror Story*) and features (*Benjamin Button, Runaway Jury*), all while the state weathered natural disasters, recession, the BP oil spill, and corruption. Up next, and still active as of 2021, is Georgia. Hollywood South, once anchored in New Orleans, has now migrated to Atlanta and along with it, thousands of jobs.

The political and economic forces that privilege one city over another drive the production of the ACL. What we see on our screens is no longer based on where a narrative was written to take place, but rather, the economics of production will now re-write narratives and locational choices so a production can generate the most revenue. Thus, if you feel you've been seeing a lot more narratives focused on Atlanta or New Orleans in your movies and TV shows for the last decade, it is no surprise. Further, this state-by-state competition is occurring on a global scale with countries vying for the same piece of runaway production. Also, we should not be quick to assume that international locations (beyond Canada) cannot be used to portray the ACL. Quite the contrary. If the economics is right, a movie like *Cold Mountain*, whose narrative takes place in Asheville, North Carolina, will film primarily in Bucharest, and use the Romanian army for American Civil War soldiers (Lukinbeal 2005a).

Many other stories of the ongoing production of the ACL abound. Centers will come and go because of the mobility of finance, the outsourcing of production, and the ongoing regional development gamesmanship of "getting a bit of Hollywood." Regional centers like LA, NYC, Tucson, New Orleans, Atlanta, and others leave a representational legacy which provides the foundation of a bricolage map that is constantly being written over, scratched out, and intertextually blended together at key locations in cities like the Nate Starkman warehouse in LA, or the Market Street Power Plant in NOLAwood (Figure 11.6).

Conclusion

An American cinematic landscape as a noun is an inscription and as a verb is always @work and becoming. A focus on text presents an inscribed landscape through a series of unfolding narratives filmed in different locations. With this approach we read the landscape through varying theoretical perspectives to uncover or explain. Amazing stories exposing the history of inscriptions, like the documentary *Reel Injuns* by Neil Diamond, show why reading ACL is so important: it allows us to share and learn from the history of representation, is a form of cultural preservation, and educates us on the politics of sociospatial representation. Critiques of the cinematic landscape span from the humanistic to the critical with a long history of interdisciplinary perspectives and approaches. Inscription defines cinematic landscape especially when the focus is on vision and the voyeur. Recognizing and understanding that landscape can reflect scopic regimes about sociospatial themes and regions allows textual analysis to uncover assumptions about what has been naturalized onto the seen/scene of the land.

Figure 11.6   Market Street Power Plant, 1601–46 S. Peters St., New Orleans. Photo by Anthony Turducken, Creative Commons Attribution 2.0 Generic license, https://creativecommons.org/licenses/by/2.0/deed.en. Image was slightly cropped.

The ACL is also not just a moving picture show inscribing or reflecting cultural meaning, but also one continually @work: it is the film production industry harvesting imagery from the land to produce a digital product for consumption. In this landscape @work, inscriptions are created at regional production centers. Regional production centers vie for the variable flow of outsourced production from LA by offering MPIs. Competition for outsourced production invariably increases a region's relevance in the ACL at least for as long as an MPI remains in effect. These regional centers and their image legacies continually write and re-write the ACL. Regional image legacies can be seen in television shows that tour Albuquerque (*Breaking Bad*), NOLA (*Tremé*), Vancouver (*Arrow*), NYC (*The Punisher*), and Atlanta (*The Walking Dead*). The ACL is a product or inscription produced by workers in regional production centers. The legacy of their work is continually removed from the mise-en-scène which allows for the suspension of disbelief that a pre-written text takes regional geographic images of space and transforms them into cinematic places. Hiding the ACL @work is the job of the scopic regime of the film production industry. Rather than the institutionalized practice of the fourth wall, where characters are unaware of the audience, this is the fifth wall: the scopic practice where the audience and actors are unaware of their production.

References

Alpers, S. 1983. *The Art of Describing: Dutch Art in the Seventeenth Century*. Chicago: The University of Chicago Press.

Anderson B. 2019. "Cultural Geography II: The Force of Representations." *Progress in Human Geography* 43 (6): 1120–1132.

Ashkenazi, O. 2020. *Anti-Heimat Cinema: The Jewish Invention of the German Landscape*. Ann Arbor: University of Michigan Press.

Bertrand, K. 2019. "Canadian Indigenous Cinema: From Alanis Obomsawin to the Wapikoni Mobile." In *The Oxford Handbook of Canadian Cinema*, edited by J. Marchessault and W. Straw, 105–124. Oxford: Oxford University Press.

Broughton, L., ed. 2020. *Reframing Cult Westerns: From The Magnificent Seven to The Hateful Eight*. New York: Bloomsbury Academic.

Casey, E. 2002. *Representing Place: Landscape Painting and Maps*. Minneapolis: Minnesota Press.

Craine, J. and S. Aitken. 2004. "Street Fighting: Placing the Crisis of Masculinity in David Finch's Fight Club." *Geojournal* 59: 289–296.

Cresswell, T. "Landscape and the Obliteration of Practice." 2003. In *Handbook of Cultural Geography*, edited by K. Anderson, D. Domosh, S. Pile, and N. Thrift, 269–283. London: Sage.

FilmLA. 2019. Feature Films: A Profile of Production. Accessed April 16, 2021. www.filmla.com/wp-content/uploads/2020/03/Feature-Films-Profile-v2-WEB.pdf.

Fleckenstein, K. 2006. "Between Iconophillia and Iconophobia: Milton's *Areopagitica* and Seventeeth-Century Visual Culture." In *Rhetorical Agendas*, edited by P. Bizzell, 69–86. Mahwah, New Jersey: Lawrence Erlbaum Associates, Inc. Publishers.

Gleich, J. 2018. *Hollywood in San Francisco: Location Shooting and the Aesthetics of Urban Decline*. Austin: University of Texas Press.

Harper, G., and J. Rayner, eds. 2010. *Cinema and Landscape*. Bristol: Intellect.

Ingold, T. 1993. "The Temporality of Landscape." *World Archeology* 25 (2): 152–174.

Katz, C. 2004. *Manhattan on Film 1: Walking Tours of Hollywood's Fabled Front Lot*. Lanham, MD: Rowman & Littlefield.

Lukinbeal, C. 2004. "The Rise of Regional Film Production Centers in North America, 1984–1997." *GeoJournal* 59 (4): 307–321.

———. 2005a. "Runaway Hollywood: Cold Mountain Romania." *Erdkunde* 60 (4): 337–345.

———. 2005b. "Cinematic Landscapes." *Journal of Cultural Geography* 23 (1): 3–22.

———. 2012. "On Location" Filming in San Diego County from 1985–2005: How a Cinematic Landscape is Formed Through Incorporative Tasks and Represented through Mapped Inscriptions." *Annals of the Association of American Geographers* 102 (1): 171–190.

———. 2019. "The Chinafication of Hollywood: Chinese Consumption and the Self-Censorship of US Films Through a Case Study of *Transformers Age of Extinction*." *Erdkunde* 73 (2): 97–110.

Mazey P. 2020. *Pastoral Music: British Film Music*. New York: Springer.

McDonald, A. 2011. "Down the Rabbit Hole: The Madness of State Film Incentives as "Solution" to Runaway Production." *The University of Pennsylvania Law School Journal of Business Law*, 101–181.

Mitchell, D. 2000. *Cultural Geography: A Critical Introduction*. Oxford: Blackwell.

———. 2003. *The Right to the City: Social Justice and the Fight for Public Space*. New York: Guilford Press.

Mitchell, W. J. T., ed. 2002. *Landscape and Power*. Chicago: University of Chicago Press.

Parmett, H. M. 2019. *Down in Treme: Race, Place, and New Orleans on Television*. Stuttgart: Franz Steiner Verlag.

Roland, Z. 2016. *Studio Labor and the Origins of Hollywood's Thirty-Mile Zone, or TMZ*. Accessed February 25, 2021. www.kcet.org/shows/lost-la/studio-labor-and-the-origins-of-hollywoods-thirty-mile-zone-or-tmz.

Scott, A. 2005. *On Hollywood: The Place, the Industry*. Princeton: Princeton University Press.

Sharp, L. and C. Lukinbeal. 2015. "Film Geography: A Review and Prospectus." In *Mediated Geographies and Geographies of Media*, edited by S. Mains, J. Cupples, and C. Lukinbeal, 21–35. Thousand Oaks, CA: Sage.

Sheil, M. 2001. "Cinema and the City in History and Theory." In *Cinema and the City: Film and Urban Societies*, edited by M. Sheil and T. Fitzmaurice, 1–18. Oxford: Blackwell.

———. 2012. *Hollywood Cinema and the Real Los Angeles*. London: Reaktion Books.

Zimmerman, S. 2000. "Landscapes of Heimat in Post-War German Cinema." In *The Geography of Cinema—A Cinematic World*, edited by C. Lukinbeal and S. Zimmerman, 171–183. Stuttgart: Steiner Verlag.

12

MUSIC AND THE AMERICAN LANDSCAPE

Rhythms of Continuity and Change

David J. Keeling and Thomas L. Bell

Music and landscapes are inextricably intertwined in the sociocultural fabric of the United States. With the expansion of economic globalization and the development of new technologies over the past 25 years, music production, distribution, and consumption have evolved in ways that have changed relationships with the cultural landscape. Geographers have long studied this interconnection between music and landscape, as it reveals symbolic meaning or represents a particular emotion or feeling about a place (Milburn 2019; Levy 2012). Landscapes reflect the dynamic interaction between specific cultural milieus and their physical settings. From humble, rural, sharecropper cabins in Mississippi to eye-catching futuristic skyscrapers in New York City, and from Kansas cornfields to regimented rows of California wine trellises, landscapes are palimpsests that reflect cultural change over time and space. Iconic music landscapes generally represent a well-known environment: places, spaces, or buildings that convey some cultural significance to the general population. Images of Graceland in Memphis, Strawberry Fields in Central Park, or the hallowed meadows of Woodstock in upstate New York, for instance, often convey specific musical meaning to certain groups of people both within the US and beyond. As American landscapes are constantly rebuilt, remodeled, renewed, and reimagined, evidence of the past remains, providing visual clues about the landscape's evolution and often inspiring musical creativity that imbues meaning and emotion to what is observed and experienced. In fact, one does not have to gaze far across the American landscape to find places, composers, and popular songs that exemplify the connection between landscape and music. Think of songs like "American Land" by Bruce Springsteen, "Pink Houses" by John Mellencamp, "New York State of Mind" by Billy Joel, "City of New Orleans" by Arlo Guthrie, and "Only N Atlanta" by Trinidad James, to name just a few. They all evoke some memory, experience, or understanding about urban, rural, or physical landscapes.

Geographers view landscapes as products of human activity, what Giddens (1991) contextualized as a social–structure–agency relationship, or the collective transformation of the environment for specific purposes. Mills (2013) posited that landscape study has moved beyond its pedagogical fossil reputation as an old-fashioned litany of landscape "characteristics" to embrace myriad conceptual, theoretical, and methodological approaches that are open to incorporating such elements as cuisine, tourism, mobility, and music. Recent research has explored the spatial value of music concerts in urban spaces (van de Hoeven and Hitters

DOI: 10.4324/9781003121800-15

A QUADRIALECTIC OF MUSIC AND THE AMERICAN LANDSCAPE

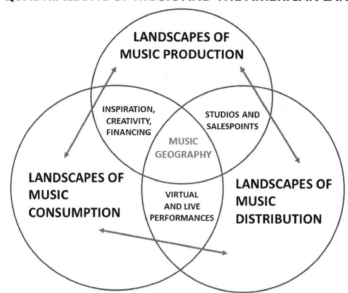

Figure 12.1 Music and the American landscape.

2020), the lyrical connections to sociopolitical structures (Post and Rhodes 2020), sonic geographies and landscapes (Gallagher et al. 2017), performance landscapes (Johansson et al. 2016), and diasporic music landscapes of blackness (Perry 2009). Indeed, the contemporary music–landscape nexus is as intellectually and creatively vibrant as at any time in the past century.

This chapter examines the music–landscape nexus by exploring how music has altered relationships between the industry, distributors, consumers, and the cultural environment. Our exploration of music landscapes engages a heuristic device called a quadrialectic framework, inspired by Soja's (1996) concept of a trialectic, that aims to tease out both evident and subtle shifts in music as a material expression of society's actions, preferences, and desires (Figure 12.1). Reciprocity and interaction between the three main elements of production–distribution–consumption are what create the fourth element, music geography, and its constituent landscapes. Each part of the quadrialectic shapes, and is shaped by, the landscapes within which it actuates and influences the myriad human–environment relationships that are experienced through and with music.

Landscapes of Music Production

American landscape influences on music production have been extensively explored in recent years (Seman 2019; Johansson and Bell 2016; Hilbruner 2015; Knight 2006). Widely admired in orchestral circles, for example, are Aaron Copland's "Appalachian Spring" that premiered in 1944, and "American Landscape" by Soon Hee Newbold, arranged for a full orchestra in 2017 and apparently inspired by train rides that revealed the full majesty of the wide-open plains and soaring mountains that stretch between America's towns and cities. Mark Templeton is a highly respected musician known for his arrangements of Amish folksongs that represent the life and landscape of the Indiana–Ohio–Pennsylvania region, and Joseph Curiale's "Prairie Hymn," from his 2001 orchestral album *The Music of Life*, was inspired by observing Nebraska farmland from the air. In the pop music genre, myriad songs have expressed the complex relationship between people and their landscapes. John Lennon and Yoko Ono wrote "Central Park Stroll" to reflect their experiences with one of New York's most beloved landscapes. Joni Mitchell in "Big Yellow Taxi" highlighted urban landscape change when she sang about paving paradise by constructing a parking lot, an urban development theme continued by Chrissie Hynde of the Pretenders. She wrote "My City was Gone" in 1982, a song about the destruction in Ohio of Akron's historic center to make way for a bland urban plaza. In 1999's "Four Wal-Marts," Baker Maultsby sang about the perceived homogenization of the American landscape, with fast-food outlets on every corner and Walmarts in every small town. This song, like others of its ilk, addressed urban sprawl in the Old South, as communities struggled to come to terms with a steady diet of new landscape excrescences such as outlet malls, superstores, theme restaurants, and traffic circles. In recent decades, rap, hip hop, country, and indie rock songs, among other genres, have touched on themes related to living in or experiencing the American landscape, from the urban experience (e.g., "Straight Outta Compton" by N.W.A.), the loss of farmland exemplified by John Mellencamp's 1985 classic "Rain on the Scarecrow," and to the isolation of rural communities represented in the 2013 musical miniseries, "The Hinterlands."

A key element in music production is the creation and recording of songs that reflect society's specific place and time. The transition from the folk, country and western, and blues rhythms that had originated in rural America to urban rock 'n' roll, rap, and hip hop in the second half of the twentieth century mirrored the changing demographic and landscape makeup of the US. By the early twenty-first century, although rural America comprised approximately 97% of the land area it only housed about 60 million people. In contrast, 3% of the land area now is considered an urbanized or peri-urban landscape, yet it contains over 270 million Americans (Ratcliffe et al. 2016). Urban environments became specific places that provided defined spaces for musical creativity and production. Cities are where record companies, recording studios, product advertisers and marketers, and music financiers agglomerate to take advantage of economies of scale, global distribution opportunities, efficient commodity chains, and affiliated linkages. Certain cities became associated with specific musical styles: New York (punk and rap), Philadelphia (soul), and Boston (alt-rock) in the Northeast; Detroit (Motown and gospel), Minneapolis (funk rock), and Chicago (blues) in the Upper Midwest; Nashville (country music), Memphis (rockabilly), Atlanta (Hip Hop), and New Orleans (jazz) in the South; and Seattle (grunge), Los Angeles (surf and glam metal), and San Francisco (psychedelic rock) in the West, among many others. Many urban landscapes have been converted into recording studios that provide the technological, creative, and material spaces that connect musicians, engineers, promoters, entrepreneurs, and industry "others." Empty warehouses, abandoned factories, garages, traditional office buildings, and houses became valuable economic, social, and cultural spaces for music creativity. A converted car garage became Sun Studios in South Memphis, as did Stax Records (originally Satellite Records) in Brunswick, northeast Memphis; a coffin showroom morphed into Muscle Shoals Studio in semi-rural Alabama; a defunct nightclub in Greenwich Village, New York, is now Electric Lady Studio; and the famous Motown Hitsville house located in New Center, Detroit, became an anchor site for this gentrified commercial and residential historic landscape located north of the CBD (Figure 12.2).

Tying together the various locations, genres, and business interests that inform a geography of music production is a challenging proposition and one fraught with generalizations and influenced by personal music preferences. However, four broad music regions can be identified as the primary drivers of the music production complex. The Northeast contains the urban megalopolis that stretches from north of Boston to south of Richmond and includes New York

Figure 12.2 Hitsville, USA: Original headquarters of Motown Records in Detroit, Michigan. Photo by authors.

City, the Jersey Shore, and points in between, housing key centers of musical productivity such as Tin Pan Alley and Broadway. The Upper Midwest links Cleveland, Detroit, Gary, Chicago, and Minneapolis musically, and are all significant in the development of musical talent and genres over the past 60 years. The Southland encapsulates the production triangle linking New Orleans, Nashville, and Memphis, famous for the blues, Music Row, and Sun Studios, respectively. Other important centers of musical creativity are also located in the South, from north Florida to Austin, Texas. Finally, the West connects the cities and hinterlands of San Diego, Los Angeles, San Francisco, and Seattle, among others, where music genres and styles reflect the cultures and landscapes of Pacific coast, Great Basin, and northern Plains environments. Measuring the overall influence of each region or urban center in the music production process can be complicated, as meaningful comparative data are not widely available. Florida (2013) analyzed the now-defunct MySpace to document the role of place in the digital music eco-system, arguing that pop songs are like widgits; they are manufactured commodities that are produced in real places by creative talent, such as composers, engineers, lyricists, publishers, etc. During the first decade of the twenty-first century, data suggested that a decline in record sales and the shift to digital in the music industry had positioned Los Angeles as the hegemon in the national pop music complex. New York came in a distant second in shaping the pop culture landscape, despite its productivity in jazz, hip hop, rap, and indie rock.

A different kind of data set developed by Yaffe (2020) for the website Musicoholics attempted to rank each of the 50 states by evaluating and linking the music genre, performer, location, and impact on the pop music landscape. In this analysis, the position of the two top cities reversed, with New York deemed to have a much broader and deeper influence on music generally, because punk and hip hop, the successors to rock and roll in terms of influence and sales, came from the city's diverse neighborhoods and spawned many groups who achieved national and international success (Ramones, Beastie Boys, Wu-Tang Clan, Blondie, et al.). No other city within the state had the influence exerted by New York City. By contrast, California was positioned in second place, in part because of the regional diversity that birthed various genres of music: the Bakersfield Sound and country music; San Francisco and psychedelic rock; Los Angeles and the surf sound, followed by metal glam and West Coast hip hop. Four southern states (Louisiana, Georgia, Tennessee, and Mississippi) were positioned from third to sixth respectively in the ranking, reflecting the influence of blues, country, hip hop, and the alt-rock/funk scene that emerged from towns like Athens and Macon, Georgia. The least influential states in this music location analysis were Wyoming, Maine, Iowa, North Dakota, and South Dakota, the bottom five with a meager level of influence in the national music scene.

Landscapes of Music Distribution

In recent decades, music distribution landscapes have changed in profound and consequential ways. Transportation and communication infrastructure have reshaped physical and digital landscapes in ways that connect music producers to consumers more efficiently, including intermediaries such as record companies, retail outlets, and a wide variety of logistics networks such as Amazon Prime or streaming services like Spotify and others. Khanna (2016) has argued that connectivity drives creativity and innovation, and that those urban centers connected to an expanding global system of critical infrastructure are more likely to dominate the nexus between music producer and consumer. He pointed out that New York City's economy alone is larger than the combined economies of 46 sub-Saharan African countries (Khanna 2016), giving it an edge in the global music industry. However, Florida (2013) has argued that Los Angeles is now the dominant center for the production of popular music, an outcome of the digital shift in the movie, television, and music industries. Their point about the rise of cities and their impact on the creative class mirrors arguments put forth by Florida and Jackson (2010) about how urban landscapes are not homogeneous and monolithic. They are comprised of culturally diverse neighborhoods, bohemian enclaves, bars, clubs, public places, and shared creative spaces that drive musical innovation. For example, in many economically and socially deprived inner-city neighborhoods of American cities, landscapes that have become all too familiar in recent years from myriad televised protests about social justice and violence against people of color, hip hop emerged as a musical form that spoke to, and about, the reality of these landscapes. Specifically, the South Bronx and Harlem neighborhoods in New York City produced a style of hip hop that had an intense territoriality about it, with its focus on life amid the landscapes that featured both real and imagined spaces (Connell and Gibson 2003).

How does the landscape, and thus the distribution of music and its associated themes, reflect this transition from rural to urban, from analog to digital, and from an industrial economy to an information-based one? Declining or disappearing industrial landscapes are one of the most visually obvious signs of a changing economy, identified so poignantly in Billy Joel's "Allentown," Hank Williams Jr.'s "Red, White, and Pink-Slip Blues," and John Prine's "Paradise," where the factories, steel mills, and coal mines closed down and jobs were lost. "Our Town," performed by James Taylor, laments the decline of small-town landscapes, identities, services, and industries as interstate freeways, beltways, and bypasses altered the site and situation of these communities. Just a few decades ago, along the rural and urban highways, byways, and railroad lines of the US, the most obvious infrastructure visually was the telephone line or utility pole— some 180 million of them still dot the landscape today. These ubiquitous telephone and power lines and their importance to the US communication and energy networks were celebrated in Glen Campbell's 1968 song, "Wichita Lineman." Today, both rural and urban landscapes are dotted with almost 350,000 cellphone towers, dishes, and their supporting infrastructure, like flagpoles, water towers, billboards, and rooftops, and another 300,000 towers are needed to support an enhanced 5G network in the years ahead (CLE 2020).

These landscape changes reflect how the music industry's relationship between producers and consumers is adapting to new ways of distributing product. Over 80% of music in 2019 was consumed via streaming services, with over one trillion individual streams (Steele 2020). Broadband internet networks, with cables often buried where telephone poles once stood, and cellphone services have replaced the ubiquitous record store as the primary means of distribution. As more music is consumed digitally today than physically, this shift is reflected in changes on the landscape that reflect the decline of brick-and-mortar music stores. At the turn of the millennium, over 4,000 independent record stores still were located in communities large and small, along with thousands of big-box retailers like Walmart and Target; now there are fewer than 1,500 independent record stores and many of these currently face extinction as a result of the COVID-19 pandemic. Record stores were ubiquitous on the retail landscape during the second half of the twentieth century, but only vinyl revival stores (Figure 12.3) seem to have survived the transition from the physical to the virtual purchase of music, as popular outlets like Tower Records and Wherehouse Music disappeared from shopping centers and strip malls (Goldman 2010). Even in the largest US cities, retail record stores are now few and far between, with New York (47), Chicago (30), and Los Angeles (29) leading the pack (Vinylives 2020). During the first decade of the twenty-first century, the music industry lost 50% of its overall sales (Smith 2020), and sales of physical albums, whether vinyl or compact disc, had fallen to just 9% of overall music consumption by 2019 (Steele 2020).

Figure 12.3 A revival vinyl store in Bowling Green, Kentucky. Photo by authors.

Apart from digital delivery, the most notable distribution impact on the landscape has been the growth of radio stations and their associated broadcast antennas. There were a little over 15,000 active radio stations in 2019, with another 15,000 FCC-licensed broadcast outlets like television stations. This is double the number of radio stations that broadcast a mix of music, news, sports, and opinion in 1970 (Watson 2020). Radio, especially listened to in cars, is still an important medium for the distribution of new and classic music. In addition, broadcast media, including commercial and subscription television channels, provide ubiquitous access to content across the American landscape, distributing music to nearly 300 million television sets via cable, satellite, and over-the-air transmission (Watson 2020). Many contemporary advertisers on radio and television are using classic rock songs to sell products to the growing baby boomer cohort, along with more recent generations, and this line of revenue from song licensing has helped to fill the gap for musicians as record sales continue to plummet.

Other important avenues of music distribution include the venues where musicians ply their trade. In urban areas across the country, many music venues started in less-desirable neighborhoods because cheap rent was an important consideration in developing the necessary space to accommodate bands and fans. One of the more famous venues in Los Angeles is the Troubadour (Figure 12.4), located in the shabby chic and historic gay district of West Hollywood, WeHo to locals, which is also home to clubs like the Viper Room and Whisky A Go Go along Sunset Strip and Santa Monica Boulevard. As well as smaller music venues in towns and cities, the American landscape hosts over 1,000 sporting arenas and concert halls that provide outlets for musical performances at a variety of scales. Football arenas are

Figure 12.4   Music venues for distribution and consumption: The famous Troubadour in West Hollywood, California. Photo by authors.

especially popular for packing in the crowds for a stadium concert. The Georgia Dome in Atlanta recorded over 73,000 for a Backstreet Boys performance in 2000; Pink Floyd packed 75,000 into the Ohio Stadium in Columbus in 1994; in 2018, Garth Brooks attracted 84,000 to a Notre Dame Stadium concert; and U2 jammed 97,000 into the Rose Bowl for a 2009 show. The largest ticketed concert in US history, however, took place in Englishtown, New Jersey, site of autoracing's Raceway Park, where the Grateful Dead drew a record crowd of 107,000 in 1977 (Rosenbaum 2020). Nationally known venues specifically hosting music performances include the Ryman Auditorium (Nashville), Stubbs (Austin, TX), Tower Theater (Philadelphia), Red Rocks (Morrison, CO), Hollywood Bowl (Los Angeles), Madison Square Garden and Brooklyn Steel (New York), among many others.

Landscapes of Music Consumption

Connell and Gibson (2003, 254) defined the landscapes of "terra digitalia" as all the ways by which music imbues places with credibility. Yet despite the significant shift to virtual platforms for the routine acquisition and consumption of music, music is still omnipresent in the myriad environments that structure the contemporary US landscape. Whether consumed at music festivals, in arenas, clubs, and cafes, at political rallies, during visits to tourist landscapes, or as backdrops to advertising and political campaigns, music remains inextricably intertwined with how society identifies and packages space and place. Many of the sites that have hosted, or still host, music festivals are in rural or peri-urban areas. Perhaps the most famous of these is the rolling farmland of Sullivan County, New York, that entered music immortality as the August, 1969, site of the Woodstock Music Festival, listed since 2017 on the National Register of Historic Places. Many other multi-day festivals have been held in rural or peri-urban locations to accommodate campers, backpackers, and day trippers, including the annual Bonnaroo Music Festival that takes place on a 700-acre farm in Manchester, TN; the Coachella Valley Music and Arts Festival in Indio Valley, California, east of Los Angeles; Burning Man held in the Black Rock Desert area of northern Nevada; the financially plagued and short-lived Lilith Fair that rotated through several small- and medium-sized cities; and the WeFest country music event held on Soo Pass Ranch in Detroit Lakes, Minnesota. Important urban-based festivals include Lollapalooza in Grant Park, Chicago; the Rolling Loud Hip-Hop Festival based in Miami, FL; and South by Southwest in Austin, Texas.

Perhaps the most landscape-visible aspect of music consumption is the tourism that has developed around specific geographic locations that share special significance because of their association with the production, distribution, or consumption process. As a supposedly ephemeral art form, music has been successful in developing and supporting a vibrant tourist economy structured around specific landscapes, lyrics, and identities. In their exploration of the Memphis music scene, Gibson and Connell (2007) wondered how contemporary culture could explain the popularity of music tourism. The status of Graceland (Figure 12.5), Elvis Presley's Memphis mansion, raises questions about how conjured memories of Elvis as musician and actor are set in a specific time and space that have, in effect, "fossilized" the landscape of an Elvis-defined America (Duffett 2013). In his analysis of Winslow, Arizona, a town made famous by a line in an Eagles song, Lashua (2018, 157) pondered why music can take an "utterly mundane landscape feature" such as a street corner and transform it into a globally popular tourist destination. Situated along Route 66, itself an important part of America's landscape mythology and culture, immortalized in an early 1960s' television series and in the Rolling Stones' 1966 cover version of the 1946 Bobby Troupe song of the same name, Winslow created a landscape of murals, signs, statues, and festivals that monetarized this artificial but physically real spot. Just a

Figure 12.5 Music tourism landscapes: Elvis Presley's Graceland in Memphis, Tennessee. Photo by authors.

year before the Rolling Stones' hit song, Bob Dylan in 1965 had released *Highway 61 Revisited*, an album that paid homage to the highway that connected his birth town of Duluth, MN, with the origin-of-the-blues cities of St Louis, Memphis, and New Orleans.

Marketing tourist landscapes for their connection, however tenuous, to music has become big business. In 1986, Cleveland, Ohio, not a city or state widely known for significant contributions to the music industry (perhaps with the exception of Chrissie Hynde and Dave Grohl), beat out New York, Memphis, Chicago, and San Francisco as the future site of the Rock and Roll Hall of Fame. Designed by architect I.M. Pei and dedicated in 1995, the venue has transformed post-industrial Cleveland, known historically for its burning Cuyahoga River and devastated industrial landscapes, into a postmodern urbanscape with a gentrified waterfront that includes a science center and sports stadium. Today, the Hall of Fame attracts hundreds of thousands of music aficionados annually and contributes significant revenue to the city's coffers, especially when the annual induction ceremony is held. A revisionist geohistorical analysis of Cleveland's role in the national music scene has rebranded the city as the birthplace of rock and roll, in no small measure due to Alan Freed's Moondog Coronation Ball in 1952, often identified as the very first rock and roll concert, and his role as a DJ playing R&B "race" records while coining the more acceptable term of rock 'n' roll for the style of music (Lashua 2018).

Urban landscapes have emerged as critical elements of music tourism studies about cultural heritage, especially as cities reimagine, regenerate, and reconstruct their environments with festivals, museums, musically relevant neighborhoods, and music heritage trails (Henshall 2020). In New York's Central Park, Strawberry Fields (Figure 12.6) has been designated as an homage to John Lennon, who lived in the nearby Dakota Building and was tragically assassinated there in 1980. The entranceway to the Dakota, along with the various Lennon-designated sites in the park, continues to attract legions of international fans of Lennon and the Beatles. In addition

Figure 12.6   Music tourism landscapes: Strawberry Fields, Central Park, New York. Photo by authors.

to those destinations already mentioned, like Graceland, Route 66, Muscle Shoals, Hitsville USA, and the Troubadour, among thousands of other sites, any location or landscape that has a connection to music production, distribution, or consumption, however tenuous, has sought to profit in some way from this relationship.

Coda

This chapter has argued that music continues to leave an indelible mark on contemporary American landscapes through music-related tourism, multiple venues for consumption, and complex logistics infrastructure, both physical and virtual, that indicate an innovative era in the music–landscape relationship. A significant unknown in this relationship is how the COVID-19 pandemic could reshape the future of the music industry and the various elements that have shaped the production–distribution–consumption nexus. Retail outlets and their associated landscapes have been changing for two decades now, and there likely will be changes in how live music performances are packaged and consumed in a post-pandemic world.

US music also continues to change and adapt to new technologies, tastes, creativity, and venues for performance and consumption, leading to a post-rock-and-roll period of music innovation that has seen songs get shorter and snappier in an attempt to boost the number of individual digital downloads. Other business models during the pandemic focus on short videos as consumers discover new music through platforms like TikTok, which is developing its own streaming service (Hall 2020). Other platforms will arise to stream live concerts to paying fans of particular artists. Indisputable is the fact that American music landscapes continue to change and adapt to globalization dynamics, to the forces of economic realignment, and to the continuing urbanization of the population.

References

CLE (Celltower Lease Experts). "2020. Industry Facts and Figures." Data updated 9/2020. Accessed March 5, 2021. www.celltowerleaseexperts.com/cell-tower-lease-news/cell-tower-industry-facts-figures-2016/.

Connell, J. and C. Gibson. 2003. *Sounds Tracks: Popular Music, Identity, and Place.* New York: Routledge.

Duffett, M. 2013. "Walking in Memphis? Redefining Mainstream Popular Music." In *Redefining Mainstream Popular Music*, edited by S. Baker, A. Bennett, and J. Taylor, 102–113. New York: Routledge.

Florida, R. 2013. "The Geography of America's Pop Music/Entertainment Complex." New York: Bloomberg City Lab, May 28, 2013. Accessed March 5, 2021. www.bloomberg.com/news/articles/2013-05-28/the-geography-of-america-s-pop-music-entertainment-complex.

Florida, R. and S. Jackson. 2010. "Sonic City: The Evolving Economic Geography of the Music Industry." *Journal of Planning Education and Research* 29 (3): 310–321, doi: 10.1177/0739456X09354453.

Gallagher, M., A. Kanngieser, and J. Prior. 2017. "Listening Geographies: Landscape, Affect and Geotechnologies." *Progress in Human Geography* 41 (5): 618–637, doi: 10.1177/0309132516652952.

Gibson, C. and J. Connell. 2007. "Music, Tourism and the Transformation of Memphis." *Tourism Geographies* 9 (2): 160–190, doi: 10.1080/14616680701278505.

Giddens, A. 1991. "Structuration Theory. Past, Present and Future." In *Giddens' Theory of Structuration. A Critical Appreciation*, edited by C. Bryant and D. Jary, 55–66. London: Routledge.

Goldman, D. 2010. "Music's Lost Decade: Sales Cut in Half." *CNN Money*, February 3, 2010. Accessed March 5, 2021. https://money.cnn.com/2010/02/02/news/companies/napster_music_industry/.

Hall, S. 2020. "This is How COVID-19 is Affecting the Music Industry." Geneva: World Economic Forum. Accessed March 5, 2021. www.weforum.org/agenda/2020/05/this-is-how-covid-19-is-affecting-the-music-industry/.

Henshall, J. C. 2020. "Clarksdale, Mississippi: Downtown Regeneration, Cultural Heritage, Tourism and Blues Music." In *Tourism, Cultural Heritage and Urban Regeneration: Changing Spaces in Historical Places*, edited by N. Wise and T. Jimura, 21–38. Cham, Switzerland: Springer.

Hilbruner, M. 2015. "'It Ain't No Cake Walk': The Influence of African American Music and Dance on the American Cultural Landscape." *Virginia Social Science Journal* 50: 73–80.

Johansson, O. and T. L. Bell, eds. 2016. *Sound, Society and the Geography of Popular Music.* New York: Routledge.

Johansson, O., M. M. Gripshover, and T. L. Bell. 2016. "Landscapes of Performance and Technological Change." In *The Production and Consumption of Music in the Digital Age*, edited by B.J. Hracs, M. Seman, and T.E. Virani, 114–129. New York: Routledge.

Khanna, P. 2016. *Connectography: Mapping the Future of Global Civilization.* New York: Random House.

Knight, D. B. 2006. *Landscapes in Music: Space, Place, and Time in the World's Great Music.* Boulder: Rowman & Littlefield Publishers.

Lashua, B. D. 2018. "Popular Music Heritage and Tourism." In *The Routledge Companion to Popular Music History and Heritage*, edited by S. Baker, C. Strong, L. Istvandity, and C. Cantillon, 153–165. New York: Routledge.

Levy, B. E. 2012. *Frontier Figures: American Music and the Mythology of the American West.* Vol. 14. Berkeley: University of California Press.

Milburn, K. 2019. "Rethinking Music Geography Through the Mainstream: A Geographical Analysis of Frank Sinatra, Music and Travel." *Social & Cultural Geography* 20 (5): 730–754, doi: 10.1080/14649365.2017.1375550.

Mills, S. F. 2013. *The American Landscape.* New York: Routledge.

Perry, M. D. 2009. "Hip Hop's Diasporic Landscapes of Blackness." In *From Toussaint to Tupac: The Black International Since the Age of Revolution*, edited by M. O. West, W. G. Martin, and F. C. Wilkins, 232–258. Chapel Hill, NC: University of North Carolina Press.

Post, C. W. and M. Rhodes. 2020. "Lyrical Geographies and the Topography of Social Resistance in Popular Music in the United States." In *Handbook of the Changing World Language Map*, edited by S. Brunn and R. Kehrein, 2535–2558. Cham: Springer.

Ratcliffe, M., B. Burd, K. Holder, and A. Fields. 2016. *Defining Rural at the US Census Bureau.* Washington, DC: US Census Bureau. Accessed March 5, 2021. www2.census.gov/geo/pdfs/reference/ua/Defining_Rural.pdf.

Rosenbaum, M. 2020. "9 of the Biggest Concerts in American History." Accessed March 5, 2021. www.radio.com/music/gallery/9-of-the-biggest-concerts-in-american-history.

Seman, M. 2019. "Punk Rock Entrepreneurship: All-Ages DIY Music Venues and the Urban Economic Landscape." In *Creative Hubs in Question: Place, Space and Work in the Creative Economy*, edited by R. Gill, A. C. Pratt, and T. E. Virani, 229–244. Cham, Switzerland: Springer.

Smith, M. D. 2020. "Ideas: Are Universities Going the Way of CDs and Cable TV?" *The Atlantic*, July 22, 2020. Accessed March 5, 2021. www.theatlantic.com/ideas/archive/2020/06/university-like-cd-streaming-age/613291.

Soja, E. W. 1996. *Thirdspace: Journeys to Los Angeles and Other Real-and-Imagined Places.* Hoboken: John Wiley.

Steele, A. 2020. "US Music Streams Topped a Trillion in 2019." *Wall Street Journal.* January 9, 2020. Accessed March 5, 2021. www.wsj.com/articles/u-s-music-streams-topped-a-trillion-in-2019-1157 8607152.

Van der Hoeven, A. and E. Hitters. 2020. "The Spatial Value of Live Music: Performing, (Re)Developing and Narrating Urban Spaces." *Geoforum* 117: 154–164, doi: 10.1016/j.geoforum.2020.09.016.

Vinylives. 2020. *FAQ.* Los Angeles: Vinylives. Accessed March 5, 2021. www.vinylives.com/faq.html.

Watson, A. 2020. *Number of Commercial Radio Stations in the United States from 1952 to 2019.* Hamburg, Germany: Statista. Accessed March 5, 2021. www.statista.com/statistics/252235/number-of-commerc ial-radio-stations-in-the-us/.

Yaffe, A. 2020. *The United States of Pop: How Each State Contributed to Popular Music.* Tel Aviv, Israel: Musicoholics. Accessed March 5, 2021. www.musicoholics.com/best-of-the-best/the-united-states-of-pop-how-each-state-contributed-to-popular-music/.

13

SPORTS AND THE AMERICAN LANDSCAPE

John Lauermann

Sports have a major impact on the American landscape. Professional sports leagues are a $75 billion per year industry (PWC 2019), with hundreds of stadiums and arenas throughout the country (Figure 13.1). Dozens of North American cities have hosted sports mega-events like the Olympics, World Cup, or Pan American Games, leveraging sport to invest billions in venues and related infrastructure (Andranovich and Burbank 2011; Kassens-Noor and Lauermann 2017). Each stadium, furthermore, anchors commercial districts and local tourism- and culture-based economies. A single stadium can generate hundreds of millions of dollars in annual economic impacts, though there is a voluminous critical economics literature arguing that sports are often financial hazards for the taxpayer, as public subsidies generate minimal return on investment (Agha and Rascher 2020; Baade and Matheson 2016; Grant Long 2013; Zimbalist 2015).

Amateur sport, likewise, is practiced in almost every American community—rural, urban, and suburban. Amateur sports landscapes are less spectacular than the professional variety, and perhaps not surprisingly they have received less attention in the research literature. But they are far more geographically widespread, practiced across a vast landscape of public parks, school ballfields, and countless miles of running and cycling trails. One recent survey estimates that 7.94 million American high school students participated in school sports programs during the 2018–2019 academic year, in particular football, track and field sports, and basketball (NFHS 2019). There is also widespread individual athleticism. For example, one sports industry association estimates that over 50 million Americans regularly practiced running in 2019 (SFIA 2020), and at least 18 million participated in formal road races ranging from community 5km runs to elite marathons (Running USA 2020).

In addition to material infrastructure, sport also shapes the representation and experience of landscape through media and popular culture. Broadcasted games are a staple of daily media consumption; for many Americans, televised aerial views of a stadium may be the most familiar representation of a place. Flagship events (e.g., the Olympics, the Super Bowl, the World Series) dominate news cycles and permeate the national conversation. Fandom and team rivalries are often components of regional identity. And the playing field is a prominent location for political performance, for instance when fans are asked to stand and salute the national anthem, or when athletes refuse and instead take a knee to protest racial injustice.

DOI: 10.4324/9781003121800-16

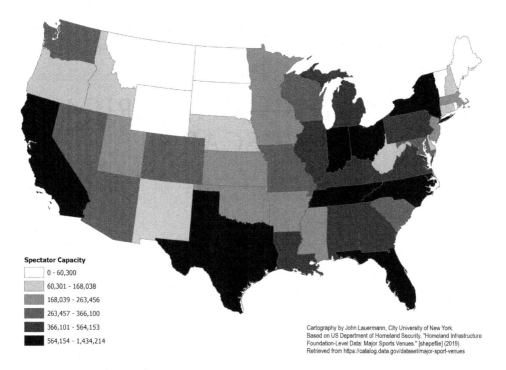

Figure 13.1 Distribution of major sports venues in the United States, by number of stadium seats. Aggregated from 691 major sports venues that host events for the Indy Racing League, Major League Baseball, Major League Soccer, NASCAR, National Basketball Association, National Hockey League, National Football League, the PGA Tour, NCAA Division 1 Football, NCAA Division 1 Basketball, and Women's National Basketball Association. Data from Department of Homeland Security (2019).

The chapter analyzes how professional and amateur sport contribute to the material development of the American landscape, and how this reflects social and political practices in American society. It begins with a review of literature on the material landscapes of sport, examining research on the landscapes of both professional and amateur sports. This includes an introduction to the sports geography subfield, as well as broader conversations on sports-themed development in geography. The chapter then uses a case study to examine the interaction between American sport and American parks, interpreting how sport shapes the landscape. While not all parks are sports-focused, most sports take place in some kind of park, and rely on the infrastructure that parks provide (e.g., related to transportation, vending, and crowd control). Parks, furthermore, are important public spaces for performing and contesting cultural practices through sport. This is illustrated with a study of one park that has a particularly long history of sports development and redevelopment: Flushing Meadows-Corona Park in New York City. The chapter concludes by drawing connections between sports landscapes and other prominent contemporary issues in the American landscape, including race, class, and populist movements.

Sports Landscapes

Research on sport geography dates to the 1970s and 80s, with early work on the relationship between sport, culture, and place (Bale 1988; Rooney 1974) and the physical landscapes of sport (Raitz 1995). For a comprehensive genealogy of the field, see Wise and Kohe (2020).

Bale's (2003) *Sports Geography* is an important milestone in formalizing the field by connecting sport to key geographical concepts like region, location, place, and globalization. More recent work has emphasized critical geographies of sport (Giulianotti 2016; Koch 2016), interpreting its intersection with power, geopolitics, and the production of space. That literature is broadly a critique of the colonial nature of sports' diffusion and globalization (Clevenger 2019; Darnell and Hayhurst 2011; Moss et al. 2018; Bale and Cronin 2020), and of "sports capitalism" (Jozsa 2017; Perelman 2012; Boykoff 2014; Collins 2013; Graeff 2019), the corporate systems that produce televised spectacles (Smart 2007) and which exploit both athlete labor and fan allegiance (Kalman-Lamb 2020).

Yet not all research on sport landscapes is framed through the lens of sports geography. Indeed, a large body of work focuses on the material landscape created by sport-themed development, with sports playing a supporting role in broader culture- and tourism-based economies. From this perspective, the spatial processes occurring outside the stadium are just as important as those happening inside it. There is a broad literature on the relationship between sports and neoliberal urban development (Koch 2018), especially focused on cities' investments in sports mega-events like the Olympics and World Cup. Over recent decades, hundreds of cities have pursued a "mega-event strategy" that leverages sports for broader economic development (Andranovich et al. 2001), or uses sports-themed investment to catalyze urban regeneration (Smith 2012). Simply bidding for these events is a major urban planning project, and it is common for cities to bid repeatedly on related events and pursue sports-related development even if their bids fail (Oliver and Lauermann 2017). Of particular concern among critical analysts is "celebration capitalism" (Boykoff 2014), the notion that private sports franchises are often heavily subsidized by the public sector despite public opposition (Kellison and Mills 2020; McGillivray et al. 2019; Hiller and Wanner 2018) and broad evidence that these subsidies generate minimal return on public investment (Baade and Matheson 2016; Long 2005; Coates 2007).

Professional sports impact the landscape primarily through their infrastructure: stadiums, arenas, racetracks, and the parks (sometimes public, sometimes private) that host them (Figure 13.1). In his analysis of stadiums and cultural landscapes, geographer Christopher Gaffney (2008, 4) points out that nearly every city in the world has a stadium and that these places provide important sites for cultural performance: "stadiums are monuments, places for community interaction, repositories of collective memory, loci of strong identities, sites for ritualized conflict, political battlefields, and nodes in global systems of sport." This cultural space is often highly commodified through sponsorships (Vuolteenaho et al. 2019) and the oligopolies of billionaire owners and professional franchises (Smart 2007). Material infrastructure is also persistently shaped by the production requirements for digital media. Indeed, while stadiums may be sites of local cultural performance, they are designed for—and earn the bulk of their revenue from—broadcast to other places. The "sports-media-business complex" (Evens et al. 2013) demands a very specific—camera friendly—aesthetic of place.

Historically, television held a pervasive influence in shaping the practice of sport and the material design of the landscape where it is practiced, both for regularly performing professional leagues and, especially, for televised mega-events (Whannel 2009; Horne and Whannel 2020). Today, the internet and social media have introduced a new (but similar) set of technical requirements for broadcasting and engaging with online fan communities in a more interactive way (Hutchins and Rowe 2012). The commercial demand for "landscapes of television" (Higson 1987) muddles the boundaries between reality and representation, transforming games into visual spectacles and implementing a rigid visual and architectural aesthetic in the stadiums and parks that host them. The International Olympic Committee, for instance, has an extensive "Look of the Games" design guide that specifies aesthetic requirements for Olympic host cities ranging from the color of the playing fields to a ban on political protests outside the stadiums (Lauermann 2016b). Yet while stadiums may be "the most global of the globalized" places (Gaffney 2008, 4), it is important to recognize that the relationship between professional sport and landscape is mutually transformative: professional sports shape the landscape, but the landscape also shapes sport. As geographer Natalie Koch (2018, 2) notes, professional sports institutions are not:

> *a priori* entities that political and economic elites simply import and deploy in the space of the city, or that urban residents choose to engage at will… [they are] sites of social and political contestation, subject to constant negotiation and re-negotiation by the various actors who participate in them and their symbolic economies.

Unlike professional sport, practiced in grand stadiums and made-for-TV landscapes, amateur sports shape the landscape through a humbler yet ubiquitous spatial form: the public park. While there is some research on the economic connection between amateur sport and tourism geographies (Lamont 2014; Hunt 2019; Cooper and Alderman 2020), its primary impact on the landscape is through local everyday practice in urban spaces (Koch 2018), especially parks. In her classic study of American urban parks, sociologist Galen Cranz (1982) identifies four general types: the "pleasure ground" predominated in the late nineteenth century, designed with rustic landscapes that bring "nature" into the city (New York's Central Park is a good example). The early twentieth century saw a boom in "reform parks," tucked into small urban spaces by Progressive-era reformers and providing playgrounds, pools, and fieldhouses for children of immigrant and working-class urban neighborhoods. By the mid-twentieth century suburbanization led to vast expansion of "recreation facility" parks, providing asphalt and grass play areas and standard playground equipment for suburban families. Today, much park building focuses on more general purpose "Open Space Systems" or "Sustainable Parks" that emphasize ecological restoration, native landscape species, and environmentally low impact sports like running (see also Cranz and Boland 2004).

Notions of what constitutes socially acceptable forms of amateur sport have shaped the material landscape of each kind of park. The "pleasure ground" was designed for bourgeois sports like bicycling and equestrianism, the "reform park" leveraged sport to assimilate immigrants and promote public health in the tenements, the "recreation facility" enabled the spread of suburban middle-class sport cultures (e.g., little league baseball and children's soccer leagues), and today's "sustainable parks" promote an interest in sports that engage with nature and have lower environmental impact. On the latter topic, there is a growing interest in reading amateur sports as part of a broader "movement space" in the city (Andrews 2017), integrated in a connected landscape of parks and other green (or gray) fitness spaces. In this sense, informal sports and fitness practices help to animate urban environments by maintaining a social infrastructure of fitness that reflects the physical infrastructure available for amateur athletes (Latham and Layton 2020).

Given the longstanding relationship between amateur sports, park design, and political agendas that use them to facilitate social goals, it is not surprising that different kinds of parks—and different kinds of sport practiced within them—reflect broader inequalities in North American society along race and class lines (Byrne and Wolch 2009; Rigolon 2016; Ngom et al. 2016). In particular, there is a growing focus on the underrepresentation of working-class athletes and athletes of color in certain kinds of sports and the kinds of parks where they are practiced. This is the case in both rural settings (e.g., nature-based sports in rural parks) (Finney

2014; Davis 2019) and urban settings (e.g., running in "green" or "sustainable" parks of gentrifying neighborhoods) (Rigolon and Németh 2018; Amorim Maia et al. 2020). It is, furthermore, not uncommon for park landscapes to be rebuilt over time as social priorities for sport change. This type of sports-themed landscape change is briefly illustrated with the following case study.

Flushing Meadows–Corona Park in New York City

Professional and amateur sports have influenced the material development of the American landscape in a wide range of settings, but have had particularly pronounced effect on urban landscapes. One park that offers a clear illustration is Flushing Meadows–Corona Park in Queens, New York. Located several miles east of Manhattan, the park reflects the more suburban landscape of the borough of Queens: it is a large space (897 acres) surrounded by medium-density commercial and mixed-use neighborhoods. Today, it is home to two major professional sports venues: Citi Field, home of the Mets Major League Baseball franchise (Figure 13.2), and the National Tennis Center, site of the US Tennis Association's annual US Open tournament (Figure 13.3). As one of the largest parks in the city, it also hosts a sizeable amount of green space, a number of community sports facilities, and major cultural institutions like the Queens Museum.

Over almost a century, the landscape of the park has been repeatedly remade through sports- and culture-themed infrastructure, reflecting broader trends in regional urbanization

Figure 13.2   Aerial view of Citi Field. Stadium viewed from the south, with Flushing Bay in the background. Surrounding parking lots serve the stadium, as does an elevated rail line visible in the bottomright corner of the image. Image courtesy of WikiCommons licensed under CC BY-SA 3.0.

Figure 13.3 National Tennis Center and World's Fair legacy infrastructure. Stadium viewed from the north. Long Island Railroad yard in the foreground. Tennis arena in the middle. Legacy infrastructure (globe and towers) from the 1964 World's Fair in the background. Image by author.

and suburbanization. The historical development (and repeated redevelopment) of this park reflects a broader history of sports in the American landscape. Originally a wetland bordering Flushing Bay, the area was used as a landfill during the nineteenth century—most famously as a dumping ground for the Brooklyn Ash Removal Company, earning the site a cameo as "the valley of ashes" in Fitzgerald's *The Great Gatsby*. It later attracted the attention of the city's most infamous builder, Robert Moses, who implemented an environmental remediation project that transformed the site into a park to host the 1939 World's Fair. At the twilight of his career he rebuilt the park once again for the 1964 World's Fair (Caro 1974). Like many of the parks Moses built, Flushing Meadows played an important role in suburban development by anchoring the construction of highway networks (technically "parkways" in local parlance, because Moses used a procedural loophole to commandeer highway funding under the guise of connecting the city to the parks). While the 1964 Fair was poorly attended and financially disastrous, it did leave a legacy of sports and culture facilities, including Shea Stadium (the original home of the Mets baseball team), the Singer Bowl stadium (used for various sporting events and concerts in the 60s and 70s), numerous courts and ballfields, and several pavilion structures.

In subsequent decades, the park was redeveloped several times (Figure 13.4). The Singer Bowl was substantially renovated into the National Tennis Center in the late 1970s. An exhibition pavilion on the west side of the park (originally constructed for the 1939 World's Fair and

Figure 13.4 Redevelopment at Flushing Meadows–Corona Park, 2009. Northern sections of the park. Shea Stadium (top left) and Citi Field underconstruction (top right) are visible in the northern end of the park. Also visible to the south of these stadiums are the National Tennis Center and Queens Museum. Imagery courtesy of US Geological Survey and Google Earth Pro.

then repurposed for the 1964 Fair) was converted into the Queens Museum in the 1970s as well. A failed bid to host the Summer Olympics during the early 2000s was used to reimagine park infrastructure, and while the Olympics never materialized the bid plan did provide an opportunity for redeveloping Shea Stadium and constructing several new amateur sport facilities (Moss 2011; Lauermann 2016a). In the 2000s Shea Stadium was demolished and replaced by Citi Field, which was constructed immediately next to the original stadium.

Elsewhere in the park, portions of World's Fair infrastructure remain, providing a tourist and amateur sports attraction. For example, each summer the New York Road Runners running club organizes a well-attended 10km race (nearly 12,000 participants in 2019), that weaves through this legacy infrastructure across the park (Figure 13.5). Thus like many American sport landscapes, the park is a palimpsest: an assemblage of historical and contemporary material culture repeatedly overwritten and repurposed as stadiums are built, renovated, and demolished. These material changes reflect the shifting role of sports-themed development in the city, as shifting sports preferences create demands for different kinds of park landscapes. While Flushing Meadows has a particularly prominent sport history, the same kind of landscape change could be traced throughout the United States as park infrastructure is repurposed to fit the changing priorities of sport culture over time.

Figure 13.5 Legacy infrastructure from the 1964 World's Fair. The central promenade of the Flushing Meadows-Corona Park, looking south. Legacy infrastructure includes the globe sculpture and wading pool (foreground) and the towers and pavilion (background). Image by author.

Conclusion

This chapter analyzed how sports contribute to the material development of the American landscape. Professional sport is a large industry anchored by hundreds of stadiums and arenas, each in turn anchoring broader tourism- and consumption-based local economies. Amateur sports have a less spectacular impact, but one that is geographically widespread as they are practiced across a vast landscape of parks, trails, and ballfields. The chapter reviewed research in sports geography, and in the broader geography literature on sports-themed development. These impacts were illustrated with a case study of a park in New York City—Flushing Meadows–Corona Park—that has been repeatedly redeveloped, over nearly a century, through sports-themed investments.

Sports landscapes matter because of their widespread economic impact, and the broad spatial footprint of stadiums, arenas, parks, and other venues. But they also matter because—as part of American cultural practice—sports reflect broader conflicts and inequalities in American society. Sports have a long and complicated relationship with American class relations: professional leagues are some of the only industries owned almost exclusively by billionaires, universities routinely exploit the labor of "amateur" student athletes who share none of the profits generated by NCAA programs, and there are economic divides in access across numerous amateur sports. Sports have a similarly complicated relationship with race, from racial disparities between the ranks of athletes, coaches, and team owners; to broad underrepresentation of communities of color in rural and suburban sport spaces.

These uncomfortable interactions with the politics of race and class position sports as a flashpoint for populist politics. The stadium has long been an arena for litigating culture war conflicts, from right-wing demands that fans stand to salute the national anthem, to left-wing protests that feature athletes taking a knee instead. But in a broader sense, the uniquely oligopolistic nature of the sports industry—with taxpayers subsidizing stadium construction on behalf of billionaire team owners—has led to a populist backlash. Mega-events like the Olympics, for instance, are now regularly countered by protest movements that oppose both the treatment of athlete labor and the impacts of the sports industry on host cities (Lauermann 2019; Boykoff 2020). The point, simply, is that like so many other aspects of the American landscape, sports are neither politically neutral nor immune to political contestation.

References

Agha, N. and Rascher, D. 2020. "Economic Development Effects of Major and Minor League Teams and Stadiums." *Journal of Sports Economics*. doi: 10.1177/1527002520975847.

Amorim Maia, A. T., F. Calcagni, J.J.T. Connolly, I. Anguelovski, and J. Langemeyer. 2020. "Hidden Drivers of Social Injustice: Uncovering Unequal Cultural Ecosystem Services Behind Green Gentrification." *Environmental Science & Policy* 112: 254–263. doi: 10.1016/j.envsci.2020.05.021.

Andranovich, G. and M. Burbank. 2011. "Contextualizing Olympic Legacies." *Urban Geography* 32 (6): 823–844. doi: 10.2747/0272-3638.32.6.823.

Andranovich, G., M. Burbank, and C. Heying. 2001. "Olympic Cities: Lessons Learned from Mega-Event Politics." *Journal of Urban Affairs* 23 (2): 113–131. doi: 10.1111/0735-2166.00079.

Andrews, G. J. 2017. "From Post-Game to Play-by-Play: Animating Sports Movement-Space." *Progress in Human Geography* 41 (6): 766–794. doi: 10.1177/0309132516660207.

Baade, R. A. and V. A. Matheson. 2016. "Going for the Gold: The Economics of the Olympics." *Journal of Economic Perspectives* 30 (2): 201–218. doi: 10.1257/jep.30.2.201.

Bale, J. 1988. "The Place of 'Place' in Cultural Studies of Sports." *Progress in Human Geography* 12 (4): 507–524.

Bale, J. 2003. *Sports Geography*. London: Routledge.

Bale, J. and M. Cronin. 2020. *Sport and Postcolonialism*. New York: Routledge.

Boykoff, J. 2014. *Celebration Capitalism and the Olympic Games*. London: Routledge.

Boykoff, J. 2020. *NOlympians: Inside the Fight Against Capitalist Mega-Sports in Los Angeles, Tokyo, and Beyond*. Winnipeg: Fernwood Publishing.

Byrne, J. and J. Wolch. 2009. "Nature, Race, and Parks: Past Research and Future Directions for Geographic Research." *Progress in Human Geography* 33 (6): 743–765. doi: 10.1177/0309132509103156.

Caro, R. A. 1974. *The Power Broker: Robert Moses and the Fall of New York*. New York: Knopf.

Clevenger, S. M. 2019. "Modernization, Colonialism, and the New Anthropology of Sport." *Reviews in Anthropology* 48 (3–4): 106–121. doi: 10.1080/00938157.2020.1743473.

Coates, D. 2007. "Stadiums and Arenas: Economic Development or Economic Distribution?" *Contemporary Economic Policy* 25 (4): 565–577. doi: 10.1111/j.1465-7287.2007.00073.x.

Collins, T. 2013. *Sport in Capitalist Society: A Short History*. New York: Routledge.

Cooper, J. A. and D. H. Alderman. 2020. "Cancelling March Madness Exposes Opportunities for a More Sustainable Sports Tourism Economy." *Tourism Geographies* 22 (3): 525–535. doi: 10.1080/14616688.2020.1759135.

Cranz, G. 1982. *The Politics of Park Design: A History of Urban Parks in America*. Cambridge, MA: MIT Press.

Cranz, G. and M. Boland. 2004. "Defining the Sustainable Park: A Fifth Model for Urban Parks." *Landscape Journal* 23 (2): 102–120.

Darnell, S. C. and L. M. C. Hayhurst. 2011. "Sport for Decolonization: Exploring a New Praxis of Sport for Development." *Progress in Development Studies* 11: 183–196. doi: 10.1177/146499341001100301.

Davis, J. 2019. "Black faces, Black Spaces: Rethinking African American Underrepresentation in Wildland Spaces and Outdoor Recreation." *Environment and Planning E: Nature and Space* 2 (1): 89–109. doi: 10.1177/2514848618817480.

Evens, T., P. Iosifidis, and P. Smith. 2013. *The Political Economy of Television Sports Rights*. New York: Palgrave Macmillan.

Finney, C. 2014. *Black Faces, White Spaces: Reimagining the Relationship of African Americans to the Great Outdoors*. Chapel Hill: University of North Carolina Press.

Gaffney, C.T. 2008. *Temples of the Earthbound Gods: Stadiums in the Cultural Landscapes of Rio de Janeiro and Buenos Aires*. Austin, TX: University of Texas Press.

Giulianotti, R. 2016. *Sport: A Critical Sociology*. 2nd ed. Cambridge: Polity.

Graeff, B. 2019. *Capitalism, Sport Mega Events and the Global South*. New York: Routledge.

Grant Long, J. 2013. *Public/Private Partnerships for Major League Sports Facilities*. New York: Routledge.

Higson, A. 1987. "The Landscapes of Television." *Landscape Research* 12: 8–13.

Hiller, H. H. and R. A. Wanner. 2018. "Public Opinion in Olympic Cities: From Bidding to Retrospection." *Urban Affairs Review* 54 (5): 962–993. doi:10.1177/1078087416684036.

Horne, J. and G. Whannel. 2020. *Understanding the Olympics*. 3rd ed. New York: Routledge.

Hunt, R. 2019. "Historical Geography, Climbing and Mountaineering: Route Setting for an Inclusive Future." *Geography Compass* 13 (4). https://compass.onlinelibrary.wiley.com/doi/abs/10.1111/gec3.12423.

Hutchins, B. and D. Rowe. 2012. *Sport Beyond Television: The Internet, Digital Media and the Rise of Networked Media Sport*. New York: Routledge.

Jozsa, F. P. 2017. *Sports Capitalism: The Foreign Business of American Professional Leagues*. London: Routledge.

Kalman-Lamb, N. 2020. "Imagined Communities of Fandom: Sport, Spectatorship, Meaning and Alienation in Late Capitalism." *Sport in Society* 24 (6): 922–936. doi:10.1080/17430437.2020.1720656.

Kassens-Noor, E. and J. Lauermann. 2017. "How to Bid Better for the Olympics: A Participatory Mega-Event Planning Strategy for Local Legacies." *Journal of the American Planning Association* 83 (4): 335–345. doi:10.1080/01944363.2017.1361857.

Kellison, T. and B. M. Mills. 2020. "Voter Intentions and Political Implications of Legislated Stadium Subsidies." *Sport Management Review* 24 (2): 181–203. doi: 10.1016/j.smr.2020.07.003.

Koch, N., ed. 2016. *Critical Geographies of Sport: Space, Power and Sport in Global Perspective*. New York: Routledge.

Koch, N. 2018. Sports and the City. *Geography Compass* 12 (3). https://compass.onlinelibrary.wiley.com/doi/abs/10.1111/gec3.12360.

Lamont, M. 2014. "Authentication in Sports Tourism." *Annals of Tourism Research* 45: 1–17. doi:10.1016/j.annals.2013.11.003.

Latham, A. and J. Layton. 2020. "Kinaesthetic Cities: Studying the Worlds of Amateur Sports and Fitness in Contemporary Urban Environments." *Progress in Human Geography* 44 (5): 852–876. doi:10.1177/0309132519859442.

Lauermann, J. 2016a. " 'The City' as Developmental Justification: Claimsmaking on the Urban through Strategic Planning." *Urban Geography* 37 (1): 77–95. doi:10.1080/02723638.2015.1055924.

———. 2016b. "Made in Transit: Mega-Events and Policy Mobilities." In *Mega-Event Mobilities: A Critical Perspective*. edited by N. B. Salazar, C. Timmerman, J. Wets, and S. van den Brouke, 90–107. London: Routledge.

———. 2019. "The Urban Politics of Mega-Events." *Environment and Society* 10: 48–62.

Long, J. G. 2005. "Full Count: The Real Cost of Public Funding for Major League Sports Facilities." *Journal of Sports Economics* 6: 119–143. doi:10.1177/1527002504264614.

McGillivray, D., J. Lauermann, and D. Turner. 2019. "Event Bidding and New Media Activism." *Leisure Studies* 1–13. doi: 10.1080/02614367.2019.1698648.

Moss, E. N., M. Hart, and L. Petherick. 2018. "Indigenous Gender Reformations: Physical Culture, Settler Colonialism and the Politics of Containment." *Sociology of Sport Journal* 36: 113.

Moss, M. 2011. *How New York City Won the Olympics*. New York: Rudin Center for Transportation Policy and Management, New York University.

NFHS (National Federation of State High School Associations). 2019. *High School Athletics Participation Survey, 2018–2019*. www.nfhs.org/sports-resource-content/high-school-participation-survey-archive/.

Ngom, R., D. Gosselin, and C. Blain. 2016. "Reduction of Disparities in Access to Green Spaces: Their Geographic Insertion and Recreational Functions Matter." *Applied Geography* 66: 35–51. doi:10.1016/j.apgeog.2015.11.008.

Oliver, R. and J. Lauermann. 2017. *Failed Olympic Bids and the Transformation of Urban Space*. London: Palgrave Macmillan.

Perelman, M. 2012. *Barbaric Sport: A Global Plague*. London: Verso.

PWC. 2019. *2019 PwC Sport Outlook: At the Gate and Beyond.* Accessed January 4, 2021. www.pwc.com/us/en/industries/tmt/assets/pwc-sports-outlook-2019.pdf.

Raitz, K., ed. 1995. *The Theater of Sport.* Baltimore, MD: John Hopkins University Press.

Rigolon, A. 2016. "A Complex Landscape of Inequity in Access to Urban Parks: A Literature Review." *Landscape and Urban Planning* 153: 160–169. doi:10.1016/j.landurbplan.2016.05.017.

Rigolon, A. and J. Németh. 2018. "'We're Not in the Business of Housing': Environmental Gentrification and the Nonprofitization of Green Infrastructure Projects." *Cities* 81: 71–80. doi:10.1016/j.cities.2018.03.016.

Rooney, J. F. 1974. *A Geography of American Sport: From Cabin Creek to Anaheim.* Reading, MA: Addison-Wesley.

Running USA 2020. *2019 US Running Trends Report.* Accessed January 4, 2021. https://runningusa.org/RUSA/News/2019/Running_USA_Releases_2019_U.S._Running_Trends_Report.aspx.

SFIA 2020. *Running/Jogging Participation Report.* Accessed January 4, 2021. www.sfia.org/reports/844_Running-Jogging-Participation-Report-2020.

Smart, B. 2007. "Not Playing Around: Global Capitalism, Modern Sport and Consumer Culture." *Global Networks* 7: 113–134. doi:10.1111/j.1471-0374.2007.00160.x.

Smith, A. 2012. *Events and Urban Regeneration: The Strategic Use of Events to Revitalise Cities.* London: Routledge.

Vuolteenaho, J., M. Wolny, and G. Puzey. 2019. "'This Venue Is Brought to You by…': The Diffusion of Sports and Entertainment Facility Name Sponsorship in Urban Europe." *Urban Geography* 40: 762–783. doi:10.1080/02723638.2018.1446586.

Whannel, G. 2009. "Television and the Transformation of Sport." *The Annals of the American Academy of Political and Social Science* 625: 205–218. doi:10.1177/0002716209339144.

Wise, N. and G. Z. Kohe. 2020. "Sports Geography: New Approaches, Perspectives and Directions." *Sport in Society* 23: 1–10. doi:10.1080/17430437.2018.1555209.

Zimbalist, A. 2015. *Circus Maximus: The Economic Gamble Behind Hosting the Olympics and the World Cup.* Washington, DC: Brookings Institution Press.

14

FROM POST-SCARCITY UTOPIA TO ZOMBIE-INFESTED HELLSCAPE

Continuity and Change in the Landscape of a Future North America

Fiona Davidson

Perhaps I see certain utopian things, space for human honor and respect, landscapes not yet offended, planets that do not exist yet, dreamed landscapes.

—*Werner Herzog (2019)*

It is Herzog's "dreamed landscapes" that form the backdrop for centuries of science fiction and fantasy stories. As readers and consumers of popular culture we are familiar with the characters, with Atwood's sullen, but resilient, Offred and Tolkien's plucky Frodo; and with the stories that have become embedded in contemporary popular narratives. A reference to "1984" speaks volumes about oppression, government control, and propaganda; to call someone a "red shirt" indicates the person concerned is expendable, disposable, and fated to die nameless and unlamented. Less familiar, however, are those "dreamed landscapes" of Herzog. The creation of new, or radically altered, landscapes is an integral part of making speculative fiction successful but has been perhaps the least celebrated aspect of the science-fiction writers' art (Miller 2016).

The *Oxford English Dictionary* (OED) definition of landscape is narrow, "all the visible features of an area of land." However, unsurprisingly, the *Oxford Dictionary of Human Geography* (Castree et al. 2013) provides a much more comprehensive definition that includes both physical and human influences and this socioeconomic, political, and cultural definition informs this chapter. Landscape, in a literary sense, is not just the physical manifestation of rivers and mountains and deserts, rather it is the entire cultural milieu; the social customs, the political systems, the economic structure, of these invented societies that create the "landscapes" of the North American future. These landscapes, created in the imaginations of writers, but often derived from deep cultural hopes and fears, run the gamut from ecological paradises to blasted heaths; from cooperative social utopias to competitive capitalist cyber-cities, all the way to terrifying savage wilderness peopled by water-stealing raiders, and zombie cannibals.

These fictions of the future, specifically the North American future, can be broken out into three subgenres best exemplified by Kim Stanley Robinson's *Three Californias* trilogy

DOI: 10.4324/9781003121800-17

(1984–1990). Set in his native Orange County, Robinson creates three futures: dystopia, cyberpunk, and utopia; setting each in a different physical and cultural landscape that is both familiar, and yet very much removed from, the Southern California of the 1980s and early 1990s. These three landscapes, the wilderness (*The Wild Shore* (Robinson 1984)), the cybercity (*The Gold Coast* (Robinson 1988)), and the garden (*Pacific Edge* (Robinson 1990)), are the archetypes for virtually all science-fiction visions of a future North America and it is these archetypes that form the organizing structure of this chapter.

Of course, these landscapes are not just archetypes for science fiction; the idea that landscapes are metaphorical representations of human achievement and aspiration is common throughout artistic, literary, and mythological communities. From Milton's *Paradise Lost* to Thomas Cole's *The Course of Empire* (Figure 14.1), the idea that the garden represents paradise, the city extravagance and indulgence which eventually collapses into the decay and peril of the wilderness is a metanarrative that has particular resonance in the wilderness-taming, anti-urban, arcadian ethos of nineteenth- and twentieth-century North America (Daniels and Cosgrove 1998; Gregory 1994; Tuan 1974, 1979).

The focus of the body of this narrative is the ways in which utopia, the cyberpunk city, and dystopia manifest in specific (if not always unique) ways in the North American landscape. Emphasizing wildness, individualism, masculinity, and technology, North American future landscapes are firmly rooted in the contemporary realities of economic, social, and racial stratification and tend to reiterate existing landscape extremes of hyperurbanization and wilderness.

Arcadia

Utopia is the oldest of the speculative fiction landscapes and is first manifest in Thomas More's *Utopia* (1516). Originally derived from the Greek οὐ τόπος (no-place), contemporary usage more commonly gives the meaning as the slightly different εὖ τόπος (well-place). In either case, while utopias might be "better" places, in the contemporary world they are also nowhere to be found, appearing only in the imagined world of a future, fictional North America.

Edward Bellamy's *Looking Backward 2000–1887* (1888) provides one of the earliest visions of what that future North America might become. Bellamy's vision of a post-capitalist utopia was the third largest seller of its time (Morgan 1944), and inspired thousands of Americans to create "Bellamy" clubs that encouraged people to work towards realizing this vision of a socially collective society where the means of production were owned by the government and everyone had access to employment, income, education, shelter, food, and medical care. Bellamy's vision is, of course, socialist. However, reluctant to use that word in case it limited the book's popularity, he labelled it "nationalism" and thereby encapsulated one of the fundamental problems of American utopian science fiction (Bowman 1958). Any realistic form of utopia must necessarily address the inequalities that arise from capitalism, leaving us with a relatively limited choice of utopian landscapes in the future North America. As Kim Stanley Robinson writes, "easier to imagine the end of the world than the end of capitalism" (2020a, 25). In the twenty-first century, with capitalism a global phenomenon, the challenge to achieve utopia appears farther away than ever.

Post-Scarcity

Nonetheless, there are utopian visions of a future North America and the most obvious workaround for the capitalist/utopia dilemma is the creation of the post-scarcity world. These are the most positive utopian futures, where the gleaming towers of *Star Trek*'s San Francisco sit

Figure 14.1 Three panels from Thomas Cole's *The Course of Empire: Arcadia, Consummation, Desolation.* Images courtesy Wikimedia Commons.

alongside a preserved natural landscape, providing a familiar link to contemporary North America. The sociopolitical landscape is equally familiar. There is no visible deprivation or discrimination, but this post-scarcity vision is one of a still ascendant North America, with a still dominant white male population.

Indeed, *Star Trek* is probably the only genuinely post-scarcity utopian view of North America. Most post-scarcity science fiction confines itself to off-Earth scenarios precisely because it is so problematic to envision a society in which everyone's basic needs are met without abandoning capitalism. Even *Star Trek*, as visionary as it is, resorts to eliding exactly how humanity transitioned from twenty-first-century global capitalism to a place where, as Picard explains in *Star Trek: First Contact* (1999), "The acquisition of wealth is no longer the driving force of our lives."

Ecotopia—In Balance with Nature

The one place where utopian views of North American futures have found a flourishing niche is in visions of ecotopia. Less overtly socialist, and more obviously reflecting culturally recognizable visions of the Garden of Eden, the concept of ecotopia gives respectability to ideas of reducing consumption, transitioning to environmentally sustainable energy and resource use, and levelling the distribution of work and reward in North American society. This is also the only vision of the future in which North America transcends contemporary gender and racial boundaries.

Ecotopia is the science-fiction articulation of paradise in the garden and, at the same time, allows for the expression of the Jeffersonian ideal of rural self-sufficiency, so critical to American social politics. Modern humans, and Americans in particular, have, in the wake of the industrial revolution and consequent urbanization, romanticized nature and rural lifestyles (Daniels and Cosgrove 1998). Where cities historically were refuges, and the wilderness savage and dangerous territory, nature is now the preserve of virtue and simplicity and real American values (Tuan 1974, 1979). These ideas of rural idyll, married to the rising environmental movement of the 1970s paved the way for an environmentally focused science fiction that presents a near-future North America in which sustainability and conservation provide a counter to consumptive capitalism, creating an entirely new social and economic society.

Among the first speculative fiction works to lay out the details of an environmentally sustainable utopia is Callenbach's *Ecotopia* (1974). The newly independent country of Ecotopia, stretching from Seattle to Carmel, flourishes with rejuvenated forests, restored wetlands and low-density city/villages, linked by renewable-energy powered public transit and interspersed with sustainable local farming enterprises and urban agriculture. The social landscape is likewise reconfigured; egalitarian, with strong grass-roots political organizations, and socially-politically active populations, much of the political power is wielded by women, with work, education, and domestic tasks undertaken on the basis of aptitude and ability, rather than by gender. In Callenbach's San Francisco a creek runs down Market Street and his car-less new towns of the South Bay evoke the nostalgia of small-town North America; places that are neither rural nor truly urban, suburbs without the sprawl.

Ecotopian ideals also inform the third novel in Robinson's *Three Californias* trilogy. Set in El Modena in 2065, *Pacific Edge* (1990) lays out a future in which the US has transitioned from resource-consumption to resource-recovery and reuse and where the protagonists first appear digging up asphalt from the abandoned I-5, for remelting into tar for roofing material.

This is still a recognizable California, the buildings, roadways, and landmarks of Orange County are intact, but often repurposed; factories and malls have become communal living spaces; parking lots are fields and orchards; roads are trails for bicycles and walkers and the occasional electric truck. There is still conflict over water appropriation and land use, but development is subordinate to long-term environmental sustainability. *Pacific Edge* is less radical than *Ecotopia* in its reimagining of California; private property is still the norm and while communal labor projects like community gardens are part of everyday life, the economy is still based on private enterprise and wage labor. However, as with Callenbach, Robinson's green future provides greater gender and racial equality, economic levelling, and political engagement with consequent landscape effects. Communities are small, walkable, and nature is integrated into the fabric of urban areas in the form of parks and urban agriculture.

This environmentally sustainable vision is imported into a high-density urban form in Newitz's *Two Scenarios for the Future of Solar Energy* (2014) where cities that are recognizably San Francisco and St. Louis are transformed through the use of solar power and bacteria to create large-scale, self-sustaining, carbon-neutral city ecosystems. In San Francisco, "The Market River cuts through downtown… and its banks are lined with everything from apples to circuit boards" (Newitz 2014, 245). This is a city where everything is locally generated and recycled, a city where gardens and nature are integrated into the urban landscape.

If it is necessary to discount post-scarcity utopias as impractical then the utopian North America of the future is primarily a landscape of distributed, small-scale, integrated urban/rural settlements. Linked by accessible public transport, fueled by renewable energy, and structured in egalitarian, democratic, socioeconomic units these societies abjure consumption and aim to balance human needs and wants with the carrying capacity of the local environment. Cities can exist but must be integrated into the environment in a way that creates sustainable landscapes. These landscapes are socially, economically, and politically radical by twenty-first-century American standards but are made desirable by their call back to rural simplicity, environmental sustainability, and the veiled invocation of paradise.

Cyberpunk and the American City

In contrast to the rural utopias of future North America, science fiction that focuses on technological change provides for a different, city-focused, vision of the future. The continuation of the present into a technologically enhanced future has always been one of the mainstays of science fiction and cyberpunk—the genre that dominated much of the 1980s and 1990s and gives us a North America that is overtly urban, crowded, capitalist, and economically divided. Cyberpunk visions of the future are focused largely on near-future scenarios, set in the early to mid-twenty-first century. In these landscapes, technology, and those who control it, primarily corporations, dominate the economic, social, and political landscape of North America.

Self-consciously a product of the rise of both computer technology and the internet in the 1980s and the ascendance of neoliberal economics in the 1990s, cyberpunk landscapes are logical extensions of many of the forces shaping contemporary North American landscapes. In the author's notes to *Virtual Light* (1993), William Gibson explicitly calls out Mike Davis' work in the urban geography of southern California as inspiration for the landscape that he creates in *The Bridge Trilogy* (1993–1997). Similarly, Neal Stephenson, in a 2014 interview, credits his training in geography for much of his ability to extrapolate the current North American world into his *Snowcrash* (1992) cyberpunk future of a franchised, fragmented America where speed, money, and technology are the only viable currencies.

The cyberpunk landscapes of a future North America are almost entirely urban. There are mentions of rural areas in some works. For example, in *Do Androids Dream of Electric Sheep?* (1967) rurality and nature are so rare and precious that people pay thousands of dollars to buy and care for fake animals, but urbanism is the abiding motif of cyberpunk. In many ways these landscapes are logical continuations of twenty-first-century American cities. The proliferation of private armies, legally independent gated communities, even extra-territorial enclaves, in the midst of US cities creates a legally and physically fragmented landscape that can already be glimpsed in Mike Davis' scorching condemnations of social, economic, and environmental problems of the Southern California metropoles of the late twentieth century (Davis 1990, 1998).

One of the most overt aspects of the cyberpunk landscape is the ubiquity of consumption, manifest in advertising (Figure 14.2). One of the earliest visualizations of cyberpunk is in Ridley Scott's *Blade Runner* (1984), where building-sized animated neon ads dominate the skyline of 2019 Los Angeles and the film is explicitly cited as influencing the world building of many of the early cyberpunk authors (Gibson 1993). The landscapes created by Gibson, Robinson, Stephenson, Suarez, and others give us a North America where the CIA is a private entity and there are mom-and-pop jails on every corner; where private security firms protect gated communities—*metacops* patrolling *burbclaves* (Stephenson 1992)—and the Chicago mob runs the largest pizza delivery service in the country. Cyberpunk delivers a landscape where a single vast economy exists to turn resources into goods that can be transported and consumed anywhere, a landscape where society is, as Neal Stephenson puts it, "chewing through the wilderness, building things and abandoning them, altering the flow of mighty rivers and then moving on because the place ain't what it used to be" (Stephenson 1992, 243).

Figure 14.2   A stylized photo of Times Square at night. Photo and stylization by author.

Movement and speed are also critical elements of the cyberpunk landscape. It is telling that the protagonists of both *Snowcrash* (1992) and *Virtual Light* (1993) are intra-city couriers, using technologically enhanced skateboard and bike to traverse the intricate roads and paths of Los Angeles and San Francisco. Hovercars are ubiquitous, from Ridley Scott's *Blade Runner* (1984), to *The Fifth Element* (2001); drone-controlled trucks and automated delivery vehicles rule the roads and the skies and urban landscapes are dominated by new smart automated highways that streamline and accelerate traffic (Robinson 1988). Space saving, car-stacking, parking garages, and the elevation of all the paraphernalia that surrounds driving, traffic lanes, signals, signage, law enforcement, into a three-dimensional landscape to facilitate aerial transportation, makes vehicles even more integrated into the urban landscape than they are in the twenty-first century.

Where utopian visions of North America resist or elide capitalism, cyberpunk futures embrace it; and with it all the social, political, and economic landscapes that go with the marketplace. Privatization, franchising, and advertising recall Allan Pred's "carousel of consumption and commodity fetishism—the merry-giddy-go-round and around of money in circulation" (Pred 1990, 145). The hyperurban landscape of speed, noise, and density also evokes the masculine landscape tropes of twentieth-century industrial and corporate power manifest in the image of the skyscraper (Domosh 1995) and contrasts with the overtly female-dominated landscapes of ecotopia.

Wasteland and Wilderness—Dystopian Futures

Cyberpunk landscapes might be dysfunctional, but it takes another level of upheaval to turn them into the true dystopias, which arise after some form of environmental, military, or political catastrophe.

While utopian literature dates to the sixteenth century, dystopian fiction is largely a feature of the twentieth century. E. M. Forster's *The Machine Stops* (1909) is often cited as the first truly dystopian story (Booker 1994) and, along with *1984* and *Brave New World*, forms the bedrock of a genre that expanded rapidly after WWII. Tied to political and social unrest, economic inequality, environmental degradation, and to fears of nuclear destruction, dystopian futures have become foundational to science fiction in the last fifty years. While the catalyst of our future destruction changes over time, the end result, the demographic collapse of humanity or its subjugation to small, powerful industrial/political or religious elites, is inevitable; hope existing only in the few individuals who are able, through luck or ability, to escape the catastrophe. In North America, dystopian futures bring us back to the wilderness, the collapse of social, political, economic, and physical infrastructure forcing the survivors into pockets of self-sufficient isolation, reminiscent of eighteenth-century frontier communities, where the quintessentially American values of self-reliance and rugged individualism are the key to survival.

The Wasteland

In the aftermath of WWII and the twin demonstrations of nuclear horror at Hiroshima and Nagasaki, it is not surprising that post-nuclear landscapes, embodying the fears of a nuclear-obsessed American public, became the most common visions of a North American future. Whether the future consisted of preparing for nuclear war as in *Foster, You're Dead* (Dick 1955) or survival in a destroyed world still at war with the Russians, as in *Second Variety* (Dick 1953), the omnipresence of the nuclear threat creates a new landscape. In many cases this devastation wrought by nuclear war is global; most famously in Philip K. Dick's *Do Androids Dream of Electric Sheep* (1967) where the entire world is blanketed by nuclear dust that shrouds the cities, and there is no escape except to leave Earth. However, stories where the nuclear devastation is more geographically contained, offer possibilities for an ex-urban sanctuary.

Robinson's *Wild Shore* (1984) presents a North America where 2,000 of the US's largest cities are destroyed in a nuclear war with the Soviets, and the country is now quarantined, forbidden from advancing beyond early industrial levels of technological development for the safety of the rest of the world. This Orange County is a landscape of restored wilderness, changed by the global cooling of a nuclear winter into a climate more like southern British Columbia. The population lives in scattered hamlets and villages. Some make a living in the ruins of the old world; the *scavengers*, sheltered and fed in the destroyed suburban wastelands of San Clemente, Irvine, and Mission Viejo; others, like the book's 17-year-old protagonist, live in small self-sufficient villages, sustained by fishing and farming, all of their technology and knowledge carefully preserved from a time before the war. These literary wastelands intersect not only with sociopolitical concerns, but also with the evolution of scientific thinking, as discussed in Box 14.1 about Martian landscapes.

Box 14.1 Landscapes of Mars: Where Science Meets Fiction

Maria Lane

Over the last two centuries, Mars has been at the center of many meeting-of-civilizations stories that conjure inhabitants and technologies capable of interstellar travel. Although these stories' landscape backdrops have changed over time, they typically reflect contemporary scientific understandings of Martian geography. At the end of the nineteenth century, however, the science of Mars was itself a reflection of many colonial assumptions drawn from the discipline of geography. Martian literary landscapes thus provide a mirror in which historical and cultural geographers can see the intertwining of science, geography, and colonial resource management.

Mars was not always perceived as having "landscapes" per se. At a minimum distance of 34 million miles from Earth, it appears to the naked eye as a mere point of light rather than as a three-dimensional sphere or a topographical surface like the moon. As fuzzy landscapes began to emerge in the telescope era, astronomers relied on visual analogy as an interpretive strategy. Throughout the nineteenth century, they compared patterns observed on Mars to known patterns in Earth's own surface. Based on these analogues, astronomers then posited specific landforms, weather events, and biomes for the red planet.

Perhaps unsurprisingly, astronomers working in Europe and North America chose landscapes close to home for comparison purposes. For example, the influential Italian astronomer Givoanni Schiaparelli determined in the 1870s that Mars was a watery world of interlinked waterways that looked much like the Mediterranean Basin. His view was then successfully challenged at the turn of the century by the American amateur astronomer Percival Lowell, who argued that the American West was a better analogue landscape.

In Lowell's interpretation, Mars was an arid, inhabited planet whose civilization depended for survival on a globe-spanning network of irrigation canals. He justified this hypothesis with frequent comparison to arid landscapes and irrigation schemes in the American West, and he convinced many members of the North American and European publics that Mars was inhabited. (Professional astronomy never subscribed to Lowell's view, although he had a few supporters among the professional ranks.)

An explosion of science fiction and popular nonfiction reflected Lowell's landscape view, often portraying Mars as a planet in crisis, with a society organized to combat the threat of planetary desertification. In these accounts, Martian landscape tropes reflected mainstream Progressive-era views about the relationship between humans and nature, and the relationship between Anglo-Americans and Indigenous peoples. Writers portrayed arid landscapes on Mars as wastelands unless they were controlled through irrigation infrastructure. They placed scientists and engineers at the top of the social hierarchy in an imagined Martian civilization. And they cast technology itself as the ultimate savior of civilization. For more on these narratives, see Lane 2011.

These same tropes animated the American settler colonial project that pushed Indigenous peoples off their lands throughout the American West. Not coincidentally, this larger project also relied on science, engineering, infrastructure, and yes, fiction, for much of its foundational success.

The Wilderness

Just as post-nuclear landscapes emphasize the destruction of urban centers, the dystopias that arise from disease create a North America returned to rural wildness. Predating post-nuclear dystopian fiction, George Stewart's *Earth Abides* (1949) creates a world where only a tiny fraction of the world's population survives a plague. In North America, isolated pockets of survivors carve out a new life as technologies fail and society returns to a pre-industrial, pre-literate, hunter-gatherer lifestyle.

Much less benign than Stewart's vision are the landscapes that arise from plagues of the living dead. The zombie genre, that first gains popular appeal with Richard Matheson's *I Am Legend* (1953) and has reached its apotheosis with AMC's *The Walking Dead*, returns North America to a frontier landscape of fortified communities and vast swathes of land inhabited by dangerous savages. These fortified landscapes hark back to colonial America and medieval Europe and symbolize the historical nature of cities as places of safety, where wildness and nature are spaces to be feared, inhabited by the uncivilized and dangerous (Tuan 1979).

An intersection of more prosaic catastrophes produces the wilderness of Octavia Butler's *Parable Series* (1993–1998); drought, authoritarian governance, corporate greed, and increased inequality create an America, where the lucky few live in secluded, guarded luxury while the mass of the population is forced into fortified neighborhoods or corporate company towns. As the neighborhoods slowly fall to drug-fueled gangs and corrupt police the only escape is to take to California's now deserted roads to find sanctuary in the isolated wilderness of the northern coastal forests (Figure 14.3). The protagonists find a temporary haven along the Lost Coast, an area of roadless forest west of Highway 1 between Leggett and Eureka; but even that cannot protect them forever as the paramilitary gangs of an authoritarian US regime eventually track them down and destroy the settlement.

Whether the catastrophe is nuclear, pandemic, or some other unspecified disaster, the isolated community, struggling to survive in the wake of the destruction of cities and infrastructure is probably the most ubiquitous vision of a North American post-cataclysm landscape. From David Brin's *The Postman* (1985) to Hugh Howey's *Silo Trilogy (Wool/Shift/Dust)* (2011–2012), by way of Cormac McCarthy's *The Road* (2006), these stories all deal with near-future catastrophes in which survival is predicated on isolation and fortification against the hazards to be found outside the walls of the known community. It is a vision of America returned to a

Figure 14.3 A photograph of a derelict motel on a California highway. Photo and stylization by author.

frontier space, with danger manifest in the unknown and unseen beyond the walls of a community fortress and where survival is predicated on self-reliance and distrust of outsiders.

Landscapes from the Twenty-First Century

It is ironic that the dystopia in Kim Stanley Robinson's *Three Californias* is post-nuclear, given his profuse writing about climate-induced dystopia in this century, and he admits that if he could go back and change anything about those early books it would be his fixation with all things Soviet (Gevers 2019). In the twenty-first century Robinson has become the most prolific writer of post-climate change science fiction, much of it very near-future—*The Climate in the Capital Trilogy* and *The Ministry for the Future*—some of it less so, *New York 2140* and *2392*. These, and a multitude of other works from Octavia Butler's *Parable Series* to Paolo Bagicalupi's *The Water Knife*, posit a North America where climate change will raise sea level, parch already dry environments to dust, and render much of the coast uninhabitable.

In the last thirty years, acknowledgement of what climate change will do to the North American landscape has become a feature of science fiction visions of the future with rising sea level, in particular, one of the new tropes of a future North American landscape. A drowned and frozen New York features as the main recognizable landscape in the scientifically implausible *Day After Tomorrow* (2012), where a warming Atlantic Ocean drives three global superstorms that drown coastal cities and leave behind a new ice age.

Much more scientifically, Robinson's *New York 2140* inhabits a world where two "surges" have raised sea level by fifty feet. Lower Manhattan is underwater but, given the sunk costs in real estate and construction, it is still inhabited. In the "New Venice" of the land below the fifty-foot watermark, the buildings are waterproofed, surrounded by canals teeming with vaporetti, their apartments occupied by permanent residents rather than serving as commercial offices or investments for the one percent. Troubles abound in this new New York; the legal status of property now caught in the new tidal zone; what to do about the millions of climate refugees fleeing their drowned coastal communities; how to deal with the disruptions of infrastructure, roads, airports, railways that hugged the low-lying real estate of coasts and river valleys. The landscape changes have wrought social change too. While the elites still go about their capital accumulation, there is a sense of community solidarity in this newly residential New York; governments have achieved the transition from fossil fuels, and wildlife is protected in vast migration corridors that cut across the North American landscape.

While most of the climate-altered landscapes of North America are relatively near future, the changes wrought by climate instability are much more extreme in far future scenarios. In *Waterworld* (1994) climate change has eradicated the entire North American continent by 2500. The ice caps have melted, raising sea level by over 25,000 feet and New York serves as a repository for valuable artifacts that Kevin Costner's "mariner" salvages for sale and trade on the floating islands that are home to the fragmented remnants of humanity. The scientific plausibility of such a change is dubious at best. Equally scientifically dubious, is Hugh Howey's *Sand Trilogy* (2011) which posits an extreme return of the Great American Desert, the front range of the Rockies buried in massive sand dunes created by a reversal of the jet stream and the entire western US reduced to Saharan-like levels of precipitation. As with *Waterworld*, neither the mechanism nor the timing of this climate-induced change is made clear, but the world-building is convincing, with Howey's sand-diver protagonists travelling from "Low-Pub" to "Danvar" in search of the same buried relics of our contemporary America, that motivate the "mariner" in *Waterworld* (1994).

Conclusion

The use of landscapes to create a literary or visual shorthand for socioeconomic continuity and/or change has a long history in human creative endeavors (Daniels and Cosgrove 1988). In film and literature, the use of iconic places and structures to embody future landscape changes, provides a visual punch to remind the audience of the magnitude of the changed future. The most memorable example is perhaps the appearance of the Statue of Liberty, half buried in a sandy shoreline, at the end of *Planet of the Apes* (1968). This is the moment in the film where the audience realizes the conceit that George Taylor is not on some distant planet, but on a future, irretrievably altered, North America.

The form of these alterations to the landscape varies in concert with the sub-genres of science fiction, with often very different landscape consequences. Utopias are a pristine mix of urban and rural, the myth of small-town community and abundant resources. Conversely, cyberpunk presents the American city on steroids with all the concomitant inequality and dysfunction of a hypercapitalist society and finally, dystopia returns America to the frontier of wilderness, isolated communities and distrust of the stranger. With the exception of the concept of ꞏꞏꞏtopia, North American science fiction tends to reinforce existing social, political, and economic realities. Whether North America of the future takes the form of a post-scarcity utopia (*Star Trek*) or a zombie-infested hellscape (*The Walking Dead*), the landscapes of the future reflect self-consciously American values of individualism, self-reliance, utilitarianism, technological prowess, and the dominance of Anglo-Saxon (usually male) society. Unlike the mystical realism of Afrofuturism which integrates nature into the landscape in innovative ways, or the socially and philosophically complex works of Russian and Continental European futurism, which rely heavily on mystically altered or entirely imaginary landscapes, North American futurism is firmly rooted in extensions of the present, visions that can be extrapolated from the cities and forests, the mountains and wilderness that currently cover the continent.

References

Bellamy, E. 1888. *Looking Backward 2000–1888*. New York: Houghton Mifflin.

Booker, K. 1994. *Dystopian Impulse in Modern Literature: Fiction as Social Criticism*. Westport CT: Praeger.

Bowman, S. E. 1958. *The Year 2000: A Critical Biography of Edward Bellamy*. New York: Bookman Associates.

Butler, O. 1993. *Parable of the Sower*. New York: Four Walls Eight Windows.

Callenbach, E. 1974. *Ecotopia*. New York: Bantam Books.

Castree, N., R. Kitchen, and A. Rogers. 2013. *A Dictionary of Human Geography*. Oxford: Oxford University Press.

Cole, T. 1836. *The Course of Empire: The Savage State*. Accessed January 12, 2021. en.wikipedia.org/wiki/Thomas_Cole#/media/File:Cole_Thomas_The_Course_of_Empire_The_Savage_State_1836.jpg.

———. 1836. *The Course of Empire: The Consummation*. Accessed January 12, 2021. en.wikipedia.org/wiki/Thomas_Cole#/media/File:Cole_Thomas_The_Consummation_The_Course_of_the_Empire_1836.jpg.

———. 1836. *The Course of Empire: Desolation*. Accessed January 12, 2021. en.wikipedia.org/wiki/Thomas_Cole#/media/File:Cole_Thomas_The_Course_of_Empire_Desolation_1836.jpg.

Daniels S. and D. Cosgrove. 1998. "Introduction." In *The Iconography of Landscape,* edited by D. Cosgrove and S. Daniels, 1–10. Cambridge: Cambridge University Press.

Davis, M. 1990. *City of Quartz*. New York: Verso.

———. 1998 *Ecology of Fear*. New York: Henry Holt and Company.

Dick, P. K. 2002. "Second Variety." In *Selected Stories of Philip K. Dick*, edited by J. Lethem, 51–95. New York: Houghton Mifflin.

———. 2002. "Foster You're Dead." In *Selected Stories of Philip K. Dick*, edited by J. Lethem, 157–176. New York: Houghton Mifflin.

———. 1968. *Do Androids Dream of Electric Sheep?* New York: Random House.

Domosh, M. 1995. "Feminism and Urban Imagery." *Urban Geography*, 1 (5): 467–472.

Gibson, W. 1993. *Virtual Light*. New York: Bantam Spectra.

Gevers, N. 2019. *Wilderness, Utopia, History; an Interview with Kim Stanley Robinson*. Accessed January 12, 2021. www.infinityplus.co.uk/nonfiction/intksr.htm.

Gregory, D. 1994. *Geographical Imaginations*. Cambridge: Blackwell.

Herzog, W. 2019. "A Conversation with Werner Herzog" *Sheffield Doc/Fest 2019*. Accessed May 11, 2021. www.youtube.com/watch?v=SXgRCI33Js0.

Howey, H. 2014. *Sand*. New York: Simon and Schuster.

Lane, K. M. D. 2011. *Geographies of Mars: Seeing and Knowing the Red Planet*. Chicago: University of Chicago Press.

Matheson, R. 1954. *I Am Legend*. New York: Gold Medal Books.

Miller, L. 2016. "Introduction." In *Literary Wonderlands*, edited by L. Miller. 3–5. New York. Black Dog & Leventhal.

Morgan, A. E. 1944. *Edward Bellamy*. New York: Columbia University Press.

Newitz, A. 2014. "Two Scenarios for the Future of Solar Energy." In *Hieroglyph,* edited by E. Finn and K. Cramer, 243–253; New York: Harper Collins.

Pred, A. 1990. *Making Histories and Constructing Human Geographies*. Boulder, CO: Westview Press.

Robinson, K. S. 1984. *The Wild Shore*. New York: Berkeley Publishing Group.

———. 1988. *The Gold Coast*. New York: St. Martin's Press.

———. 1990. *Pacific Edge*. New York: Tom Doherty Associates.

———. 2017. *New York 2140*. London: Orbit Publishing.

———. 2020a. *The Ministry for the Future*. London: Orbit Publishing.

———. 2020b. *Three Californias Trilogy*. New York: Tor Books.

Shiau, Y. 2017. *The Rise of Dystopian Fiction*. Accessed January 11, 2021. electricliterature.com/the-rise-of-dystopian-fiction-from-soviet-dissidents-to-70s-paranoia-to-murakami/.

Stephenson, N. 1992. *Snowcrash*. New York: Del Rey Books.

Stewart, G R. 1949. *Earth Abides*. New York: Random House.

Tuan, Y-F. 1974. *Topophilia*. New York: Columbia University Press.

———. 1979. *Landscapes of Fear*. Minneapolis: University of Minnesota Press.

SPECIAL ESSAY
BRIDGING SOCIAL AND POLITICAL LANDSCAPES

American Landscapes Under Siege: A Provocation

Joshua Inwood and Derek H. Alderman

As we write this chapter in the summer of 2020, the United States is in crisis, and by extension, America's landscapes are under siege. The COVID-19 virus is ravaging communities across the country. Many of the most vulnerable are most exposed to the contagion, and a US healthcare system that is underfunded, especially as it relates to economically distressed communities, and oriented towards profit instead of care, is stretched to meet the needs of the sick and the dying (Gaffney et al. 2020). Because of shutdowns and quarantine orders, millions of Americans have lost jobs, and the anemic—if not callous—federal response of the Trump Administration to the suffering has exposed and exacerbated existing wealth gaps and exposed wage workers to increased poverty and precarity (Badger 2020). Accompanying the coronavirus has been an "infodemic" of widely disseminated misinformation, often spread by internet rumor and high-ranking conservative public officials denying the very realities of science and public health (Richtel 2020). Not surprisingly, response to the pandemic became politicized as armed militias, some of whom are associated with far-right extremist groups, protested stay-at-home orders and the wearing of masks in public by taking over state capitals and holding "open up" rallies across the country. In at least one case, the Michigan legislature decided to cut short its legislative session because lawmakers did not feel safe enough to continue (Beckett 2020). These protests have undermined confidence in public health precautions to the outbreak and have exacerbated the problems associated with the coronavirus.

Also gripping America is the no less lethal but much older pandemic of police brutality against people of color. Catalyzed by the lynching of George Floyd by Minneapolis police officers who choked and abused him in front of the public, tens of thousands of people in dozens of cities took to the streets and marched in the face of militarized state security forces gassing peaceful protestors and shooting "non-lethal" projectiles into crowds (Taylor 2021). Perhaps most famously, in front of the White House, the US Park Police were accused at the time of violently clearing Lafayette Square of peaceful protestors on June 1, 2020 so President Donald Trump could have a photo of him holding a Bible in front of St. John's Church, which sits across the street from the White

DOI: 10.4324/9781003121800-18

Figure SE.1 Fence in Lafayette Square across from the White House on Juneteenth, a few days after the Black Lives Matter Protest in Washington, D.C., June 19, 2020. Library of Congress, Prints & Photographs Division, photograph by Carol M. Highsmith [reproduction number, LC-DIG-highsm-63641]. Public domain image available at https://www.loc.gov/pictures/item/2020720096/

House. This photo-op set off a nationwide debate about the expansion of militarized policing that further fueled calls for defunding police departments across the country (Gjelten 2020; Stockman and Eligon 2020). Although a Department of Interior report issued a year later would purport that Park Police and other law enforcement agencies cleared Lafayette Square to install fencing around the White House and not expressly for Trump, vocal critics of the report suggest that the timing of the President's photo-op was more than just coincidental and there is little disagreement that police used excessive force against protesters (Cooper 2021). The almost ten-foot black fencing encircling the White House visually and physically cut off the Presidential mansion from the public while also creating a canvas for demonstrators to display images of George Floyd and other murder victims and signs asserting that "Black Lives Matter" and calling for immediate police reform (Figure SE.1). Later in June 2020, authorities placed additional fencing in Lafayette Square around the statue of Andrew Jackson after demonstrators sought to use ropes and chains to pull down the memorial in protest of the president's role in owning slaves and as a purveyor of white settler colonialism and the forced removal of Native Americans east of the Mississippi River (Figure SE.2). In response to these fortification efforts, Washington DC's mayor changed the name of the square in front of the White House to Black Lives Matter Plaza and had the asphalt of the road that runs up to the White House painted with the slogan "Black Lives Matter" in letters large enough to be seen by overflying aircraft. These pavement murals grew in popularity on streets in towns across the country. While communities of activists engaged in this place marking to articulate insurgent urban messages against systemic racism, public officials also got into the act and appropriated a "black aesthetics" to rebrand themselves as supportive of the struggle without really making structural changes in the lives of Black communities (Summers 2020).

Figure SE.2   Fortifications placed around Andrew Jackson statue in Lafayette Square across from the White House, Washington, D.C., June 19, 2020. Library of Congress, Prints & Photographs Division, photograph by Carol M. Highsmith [reproduction number, LC-DIG-highsm-63647]. Public domain image available at https://www.loc.gov/pictures/item/2020720102/

Box SE.1 The History of Race and Landscape in the United States

Aretina R. Hamilton

The history of race and landscape in the United States is long and rich. While cultural geographers examine contemporary issues of race and landscape ranging from settler colonialism to gentrification to residential segregation to gerrymandering, the true origins of this history date back to before the founding of the nation, when the Americas were viewed as a pure, uninhabited, pastoral landscape. As numerous scholars have documented, the Great Woods of the Americas were seen as pure and untapped resources despite the fact that they were already inhabited by Indigenous peoples. However, since Indigenous peoples were viewed as heathens, barbarians, and savages, neither their stake in the land nor their very humanity were recognized by the European colonists. Thus, even before the founding of the colonies, a racialized caste system was already baked into the structure of early America.

Between the seventeenth and eighteenth centuries, colonial powers entered the Americas in search of economic opportunity, but also in search of a white Anglo-Saxon place that mirrored their cultural homelands in Spain, Britain, and France. Just as the European Anglo-Saxon Protestant colonists perceived Indigenous peoples as savage and barriers to progress, they viewed themselves as civilized and thus the only rightful inheritors of the land. As the most dominant European power, the British in particular changed the face of the physical terrain and the cultural landscape that would become the United States.

The second stage in the history of race and space in the US came with the arrival of kidnapped Africans from the continent during the transatlantic slave trade, with the earliest records of enslaved Africans arriving on our shores in 1619. Here we witnessed a secondary and more permanent stage in the racialization of space. While the racial caste system began with the genocide of Indigenous peoples, it would become permanent with the kidnapping, enslavement, and subjugation of African Americans. As McKittrick writes, "The plantation evidences an uneven colonial–racial economy that, while differently articulated across time and place, legalized black servitude while simultaneously sanctioning black placelessness and constraint" (McKittrick 2011, 948). This continuous sense of placelessness that Black Americans experienced, as well as the societal beliefs that non-European bodies would contaminate any spaces they occupied, became a dominant ideology that shaped early Americans' view of the landscape and the communities they called home. In many ways, these racial conquests in early America set the stage for our contemporary understandings of race and the landscape.

Later in the nineteenth century, ideas of nationhood became even more complex as the US experienced a wave of immigration largely from southern and eastern Europe. Much like black and brown peoples, these immigrants were viewed as threats to the racial and purity and moral order of the country. While southern and eastern Europeans were closer in proximity to the white gentry who controlled social and political life in the early Americas, they were racialized based on ethnicity and seen as threats to be contained in urban centers. In early twentieth-century descriptions of these communities, ethnic enclaves comprised of Jews, Catholics, Italians, and others were depicted as non-white and second-class citizens. While their subjugation and containment was neither as brutal nor as complete as that of Indigenous peoples and African Americans, the physical spaces they occupied were limited to those that were deemed outside of the paradigm of civilization and respectability.

Between 1866 and 1877, an era known as the Reconstruction, newly freed African Americans successfully claimed space both politically and economically. Across the South, Black citizens were elected to public office, they created Black businesses, formed Freedmen's Bureaus, founded Black colleges and universities, and developed Black towns in which African Americans claimed spaces of freedom and opportunity to which they had been long denied access. During this time, cities such as Wilmington, North Carolina, Tulsa, Oklahoma, Atlanta, Georgia, and Charleston, South Carolina were racially integrated, and Black Americans enjoyed unprecedented opportunities for education and economic mobility, political influence, and social acceptance. However, this period ended abruptly with the rise of white supremacy led by not only the Ku Klux Klan, but white Citizens' Councils composed of everyday citizens (Camarillo 2013). Lynching, mob violence, and other forms of white racial terror were weaponized against African Americans, resulting in murders and the destruction of Black towns and Black businesses. This led to an exodus of over 6 million Black Americans to northern and midwestern cities. Known as the Great Migration, this period lasted from 1910 to 1940.

However, this optimism and opportunity for Black Americans was short-lived. The experience of Black marginalization, which had long locked Black Americans into place in the South, thus continued in the north. By the mid-twentieth century, public spaces purposely designated for Black Americans were increasingly constricted by public policies such as redlining, restrictive covenants, urban renewal and highway construction projects. This intentional campaign by government officials at the federal, state, and local levels demolished prospering Black communities, destroyed

Black businesses, homes, and wealth, and condemned many Black Americans to the bottom of the social and economic ladder (Lipsitz 2007). Simultaneously, the rise of sundown towns drove out BIPOC residents and ethnic "others" from small-town America, consigning these groups to urban centers and naturalizing their subordinate position as one of poverty, destitution, and segregation. Much like Southern plantations, Black landscapes and all that resided in or near them—now called "ghettos"—were viewed as corruptive influences to be contained (Anderson 2015). And Black people's containment in these spaces was seen as a choice, not as a purposeful campaign championed by government officials, corporate executives, policymakers, white homeowners, and others in power.

Consequently, the desire to maintain pure white spaces resulted in many of the issues that continue to plague Black communities today. Recently, we have seen a resurgence in protest movements such as Black Lives Matter that call attention to police violence, access to affordable housing, wealth inequality, disparities in access to healthy food, and educational inequalities (Hamilton and Foote 2018; McKittrick 2011; Woods 2002). These inequalities are not coincidental. They are baked into the American landscape by design.

References

Anderson, E. 2015. "'The White Space.'" *Sociology of Race and Ethnicity* 1 (1): 10–21. doi:10.1177/2332649214561306.

Camarillo, A. M. 2013. "Navigating Segregated Life in America's Racial Borderhoods, 1910s—1950s." *The Journal of American History* 100 (3): 645–62. www.jstor.org/stable/44308757.

Hamilton, A. R., and K. Foote. 2018. "Police Torture in Chicago: Theorizing Violence and Social Justice in a Racialized City." *Annals of the American Association of Geographers.* doi:10.1080/24694452.2017.1402671.

Lipsitz, G. 2007. "The Racialization of Space and the Spatialization of Race: Theorizing the Hidden Architecture of Landscape." *Landscape Journal* 26 (1): 10–23. www.jstor.org/stable/43323751.

McKittrick, K. 2011. "On Plantations, Prisons, and a Black Sense of Place." *Social & Cultural Geography.* doi:10.1080/14649365.2011.624280.

Woods, C. 2002. "Life After Death." *Professional Geographer* 54 (1): 62–66. doi:.1111/0033-0124.00315.

As protests over George Floyd's death increased, they stretched across the country and further galvanized the post-Charleston and post-Charlottesville movement to topple and dismantle long-standing statues, memorials, and the names of public institutions and places that valorize Confederate soldiers, slave owners, segregationists, leaders of the KKK, and other purveyors of racial inequality (Kelly 2020). As these monuments to white supremacy have fallen, calls for the removal of other figures, including George Washington, Ulysses S. Grant, Andrew Jackson, and Thomas Jefferson, have proliferated—provoking a wider nationwide reckoning with the fact that anti-Black discrimination and violence is foundational to the American project rather than merely an uncomfortable aberration (Gowen 2020). These and other campaigns indicate the central place that landscape plays within our lives, not just as the mere settings or stages for the ongoing American crises but as active participants in shaping the form, meaning, and stakes of recent struggles. However, the country's landscapes have always—from their inception—been deeply involved in constituting and structuring deep-seated fights over identity, over economics, and the political and cultural structures that come to define and dominate our lives (and the lives of others). In this respect, US places and spaces have always been under siege, even if those contests and their consequences do not capture major headlines or provoke demonstrations in the streets.

Importantly, the American landscape is not simply the product or result of social power relations; rather, it is also, knowingly and unwittingly, productive and generative of the ideologies, social practices, and tensions that have come to define the nation. In particular, over the past few months, the American landscape has revealed a nation riven with structural inequalities and racism. A nation that has hollowed out and devastated social services and the welfare state in a broader neoliberal effort to concentrate wealth in the hands of the few at the expense of the many. These realities raise a broad array of questions for scholars and practitioners of landscape studies. Important among those questions is: *What role can critical scrutiny of the landscape and its revelatory power and meaning play in exposing and challenging the conditions of poverty and racism, violence and militarism, sexism, and patriarchy that characterize life and death in the United States?* Such a question does not just elevate the stakes of studying the American landscape. Rather, it is also suggestive of the need for scholars to leverage their writings, voices, and other practices to make ethical interventions that advance public understanding of how landscapes are formed, how they operate, and for whom they serve (and don't serve). Landscape intervention is also about scholars working in solidarity with and support of historically marginalized groups put at risk by dominant framings of American culture and politics (Alderman and Inwood 2019). In particular, we join Kellner (1997) and Kurtz (2019) in believing these interventions should follow a model of "democratic public intellectualism." Rather than a lone, elite authority, such an intellectual works to widen the circles of dialogue and engagement with wider publics to generate a variety of knowledge innovations and communication channels with the capacity to do social work.

Framing this essay less as a standard research piece and more as a provocation, we suggest that landscape studies must be centrally focused on addressing and exposing social tensions and inequalities, transforming those moments in which the American landscape is under siege into critical opportunities for critique and intervention. Doing so requires a revolution in theorizing a landscape approach, one that values the social theory and knowledge production of oppressed communities and brings those ideas in conversation with traditionally privileged academic epistemologies. For this discussion, we chart a future path of analysis and praxis that joins the Black Freedom Struggle's emphasis on the necessity of exposing the "black box" of inequalities operating within communities and a postcolonial "contrapuntal" method of analysis that makes visible the absences, tensions, and inequalities that always underlie and circulate through Americas' landscapes.

Exposing the Black Box of Community

An important principle in this project of understanding how landscapes are created, lived, experienced, and fought over is engaging the full range of social theory. Scholars must avail themselves not only to the established and emerging academic theories but also those frameworks of sense-making and strategizing that arise from the organic intellectualism of everyday activists and citizens knee-deep in the grounded struggles for survivability and equality. Wright (2019, 172) makes this point powerfully: "scholars [should] take seriously the questions, research methods, and analyses emerging from the general public, particularly from within communal and political organizations." Such an approach arises from the hard-learned lessons of oppressed peoples who engage in struggle and who, in many cases, place their lives and the lives of their colleagues into each other's hands as they organize to take on oppression. Such a grounded and materially constituted reality forged through lived, activist lessons, holds the potential to open up academic understandings of landscapes in new ways that push the boundaries of what traditionally counts as knowledge and how we understand the politics of knowledge-making. This melds praxis and academic practice in ways that can challenge traditional western forms of knowledge creation and elevate the significant and important contributions of peoples whose work, because of the politics of who is considered an intellectual, has been erased or ignored.

The history of America's Civil Rights Movement often contains important alternative landscape epistemologies not adequately acknowledged (McCutcheon 2019). While traditional Anglo-American Geography spent much of the 1950s and 1960s debating whether the discipline should be idiographic or nomothetic, activists were engaged in struggles that literally remade and refashioned the US landscape, opened up public spaces, expanded the right to vote, and took on the foundational role of white supremacy in the United States. As civil rights groups engaged in these processes, they also revealed deeper truths about how to understand the landscape and the central role that it plays in social justice work. For example, in our ongoing examination of the Student Non-Violent Coordinating Committee (SNCC), we are reminded of how the 1960s civil rights group made critical use of social and spatial data and mapping while conducting landscape analysis in their struggle to topple white supremacy (Inwood and Alderman 2020). SNCC's story not only expands our understanding of the organic intellectual labor and planning behind the Civil Rights Movement, but it also helps define more generally new theoretical ways to think about the relationship between space and society and the role of the landscape in the processes of oppression and resistance.

For those not familiar with SNCC, they were a group of mostly young organizers who worked in the deep southeastern United States, undertaking a grassroots, community-centered approach to organizing inspired by noted civil rights leader Ella Baker, who organized the founding conference of SNCC. Baker and other SNCC workers eschewed the idea of charismatic civil rights leaders for empowering marginalized communities to understand the nature of their oppression and advocate for themselves. Working in rural Southern communities to register voters and address economic precarity, SNCC organizers risked life and limb to secure Black freedom dreams and directly challenged anti-Black racism. Because SNCC worked in local communities, they developed a complex and multifaceted research department that would provide local organizers with actionable information, communication, and education—much of it geographic in nature. That research was carried out with the goal of helping activists in the field understand and subvert racialized landscapes, raising the consciousness of communities, and encouraging those communities to demand and secure basic political and economic rights and to envision and realize a different and more just sociospatial future (Inwood and Alderman 2020).

Charles Sherrod (n.d.), a prominent civil rights organizer in Albany, Georgia, described SNCC's process as "exposing the black box of community." He wrote that every person lives within a box that leads them to understand their communities and their place within that community. Because of racism and dominant understandings of how the economy and politics work, many people become confined to their box and come to accept conditions of exploitation and inhumanity that structure their daily lives. The black box of community holds academic purchase for scholars of landscape studies. The purpose of SNCC organizers when they went into a community was first to ascertain the nature and dimensions of that box—to research and understand the local and on-the-ground spatial and social conditions of race and inequality. From that landscape knowledge, field activists would create political, cultural, and educational programs that would allow people to see who they truly were, who they could be, and that they had the power to change their conditions. Confronting the black box of the community was central to how SNCC operated and the kinds of work in which they engaged. Although SNCC workers would not identify themselves as landscape analysts or geographers, an examination of the organization's archives clearly shows evidence of a Black spatial imaginary and a keen understanding that dismantling the box of racism that enveloped communities required understanding how these communities were situated with and shaped by a broader historical geography of racialized social relations.

We suggest that practitioners of landscape studies examine how the black box of normalized inequality comes to define the US landscape and how the landscape, as an extension of that box, contributes to the obscuring and hiding of injustice in plain sight. In other words, the study of the landscape must undertake a broader political project to expose the structures of inequality that come to define the social, political, and economic realities of life in the United States. This project is not only necessary academically but also relevant to myriad communities inhabiting landscapes of trauma, suffering under the yoke of unfettered capital accumulation and rampant racism, the violence of militarism, and the tyranny of sexism and gender discrimination. These boxes and their attendant landscapes come to define the lives of millions of people, lie at the heart of the current crisis in America, and heavily shape the lived futures and well-being of marginalized communities. The study and understanding of landscape must be at the forefront of efforts at exposing these boxes, not only for highlighting their existence but also in providing a foundational understanding of the United States that can facilitate social justice movements and abolition geographies (Gilmore 2017). As we suggest in the next section, a "contrapuntal" approach is one potentially useful means of exposing the black box of inequalities that has for so long become naturalized and projected through the American landscape.

A Contrapuntal Approach

Within the context of the black box, there are conditions that undergird the unfolding traumas that exist within and perpetuate processes of oppression. That oppression is central to forming the broad contours of American landscapes, even as those landscapes are sites of uprisings, resistance, and freedom movements. We confront a central question in this chapter that should be at the forefront of many scholars' minds: *How can we approach the landscape to expose and cast critical scrutiny on the underlying structures and hidden processes that support the broader project of white supremacism and racial capitalism?* We suggest that postcolonial theory offers a viable method of landscape analysis capable of uncovering and de-naturalizing the seemingly impenetrable black box of inequality surrounding communities and shed light on the broader social, psychological, and geographic drivers of oppression. Specifically, inspired by the work of Edward Said, there is promise in developing the idea of a "contrapuntal reading" of the American landscape.

Contrapuntal reading was an idea that Said developed in his book *Culture and Imperialism* (1993). Written after his more widely cited book *Orientalism, Culture and Imperialism* was Said's attempt to address critiques of *Orientalism* and those who argued he focused too much on a kind of passive acquiescence to European domination by colonized peoples (Mortimer 2005). Focused on what Mortimer (2005, 55) describes as "indigenous [sic] struggles against colonial domination and anticolonial resistance," *Culture and Imperialism* introduces the idea of contrapuntal reading; "a form of reading back from the point of view of the colonized that brings to light the hidden colonial history that permeates" European culture and history (Mortimer 2005,

55). Important to Said, who drew from the idea of counterpoint within western classical music, was developing the capacity to read texts in ways that took into account the processes of both imperialism and resistance, to make visible and heard those voices and histories of marginalized groups typically excluded from dominant, naturalized stories within society. For him, the role of the intellectual is to oppose consensus and orthodoxy, to interpret and question the consolidation of authority, and to recall what (and who) is forgotten and ignored (Chowdhry 2007). This emphasis on challenging what is naturalized and unquestioned harkens to Charles Sherrod's admonition that an anti-racist epistemology or knowledge system must question and unpack the black box of institutionalized patterns of inequality and expose the broader community conditions and power relations that oppress people.

A contrapuntal lens, with the stress it places on challenging cemented, common-sense understandings of social life, holds great potential to move away from elite, univocal readings of the American landscape and make visible the intertwined and overlapping histories of the dominant and the subaltern. Lachman (2010, 164) notes that the idea of contrapuntal reading "enables us to think through and interpret together experiences that are discrepant, each with its particular agenda and pace of development, its own internal formations, its internal coherence and system of external relationships, all of them coexisting and interacting with others." In other words, a contrapuntal reading focuses on the interactions and tensions that exist between a range of contradictory power-laden juxtapositions written into and through the landscape as both a thing, process, and set of practices. Importantly, a contrapuntal methodology, according to Said, would strive to view the landscape through the eyes of those social actors and groups normally exiled from or marginalized within those places and spaces, while also interrogating the historical and contemporary assumptions and relations that uphold this exclusion (Bilgin 2016). A contrapuntal approach, when married with SNCC's mobilizing against and within the black box of racism, pushes us to recognize the necessity of landscape studies to read and juxtapose the realities and histories of inequality and white supremacism against the experiences and histories of Black dispossession, resilience, and resistance. Such an analysis has always been needed but especially so during the current siege. A contrapuntal reading that actively recovers and acknowledges oftentimes forgotten subjects and worldviews is a way of highlighting how the erasure of history, people, and culture manifests through the cultural landscape. Rather than remaining silent, traditionally excluded identities and perspectives can occupy and thus be read as having an actual presence within the landscape itself.

It is this leveraging of juxtaposition—the richly detailed story of those in power, on the one hand, and the silences and resistances of the marginalized, on the other hand—that makes contrapuntal reading a potent analysis for interpreting landscapes in the United States. The analytical potency comes from how a contrapuntal approach exposes how America's landscapes—rather than mere reflections or bystanders of inequality—are built into and through the processes of racial capitalism and settler colonialism. Take, for example, the point that Said (1993) makes in arguing that the very absence of discussions of slavery and colonialism in nineteenth-century British fiction belie the role that these exploitative processes played in forming the landscapes and lifestyles of white elites. The US is especially in need of landscape analysis that makes the lives and histories of the enslaved a counterpoint to the story and legacies of those benefitting from and complicit in slavery. As Inwood and Brand (2021) argue, the everyday spaces of many American cities still in use today—from buildings and roads to ports and rail lines—were built by enslaved Black workers, but these pieces of infrastructure are seldom discussed as "landscapes of race." By failing to read, in a contrapuntal fashion, the engineering of US landscapes against the history of how slavery is built into landscapes and continues to generate wealth, we lose an opportunity to expose how racism operates within the black box of communities and we actively disinherit people of color from their roles as the builders of the nation. Contrapuntal reading opens space to see and challenge these silences around race and racism. Contrapuntal readings also push against oblique mentions that tend to bracket racism within a particular era or region, when actually all landscapes within the United States are built upon a foundation of race. The very absence of discussions around the violent dispossession of Indigenous peoples as well as the institution of regimes of racial capital are the central stories that exist throughout the US landscape. A contrapuntal future for American landscape studies must be full-throated in challenging the assumed coherence of those landscape stories, making visible those silences about race and power, and centering the exiled histories and voices of BIPOC (Black Indigenous People of Color). As a way of brief illustration, we focus on a contrapuntal reading of Black Lives Matter plaza in Washington, DC where we can begin to outline an agenda that highlights how these processes play out.

Reading Black Lives Matter in Lafayette Square

Lafayette Square in Washington, DC is arguably one of the most famous parks in the United States. Located on the north side of the White House, it is the park used by Presidents in viewing their inaugural parades. And on most days, Lafayette is crowded with tourists and others hoping to get a glimpse of the White House and the Old Executive Office Building located near the West Wing of the White House Complex. In June of 2020 and in the face of rising tensions over the killing of Black men and women by police forces in the United States, hundreds of non-violent

Figure SE.3   The words Black Lives Matter painted in large yellow capital letters on 16 Street NW near the White House, June 19, 2020. The mural marked the recently named Black Lives Matter Plaza and became a powerful icon and gathering place for demonstrations. Library of Congress, Prints & Photographs Division, photograph by Carol M. Highsmith [reproduction number, LC-DIG-highsm-63668]. Public domain image available at https://www.loc.gov/pictures/item/2020720123/

protestors gathered in this historic square to protest police brutality. These protests wanted to draw attention to the brutal and discriminatory realities of police violence and the seemingly endless accounts of extreme violence that characterize modern police interventions in mostly minority communities. In response to these protests, the square was cleared of demonstrators and fortified with fences that would remain in place for almost a year. As mounted and armed police stormed non-violent protestors in June 2020, they used tear gas, billy clubs, and shields to push the crowd back. As we detailed in our introduction, following the efforts of the police, President Trump walked across Lafayette Square in order to get a photograph in front of St. John's Episcopal Church holding a Bible. As a strong rebuke to this clearing of the square, Mayor of Washington, DC, Muriel Bowser, ordered that the words "Black Lives Matter" be painted in large yellow capital letters on 16th Street on the north of Lafayette Square in front of the White House while also declaring that the District welcomes all peaceful protesters (Figure SE.3). Mayor Bowser also officially renamed a pedestrian section of that part of the street outside of the White House as Black Lives Matter Plaza and marked that designation with a road sign.

Box SE.2 The Roots and Routes of a Black City

Brandi T. Summers

Washington, DC has had a long and unique Black cultural and political history. In the early nineteenth century, Washington welcomed runaway slaves and freedmen and women seeking refuge. Because federal legislators preferred to stay in Washington only a few months out of the year, Black people had more freedom and economic opportunities. Throughout the nineteenth and twentieth centuries, the District was the center for anti-lynching, anti-segregation, and voting rights campaigns. Since that time, Black people have continued to build vibrant neighborhoods complemented by thriving commercial districts.

In many ways, DC is and has always been a battleground. It has been the location of so many social, economic, and political battles that have not only impacted District residents, but most American citizens. At the same time, DC is "chocolate," a Black city, because of the sights (and sites), sounds, decadent tastes, politics, landmarks, and its history. In a city that has only been led by Black mayors, DC also boasts its unique style of music (go-go) and celebration. These cultural and political attributes derive their roots from the routes Black people have traveled to be in DC.

What often goes unsaid is that DC residents live under some of the least democratic conditions in the country. Washington (the federal city) and DC (the "Chocolate City") have operated as two separate spaces, but intertwined in significant ways. DC has had a particularly complex relationship to the federal government. Beleaguered by an intricate history between the federal government and its Black constituents, DC only received limited home-rule after the enfranchisement of Blacks in other locations in the United States. The DC Home Rule Act of 1973 allowed the city to elect a mayor and city council members and designated an annual payment for the District, but final say over the District's budget and legislative power over city matters still remains with the House Committee on DC, which also has the veto right on any city legislation. Home rule meant that DC could, for the first time, operate like an independent municipality, with several stringent rules attached. New provisions afforded many Black residents with unprecedented opportunity. Nevertheless, DC remained a contentious environment for its high concentration of (Black) poverty (since white poverty in the District is virtually non-existent), unemployment, and class and racial stratification. The federal government continues to have distinct power over and within the

District politically. Residents have no voting congressional representation, despite paying federal taxes, and have only been able to vote in presidential elections since 1964.

Home rule, the fight to get it, and the federal government's unique oversight (or lack thereof) of DC, puts Black Washingtonians at a distinct disadvantage. The District's legal disfranchisement is a major metaphor for the long-term disfranchisement of Black people across the United States at the hands of federally instituted and supported white supremacy; Black residents of DC are up close and personal with this disfranchisement.

Today, DC remains largely segregated as the population on the eastern side of the city remains almost exclusively Black, and mostly white on the western end. Nevertheless, with its Black population dipping below 50 percent in 2010, DC joins a collection of US cities like Oakland, Chicago, and St. Louis, where the population of Black residents has been in steady decline, largely due to economic factors (primarily income inequality and a dearth of economic opportunities) that have led to higher unemployment rates, declining wealth accumulation, and lower educational attainment in comparison to their white counterparts.

Despite the trenchant ubiquity of gentrification in DC, there are an array of community groups, initiatives, policy, and ongoing cultural activities that address the sociocultural and socio-spatial shift in the city's landscape. Three particular levels reflect activities in the current political struggle against gentrification and displacement in the District: policy, grassroots organizing, and artistic/cultural activities. Each of these groups and initiatives uniquely address structural inequalities that proliferate and bolster gentrification and displacement. All of them represent, in their own way, the importance of advocating longtime DC residents' right to stay and how unnatural and violent gentrification is as a process.

We are now experiencing unprecedented conditions—the threat of a global pandemic intertwined with a global fight against racism, white supremacy, and state-sanctioned violence against Black people, specifically in the US. What the spread and morbid effects of COVID-19 have revealed is the co-evolution of Black inequity in various parts of life. What COVID-19 and the "Black Spring" revolution have revealed are the cracks in the system that require us to think differently about how we use our power to enact change. In many ways, Washington DC must reckon with the ways that freedom is built on the backs of Black Washingtonians, but more specifically, the blackness that the bodies of these people register.

The importance of a place like DC is not simply a migration story, or even the diverse experiences of Black people, but that Black people *built* the city in meaningful ways. Calling out the ways the violence of gentrification and displacement disproportionately impacts Black Washingtonians allows them to imagine new ways to take up space, to expand, and potentially grow.

The renaming of the plaza and the painting of the street is a visible reminder of how the American landscape is a site of struggle, a site where ideas about belonging and race play out in a series of sometimes spectacular struggles to redefine for whom and in what way the landscape should speak to broader ideas about the nation. In this sense and as has been written (Lewis 1979), the landscape is like an "unwitting biography." But as we have alluded to in this essay, the landscape can conceal and hide, obscure and obfuscate as much as it can reveal about who we are and what the foundations of the nation are about. The renaming of Black Lives Matter Plaza is significant not only for the symbolism, but also for the ways we can begin to think about how to read the US landscape contrapuntally.

The reality is that Black lives have always mattered to the square and that Black lives and destinies have long been intertwined with the very seat of US government. Recall from the section on contrapuntal reading that we cannot ignore the fact that the West has long enjoyed a life of comfort furnished through the labors of colonized others or the enslaved and Lafayette Square is a strong manifestation of this reality. Named after the Marquis de Lafayette, a French army officer who served with George Washington in the American Revolution, Lafayette at various times in his life owned enslaved Black workers. This reality speaks to broader ways that the enslaved were central to the story of the American empire and speaks to a central contradiction in ideas of freedom and liberty that animate not just the American origin story but the very conceptions of purity and authority that we attach to national memorial spaces. More to the point, Lafayette Square was at one time a large slave market in the nation's capital (Brown 2000). The park, located mere feet from the entrance to the White House, saw men, women, and children paraded up Pennsylvania Avenue and auctioned off as the tobacco economy of the upper South collapsed and as the needs for enslaved labor on the massive cotton plantations in the Deep South expanded.

This history, written out of mainstream understandings of the nation's development, points to the reality that the square has not only always been a site of contestation, but when read contrapuntally it points to the foundational role that slavery played in the making of America. This landscape is now and has always been a site central to the struggles of Black people. Reading this landscape contrapuntally means putting the efforts to erase, remove, and push out Black people—made most visible during the BLM protests—into a broader perspective that highlights the central role that slavery and the Black experience are to the American landscape. Far from placing Lafayette Square into a black box that obscures or obfuscates these histories, we instead need to see this, and other landscapes within a framework of racial capital and struggle. In so doing we can reveal the deeper and more sustained struggles that animate not just this, but all landscapes within the United States.

References

Alderman, D. H. and J. F. J. Inwood. 2019. "The Need for Public Intellectuals in the Trump Era and Beyond: Strategies for Communication, Engagement, and Advocacy." *The Professional Geographer* 71 (1): 145–151.

Badger, E. 2020. "The Year Inequality Became Less Visible, and More Visible Than Ever." *New York Times*, December 28, 2020. www.nytimes.com/2020/12/28/upshot/income-inequality-visible.html.

Beckett, L. 2020. "Armed Protesters Demonstrate Against Covid-19 Lockdown at Michigan Capitol." *The Guardian*, April 30, 2020. www.theguardian.com/us-news/2020/apr/30/michigan-protests-coronavirus-lockdown-armed-capitol.

Bilgin, P. 2016. "'Contrapuntal Reading'" as a Method, an Ethos, and a Metaphor for Global IR." *International Studies Review* 18 (1): 134–146.

Brown, D. L. 2020. "A History Lesson for Trump: Lafayette Square was Once Bordered by 'Slave Pens.'" *The Washington Post* June 5, 2020. www.washingtonpost.com/history/2020/06/05/lafayette-square-slave-market-dc-protests/.

Chowdhry, G. 2007. "Edward Said and Contrapuntal Reading: Implications for Critical Interventions in International Relations." *Millennium* 36 (1): 101–116.

Cooper, R. 2021. "Trump's False Lafayette Square Exoneration." *Yahoo News*, June 11, 2021. news.yahoo.com/trumps-false-lafayette-square-exoneration-095509572.html.

Gaffney, A., D. U. Himmelstein, and S. Woolhandler. 2020. "COVID-19 and US Health Financing: Perils and Possibilities." *International Journal of Health Services* 50 (4): 396–407.

Gilmore R. W. 2017. "Abolition Geography and the Problem of Innocence." In *Futures of Black Radicalism*, edited by G.T. Johnson and A. Lubin, 225–240. New York: Verso.

Gjelten, T. 2020. "Peaceful Protesters Tear-Gassed to Clear Way For Trump Church Photo-Op." *NPR. org*, June 1, 2020. www.npr.org/2020/06/01/867532070/trumps-unannounced-church-visit-angers-church-officials.

Gowen, A. 2020. "As Statues of Founding Fathers Topple, Debate Rages Over Where Protesters Should Draw the Line." *Washington Post*, July 7, 2020. www.washingtonpost.com/national/as-statues-of-found ing-fathers-topple-debate-rages-over-where-protesters-should-draw-the-line/2020/07/07/5de7c 956-bfb7-11ea-b4f6-cb39cd8940fb_story.html.

Inwood, J. F. J. and D. H. Alderman. 2020. "'The Care and Feeding of Power Structures': Reconceptualizing Geospatial-Intelligence through the Countermapping Efforts of the Student Nonviolent Coordinating Committee." *Annals of the American Association of Geographers* 110 (3): 705–723.

Inwood, J. F. J. and A. L. Brand. 2021. "Slave-Built Infrastructure Still Creates Wealth in US." *The Conversation*, February 5, 2021. theconversation.com/slave-built-infrastructure-still-creates-wealth-in-us-suggesting-reparations-should-cover-past-harms-and-current-value-of-slavery-153969.

Kellner, D. 1997. "Intellectuals, the New Public Spheres, and Techno-Politics." *New Political Science* 41–42: 169–188.

Kelly, C. 2020. "Confederate Monuments Toppled, Burned as Protests over George Floyd's Death Continue." *USA Today*, June 1, 2020. www.usatoday.com/story/news/nation/2020/06/01/george-flo yds-death-causes-confederate-monuments-targeted/5310736002/.

Kurtz, H. E. 2019. "Public Intellectualism as Assemblage." *The Professional Geographer* 71 (1): 179–183.

Lachman, K. 2010. "The allure of counterpoint: History and reconciliation in the writing of Edward Said and Assia Djebar." *Research in African Literatures* 41 (4): 162–186.

Lewis, P. F. 1979. "Axioms for Reading the Landscape: Some Guides to the American Scene." In *The Interpretation of Ordinary Landscapes*, edited by D. W. Meinig, 11–32, New York: Oxford University Press.

McCutcheon, P. 2019. "Fannie Lou Hamer's Freedom Farms and Black Agrarian Geographies." *Antipode* 51 (1): 207–224.

Mortimer, M. 2005. "Edward Said and Assia Djebar: A Contrapuntal Reading." *Research in African Literatures* 36 (3): 53–67.

Richtel, M. 2020. "W.H.O. Fights a Pandemic Besides Coronavirus: An 'Infodemic'." *New York Times*, February 6, 2020. www.nytimes.com/2020/02/06/health/coronavirus-misinformation-social-media.html.

Said, E. 1993. *Culture and Imperialism*. London: Vintage Press.

Sherrod, C. n.d. "Non-violence." Lawyer's guild archive. The Bancroft Library. The University of California Berkeley, Call Number: 99/280 Ctr. 60.

Stockman, F. and J. Eligon. 2020. "Cities Ask if It's Time to Defund Police and 'Reimagine' Public Safety." *New York Times*, June 5, 2020. www.nytimes.com/2020/06/05/us/defund-police-floyd-protests.html.

Summers, B. 2020. "We Need Action to Accompany Art." *Boston Globe*, June 11, 2020. www.bostongl obe.com/2020/06/11/opinion/we-need-action-accompany-art/.

Taylor, D. B. 2021. "George Floyd Protests: A Timeline." *New York Times*, March 9, 2021. www.nytimes. com/article/george-floyd-protests-timeline.html.

Wright, W. J. 2019. "The Public is Intellectual." *The Professional Geographer* 71 (1): 172–178.

PART III

US Political Landscapes

Chris W. Post

We witness political geographies at all scales, from international geopolitics to the body politic. What these landscapes and places all hold in common is the attainment, attempted or achieved, and maintenance of power. Be it economic, social, environmental, local or international, political landscapes stand omnipresent as a means to direct our lives, or simply tell us what to do. Some suggest that politics is necessary to prevent chaos. Others battle the expression of power at every turn they see its manifestation. Already in this book we've seen the impact of politics in other landscapes. Rich Schein's foreword on the very notion of landscape brings to light the aspect of power over place and how it negotiates the wielding of power over where we can live, how we make a living, and who possesses the power to order the landscape to make it more inclusive for all Americans. This section's chapters represent mostly national and regional scale political landscapes that present themselves frequently in our lives.

Barney Warf starts this section with an overview of America's imperial position in the world. This chapter lays out the history of the United States' ascent in global politics from British colony to global superpower. Though it focuses less on any particular landscape, this chapter provides the necessary foundation upon which following chapters—especially those on the border landscape, the public land survey system, and electoral landscapes—build their analyses.

After establishing America's position in the world and how it developed over time, Kenneth Madsen sets America's border landscape as a stark contrast between our northern and southern borders. Keeping with this volume's focus on recent changes, Madsen focuses mostly on the changes and challenges surrounding the US border with Mexico and the role of former President Trump's attempts to fortify and militarize that line, regardless of environment, economics, or need for such a system.

Within our borders, the lines that impact us most are those of the United States Public Land Survey System—or PLSS. Timothy Anderson explores the American grid, from which most of the country's land is defined and planned, in a new way that incorporates spatial theory and the concept of settler colonialism that understands colonization, in the American sense, as the daily work of white settlers at a more local scale.

As we set our farms and cities on this grid, we vote in precincts also established by similar grid systems. Fred Shelley and Heather Hollen discuss the impacts of electoral politics on our landscapes in various forms, with a focus on the 2020 Presidential election. Amidst a pandemic that saw increased absentee voting, the Republican and Democratic Parties also held

DOI: 10.4324/9781003121800-19

different convention formats—virtual (Democrat) and in person (Republican). The election also manifested many of the social tensions in the US, stemming from concerns over immigration, policing, and the racialization of Chinese Americans during the COVID-19 pandemic. Ultimately, in January of 2021 as the senate approved the electoral votes, the landscape of Washington, DC, turned into a site of violence, attempted coup, and even death.

The next two chapters of this section dovetail with one another. Erika Doss works at the intersection of emotional affect and memorialization. Building on the cultural politics covered by Shelley and Hollen, Doss explores the work of artist Judy Baca in Los Angeles and how her art spurs a reconsideration of colonial history and modern inclusion in that city to produce a resonant emotional response to her work and larger perspective. In doing so, Doss evidences how memorial landscapes become real sites of social mediation (see Schein's chapter in this volume) through particular emotions, specifically anger. Jacque Micieli-Voutsinas adds to Doss' chapter by exploring more-than-representational understandings of heritage landscapes and how such a methodology reveals the dynamism of such places.

Naming our places also remembers our pasts. In his chapter on place naming practices, Jordan Brasher assesses how we increasingly recognize the memorial power of what we name our streets, towns, and even college campus buildings. Such naming conventions indelibly empower some community members over others. Those persons with more social, economic, or political capital receive the privilege of naming places for themselves or for others whom they feel deserve such material legacies. Brasher analyzes toponymic changes on the campus of the University of North Carolina stemming from the many buildings named after figures affiliated with slavery, the Confederacy, the Ku Klux Klan, and other white supremacist movements (also witnessed at many other southern colleges).

Finally, in America, imprisonment is capital and Matt Mitchelson details the development of the incarceration landscape, which has seen increased focus in landscape studies over the past twenty years. Prisons blur the lines between politics and economics. Many states now outsource their prisons to private corporations who seek to profit on increased incarceration rates. Such influence not only jeopardizes the integrity of a state's policing system, but also impacts its economics and, as Mitchelson details in Virginia, local development.

As a collection, these essays investigate how society negotiates power through the landscape at various scales. These examples also highlight the ability of politics to intersect with multiple other themes—race, economics, memory, ecologies, and naming, to state a few from these examples. Truth is that any landscape exists as an expression of power over place at some scale and point in time. These chapters show how this is accomplished and how we can better identify these spaces and then counter such movements of exclusivity and division.

15

AMERICAN IMPERIALISM

Barney Warf

Although many Americans are uncomfortable with the fact, the US is an empire. Indeed, it is arguably the largest and most powerful empire in world history. The topic has received significant scholarly scrutiny (e.g., Ferguson 2004; Bacevich 2004; Berman 2006; Chomsky 2003). Yet, to many Americans, the reality of the empire is obscured; it is a reality that does not mesh well with prevailing mythologies of the US as a fountainhead of liberty and democracy. When Americans do admit to being an empire, it is typically portrayed as a force for good, one that spreads democracy across the world (Immerman 2010).

This chapter examines American imperialism in several steps. It begins by contextualizing it historically, noting its roots in the Euro-American conquest of North America and the first significant overseas conquests in Hawaii and the 1898 Spanish–American War. The second part examines American hegemony during the post-World War II boom even as it was constrained by the Cold War. The third section turns to the late twentieth century, when the empire began to exhibit serious symptoms of decay following the petrocrises and the rise of foreign competitors. The chapter notes that shifts in the US position internationally were mirrored in its changing landscapes, such as the decline of the Manufacturing Belt. Fourth, it focuses on contemporary American imperialism in the twenty-first century; in the wake of the collapse of Communism and the fall of the Soviet Union, the US once again enjoys a period of unsurpassed dominance. Next, it addresses the cultural and ideological dimensions of US hegemony, which are as important as the economic, political, and military ones. The conclusion highlights major themes and touches on a persistent opposition to the American empire.

The Origins of Empire

The US is both a product of colonialism, largely British, and a producer of colonies in its own right (Doolen 2005). The country's political landscape was forged through the annexation of several territories over time (Burns 2017), including the Louisiana Purchase of 1803 and the acquisition of Florida from Spain (Figure 15.1). The waves of expansion of European-Americans across North America in the eighteenth and nineteenth centuries laid the basis for the subsequent overseas expansionism in the twentieth. Guided by "Manifest Destiny," the notion that the US is an exceptional nation with the right, even the obligation, to conquer other peoples, white settlers effectively exterminated the Native population. Additionally, in much of the country, slaves from Africa provided a cheap labor force that worked the cotton

DOI: 10.4324/9781003121800-20

Figure 15.1 Territorial Acquisitions of the United States since the colonial period. Source: Wikimedia Commons (Public Domain): https://upload.wikimedia.org/wikipedia/commons/9/94/U.S._ Territorial_Acquisitions.png

and tobacco farms of the Southern and Middle Colonies and forged many of the colonial and early republic's landscapes. In the 1840s, wars with Mexico led to the seizure of many territories that later became states: Texas, Arizona, New Mexico, and California. In short, territorial seizures were central to the formation of the US as a political and spatial entity.

By the late nineteenth century, as noted by the historian Frederick Jackson Turner, the frontier reached the Pacific Coast (Turner 1920). In Turner's view, the frontier—that border separating productive, white, Christian culture from the brown, heathen Natives—proved instrumental in the formation of the US as a democracy. In many ways, the subsequent expansion of the United States overseas may be viewed as an extension of the frontier on an international scale.

In the 1890s, the encroachment of the US overseas began in earnest. In 1893, the US overthrew the independent Kingdom of Hawaii (Merry 2000; Nasser 2021). The islands were vital to the growing trans-Pacific trade networks and were lucrative for the influential sugar industry. In 1898 the US annexed the kingdom, in 1900 it was made a territory, and in 1959 it became the 50th state, extending the country's landscape into the Pacific Ocean.

The most obvious sign that the US was becoming an imperial power was the 1898 Spanish–American War, which ended the Spanish Empire (Foner 1972). In this conflict, the US seized Cuba, which became independent in 1902, and Puerto Rico, which remains an American colony to this day. In the Pacific Ocean, the US seized Guam and the Philippines, entering into a bloody conflict with independence activists there. The victories were celebrated as part of a new empire stretching over 10,000 miles in length (Figure 15.2). The Philippines eventually achieved nominal freedom from US control in 1946, following the Japanese occupation during World War II. Nonetheless, the Spanish–American War established the US as the foremost power of the Caribbean and a growing presence in East Asia.

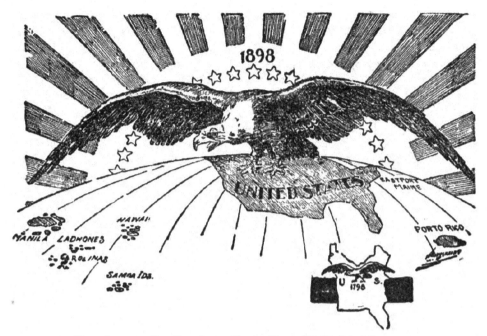

Ten thousand miles from tip to tip.—Philadelphia Press.

Figure 15.2 1898 Cartoon in the Philadelphia Press. Source: Wikimedia Commons (Public Domain), https://commons.wikimedia.org/w/index.php?search=spanish+american+war+cartoon&title=Specia l:Search&profile=advanced&fulltext=1&advancedSearch-current=%7B%7D&ns0=1&ns6=1&ns12=1 &ns14=1&ns100=1&ns106=1#/media/File:10kMiles.JPG

In Latin America, the Monroe Doctrine guided US policy for more than a century (Sexton 2011). It held that any foreign intervention there was to be considered a hostile act and effectively served to legitimate American intervention there, and became a foundational principle of American foreign policy. The doctrine established the Western Hemisphere as the rightful "sphere of influence" of the US and was invoked repeatedly by numerous presidents. It was used to justify armed interventions by American troops in Venezuela (1895), Panama (1903), the Dominican Republic (1905), Nicaragua (1909), Mexico (1914), Haiti (1915), and the Virgin Islands (1917).

The Height of Empire

World War II marked a decisive turning point in the history of American imperialism. Globally, the US assumed the hegemonic role once performed by Britain. Whereas its allies and enemies lay in smoking ruins, the US emerged unscathed, the landscape of its cities and infrastructure intact, its population largely free of the trauma visited on other countries. Domestically the post-war boom was reflected in rapidly rising standards of living, low unemployment, growing wages, and massive suburbanization. The Manufacturing Belt was the largest and most productive cluster of manufacturing firms on the planet.

New US industrial and financial landscapes reinforced at home this new international dominance while the country urbanized faster than any other in the world. Subsequently, the material skylines of cities throughout the country—particularly in the Northeast and Great Lakes regions, such as New York, Philadelphia, Chicago, Cleveland, and Detroit—became the manifestations of political and economic power and global capital circulation (Ford 1992).

During this period, with little foreign competition, the US enjoyed unparalleled political hegemony. *Time Magazine* Publisher Henry Luce triumphantly declared in *Life* the twentieth century the era of the Pax Americana or the "American Century" (Luce 1941).

To cement its position of leadership, the US established a number of organizations and treaties. These included the Bretton Woods system that established the World Bank, International Monetary Fund, and a system of fixed exchange rates in which most currencies were pegged to the US dollar, which was in turn pegged to gold at the rate of $35/ounce, a system that lasted until 1973. Similarly, the United Nations was founded in San Francisco in October 1945 as a means of fostering diplomacy and mobilizing world opinion. Many viewed it as a tool of US foreign policy.

In 1947, the US helped to form the General Agreement on Trade and Tariffs (GATT), a system designed to reduce protectionism worldwide, and, not coincidentally, open markets to US firms; GATT lasted until its mutation into the World Trade Organization in 1995. In 1948, the European Recovery Program, better known as the Marshall Plan, provided $12 billion in loans and grants to help rebuild the shattered economies of that continent and to prevent them from falling under the sway of the Soviets. In 1949, the US founded the North Atlantic Treaty Organization (NATO; Figure 15.3). In 1954, it created the Southeast Asian Treaty Organization (SEATO), which was dissolved in 1977. The US also signed a series of bilateral military treaties with the Philippines (1951), Australia and New Zealand (1951), South Korea (1953), Japan (1960), and Thailand (1966). In the Middle East, the US walked a fine line between support for Israel and alliances with various Arab regimes.

Two dimensions are fundamental to understanding late-twentieth-century American imperialism: neocolonialism and the Cold War. Neocolonialism differs from conventional

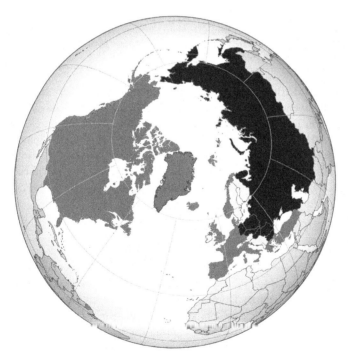

Figure 15.3   Border of NATO and Warsaw Pact in contrast to each other from 1949 (formation of NATO) to 1990 (withdrawal of East Germany). Source: Created from Natural Earth Data using gringer's perlshaper script. Author: Heitor Carvalho Jorge. Wikimedia Commons (Public Domain): https://commons.wikimedia.org/w/index.php?curid=88890272

colonialism in that it entails indirect economic, but not formal political, control. Into this scene stepped cash-rich transnational corporations, which exploded in size, number, and power after World War II. American transnational corporations were among the largest in the world. Much of American foreign policy during this era may be seen as a strategy to keep the world safe for American corporations. Thus, if old-fashioned colonialism was decisively European in nature, neocolonialism was in large part American by design.

Simultaneously, the US entered a rivalry with the world's other superpower, the Soviet Union, known informally as the Cold War. In 1946, George Kennan, the American *charge d'affaires* in Moscow, sent a famous 8,000-word telegram message to the government in Washington, DC warning about the dangers of the Soviet Union. In 1947, President Truman declared that US foreign policy would be centered on anti-communism, sketching a vision that became known as the Truman Doctrine. At the heart of the Cold War was the nuclear weapons race; the two superpowers controlled 98% of the world's most devastating weapons, enough to ensure "Mutual Assured Destruction."

The superpowers jockeyed for influence in the developing world, cultivating ties with diplomacy, foreign aid, trade programs, and offers of alliances as well as a protracted series of struggles to win allies among the newly independent states in Africa and Asia. While many states aligned themselves with one or the other, some countries skillfully played the two powers off against each other (e.g., Egypt, India, Ethiopia).

Although the Cold War never involved military conflicts between American and Soviet troops, it was manifested in a series of violent regional conflicts in which the superpowers supported opposing sides. The Korean War (1950–1953) was the first of these, although it ended in a stalemate. So too was the Vietnam conflict (1960–1975), which ended with the defeat of the US, its most important loss during the Cold War. The American landscape simultaneously became ever-more dotted with military bases as the military–industrial complex grew during this time, particularly in the American South where rural land remained predominantly Black-owned and therefore, due to racist land and property policies, less expensive and easier to claim as eminent domain.

Moreover, the US long supported a series of unsavory dictatorships around the world, many of which were highly militarized and anti-democratic in nature (Schmitz 1999), including Spain, Congo, Haiti, Bolivia, Nicaragua, Paraguay, Turkey, Greece, the Philippines, Indonesia, South Korea, Taiwan, South Vietnam, and Iran. Such regimes often engaged in brutal suppression of their own populations, censorship, and human rights abuses. Yet their sins were generally overlooked in the name of anti-communism.

If governments were not receptive to American advances, they could be overthrown by the CIA, as happened in Iran (1953), Guatemala (1954), Iraq (1963), Brazil (1964), Indonesia (1965), the Dominican Republic (1966), Cambodia (1970), Bolivia (1971), Chile (1973), Chad (1982), Grenada (1983), and Panama (1989). Indeed, regime change has been a foundation of American foreign policy for decades (Kinzer 2006). In the 1980s, the US intervened in civil wars in El Salvador and Nicaragua. This long series of interventions and regime changes earned the US enormous enmity around the world, undermining its self-proclaimed image as the bastion of democracy.

Militarily, the US in the twentieth century became the strongest geopolitical force in world history. To sustain its empire, it devoted a significant share of its public resources to the military. The budget of the US Defense Department—now more than $700 billion annually—exceeds that of the next 10 countries combined (it comprises one quarter of the world's total expenditures).

The Decline of Empire

The decline of American imperialism has been a long, gradual, and drawn-out process (Bello 2005). Some date the beginning of the decline of American imperialism to its defeat in Vietnam. The Bretton Woods Agreement came under increasing strain, largely due to the depletion of US gold reserves. In 1971, the US uncoupled the dollar from gold, and in 1973 the entire system collapsed.

Simultaneously, international competition began to grow. Japan became, briefly, the second largest economy in the world. Soon thereafter, the "Tigers" of South Korea, Taiwan, Hong Kong, and Singapore joined the fray. Numerous industries in the US began to move to East Asia, attracted by lower labor costs. In the late 1970s, the petrocrises witnessed the price of petroleum increase markedly, accelerating the process of deindustrialization in the US. By the 1980s, a new generation of tigers attracted American firms, including Thailand, Malaysia, and Indonesia. And of course, hovering behind them all, was China, which began eagerly to court foreign capital and companies.

The domestic repercussions of these changes were devastating, greatly altering the fabric of the American landscape. These changes were seen in widespread closures of factories, which turned the Manufacturing Belt into the Rustbelt in a remarkably short period of time. Once-thriving communities in the Northeast and Midwest witnessed a surge in unemployment, declining incomes, and falling property values. The deindustrialization of the US signaled the beginning of a prolonged period of wage stagnation and declining social mobility, leading some to question the vitality of the "American Dream." The middle class increasingly felt squeezed by stagnant incomes and rising prices for housing, education, and health care.

In the 1980s, under the Reagan Administration, the US (with Britain) pioneered the global triumph of neoliberalism, which has become arguably the most potent political force in the globe today. The central elements of neoliberalism include "free" markets; privatization and deregulation; and reductions in social services. In the US neoliberal policies led to significant rounds of tax cuts, which greatly benefitted the wealthy. As government spending on housing, education, and transportation declined, the social safety net upon which many poor and disadvantaged people relied became increasingly tattered (Gritter 2018; Dickinson 2019). The country has the highest level of income inequality in the economically developed world (Filauro and Parolin 2019). Class inequalities were magnified by racial and ethnic inequalities, exacerbating the wealth division between whites and people of color (Bruch et al. 2019). Another result was soaring federal budget deficits: by the late 1980s, the US had become the largest debtor in the world, heavily reliant on foreign funds (notably Japanese and Chinese) to finance its debt.

Because American imperialism is structured around the interests of corporations, trade agreements were a vital part of US hegemony. In 1994, the North American Free Trade Agreement went into effect, largely turning North America into one giant free trade zone. Economically, it merged the US landscape with that of its neighbors.

A dramatic turn of events in the late 1980s ushered in a new stage of American imperialism. In 1989, the Berlin Wall came down, signaling the sudden and unexpected collapse of global communism and the end of the Cold War. The collapse of Communism removed the last obstacle to American hegemony, leaving the US unquestionably the world's only superpower. Moreover, it served to legitimize neoliberal policies across the globe, offering "proof" that the American mode of organizing the economy was successful and inevitable.

Contemporary American Imperialism

In the twenty-first century, American imperialism assumed a new form: globalized, digitized, and financialized. The internet led to an historically unprecedented surge in global connectivity. The rise of China as the world's second-largest economy challenged the US role as the sole superpower. Corporate supply chains became increasingly complex and globalized, so that disruptions in one place reverberated throughout many others. Numerous non-state actors emerged on the global stage, including transnational drug cartels, Muslim fundamentalists (e.g., al Qaeda, Islamic State) non-governmental organizations, and ethnonationalist movements.

The September 11, 2001, attacks on the World Trade Center and the Pentagon revealed that opposition to US imperialism was widespread, intense, and organized. The American response was the "war on terror" (Fouskas and Gökay 2005), including the overthrow of the Taliban in Afghanistan in 2001 and the invasion of Iraq in 2003. The Iraq War cemented US hegemony in the Middle East, much as the conquest of the Philippines made the US an Asian power in 1898. Both the Iraq and Afghan wars proved to be diplomatic, military, and humanitarian disasters, earning the US widespread enmity across the globe (Burbach and Tarbell 2004).

American foreign policy centered on eliminating barriers to trade and capital movements. To accomplish these goals, foreign policy became heavily militarized. Diplomacy, multilateral coordination, and soft power all took a back seat to armed intrusions. Cyberwarfare became an integral part of the control of battle spaces across the globe. Reliant on sophisticated technologies such as drones, military interventions could be accomplished with few American casualties, as strikes in Pakistan, Syria, Yemen, Libya, and Somalia indicate. Such moves were often justified as pre-emptive, humanitarian, and strategically necessary (Sakellaropoulos and Sotiris 2008). A foreign policy based on military strikes reflected deeply embedded American hubris and led much of the world to view the US as a rogue state unbound by the normal rules of global geopolitics (Blum 2005b).

The status of American imperialism was greatly complicated under the presidency of Donald Trump (2017–2021). On the one hand, Trump espoused an isolationism antithetical to prevailing conservative orthodoxy, promoting an "America First" agenda, withdrawing from the Trans-Pacific Partnership, UNESCO, the Paris Climate Accords, the nuclear deal with Iran forged by President Obama, and the World Health Organization (in the midst of the coronavirus pandemic). He engaged in trade wars with China, Canada, and Europe. On the other hand, Trump did not engage in significant military interventions and withdrew many US troops from Iraq and Afghanistan. Trump was often contemptuous of allies and supportive of foreign autocrats, notably Russia's Vladimir Putin. He was particularly hostile to immigrants and refugees, and sought to build a wall along the border with Mexico.

Finally, it must be emphasized that the status of the American economy is foundational to its ability and incentive to maintain an empire. With 4% of the world's population, the US produces one quarter of world GDP. It is the only populous country that is also very wealthy. Sustained by well-capitalized corporations, business-friendly governments, a well-developed technology sector, and world-class universities, US productivity remains among the highest in the world. While it may be a wounded giant, the American economy remains the envy of much of the world.

The Cultural Dimensions of American Imperialism

One critical dimension of US hegemony lies within the broad domain of culture. As Nye (1990, 2004) famously noted, "soft power" is an important part of political rule. Soft power involves the ability to inspire admiration, to persuade, and to attract a loose following or allegiance. Because culture and power are inseparable, the role of American culture as a form of imperialism is important.

Racism was also an integral part of American imperialism. In the nineteenth century, racism essentially equated territorial expansionism (both domestic and foreign) with the benefits of white superiority and the spread of civilization (Love 2004; Doolen 2005). The conquest of non-white peoples abroad was mirrored in the suppression of people of color at home, particularly African Americans, and the landscapes of inequality that they inhabit. Although overt racism began to dissipate in the late twentieth century, it was mirrored in the intense Islamophobia that gripped the country in the twenty-first (Kumar 2012).

As American-style capitalism has spread throughout the world, it has had profound impacts on cultures everywhere. Barber (2003) notes American dominance in the "infotainment telesector," or the media, Hollywood movies, fashion, and other symbolically laden products and services. Among the mechanisms that allow this culture to spread globally are various aspects of the mass media and consumer culture—newspapers, magazines, the internet, music, television, films, fast-food franchises, and fashions. Thus, Hollywood movies and television shows are popular the world over; billions of people dress like American teenagers and eat at McDonald's; music from the United States can be heard all over the planet. American sports, such as baseball and basketball, have become popular in Latin America and Asia. The globalization of American culture to the global landscape has thus been a powerful product of the country's imperialism.

As the world's largest economic, military, and political power, the United States is simultaneously envied, imitated, and despised. Admiration for American culture is typically strongest among the young, who often associate Western culture with status, fun, sex, and hope; the elderly tend to be more traditional, so globalization creates a generation gap in terms of outlook and preferences.

Concluding Thoughts

Many Americans are reluctant to think of the US as an imperial country. Yet an empire it is. American imperialism has taken varying forms over time, depending on domestic and global political and economic circumstances. In the nineteenth and early twentieth centuries, it resembled European colonial powers with its conquests of Hawaii, Cuba, Puerto Rico, and the Philippines. After World War II, the US paved the way for neocolonial rule, in which overt political control is eschewed in favor of behind-the-scenes economic control, a strategy made necessary in part by the collapse of European empires and the Cold War with the USSR. In the late twentieth century, as the oil crises, financialization, and the microelectronics revolution reshaped the contours of the world-system, American imperialism reinvented itself yet again in the form of a digitized, neoliberal network. All of this was accompanied by a command performance in the domains of culture and ideology, giving the country enormous "soft power" that won the hearts and minds of billions, led to widespread imitation, and helped lubricate markets for American goods and services. While US foreign policy is often despised, American culture is frequently celebrated, often by the very same people.

The changing US position internationally was inescapably mirrored in changes in its domestic landscapes. These changes included territorial acquisitions either through wars or treaties. The rise and fall of the Manufacturing Belt paralleled the country's dominance after World War II and decline starting in the 1970s. Suburbanization reflected the nation's high standards of living, shift to a service-focused economy, and furthered a history of racial segregation. Finally, and a focus of several chapters in this text, areas of environmental alteration, borders landscapes, and economic vitality reinforce the intense social inequalities exacerbated by these neoliberal expansions.

References

Bacevich, A. 2004. *American Empire. The Realities and Consequences of US Diplomacy.* Cambridge, MA: Harvard University Press.

Balibar, É., Brandes, S., Brown, W., Cooper, M., Elyachar, J., Feher, M., Moodie, M., Newfield, C., Plehwe, D., Rofel, L., and Salzinger, L. 2019. *Mutant Neoliberalism: Market Rule and Political Rupture.* New York: Fordham University Press.

Barber, B. 2003. *Jihad vs. McWorld.* New York: Random House.

Bello, W. 2005. *Dilemmas of Domination: The Unmaking of the American Empire.* London: Zed.

Berman, M. 2006. *Dark Ages America: The Final Phase of Empire.* New York: W.W. Norton.

Blum, W. 2005a. *Freeing the World to Death: Essays on the American Empire.* Monroe, ME: Common Courage Press.

———. 2005b. *Rogue State: A Guide to the World's Only Superpower.* Monroe, ME: Common Courage Press.

Bruch, S., A. Rosenthal, and J. Soss. 2019. "Unequal Positions: A Relational Approach to Racial Inequality Trends in the US States, 1940–2010." *Social Science History* 43 (1): 159–184.

Burbach, R. and J. Tarbell. 2004. *Imperial Overstretch: George W. Bush and the Hubris of Empire.* London: Zed Books.

Burns, A. 2017. *American Imperialism: The Territorial Expansion of the United States, 1783–2013.* Edinburgh: Edinburgh University Press.

Chomsky, N. 2003. *Hegemony or Survival: America's Quest for Global Dominance.* New York: Henry Holt.

Dickinson, M. 2019. *Feeding the Crisis: Care and Abandonment in America's Food Safety Net.* Berkeley: University of California Press.

Doolen, A. 2005. *Fugitive Empire: Locating Early American Imperialism.* Minneapolis: University of Minnesota Press.

Ferguson, N. 2004. *Colossus: The Rise and Fall of the American Empire.* New York: Penguin.

Filauro, S. and Z. Parolin. 2019. "Unequal unions? A comparative decomposition of income inequality in the European Union and United States." *Journal of European Social Policy* 29 (4): 545–563. doi:10.1177/0958928718807332.

Foner, P. 1972. *The Spanish-Cuban-American War and the Birth of American Imperialism: 1898–1902.* New York: New York University Press.

Ford, L. 1992. "Reading the Skylines of American Cities." *Geographical Review* 82 (2): 180–200.

Fouskas, V. and Gökay, B. 2005. *The New American Imperialism: Bush's War on Terror and Blood for Oil.* Westport, CT: Greenwood.

Gritter, M. 2018. *Republican Presidents and the Safety Net: From Moderation to Backlash.* Lanham, MD: Rowman & Littlefield.

Hazarika, O. and V. Mishra. 2016. "South Asia as a Battleground." *World Affairs* 20 (3): 112–129.

Ignatieff, M. 2003. The Burden. *New York Times Magazine* (January 5). www.nytimes.com/2003/01/05/magazine/the-american-empire-the-burden.html

Immerman, R. 2010. *Empire for Liberty: A History of American Imperialism from Benjamin Franklin to Paul Wolfowitz.* Princeton, NJ: Princeton University Press.

Johnson, C. 2006. *Nemesis: The Last Days of the American Republic.* New York: Henry Holt.

Kinzer, S. 2006. *Overthrow: America's Century of Regime Change from Hawaii to Iraq.* New York: Henry Holt.

Kohut, A. and Stokes, B. 2006. *America Against the World.* New York: Henry Holt.

Kovalik, D. 2018. *The Plot to Control the World: How the US Spent Billions to Change the Outcome of Elections Around the World.* New York: Hot Books.

Kumar, D. 2012. *Islamophobia and the Politics of Empire.* Chicago: Haymarket Books.

Laidi, Z. 1990. *The Superpowers and Africa: The Constraints of a Rivalry, 1960–1990.* Chicago: University of Chicago Press.

Love, E. 2004. *Race over Empire: Racism and US Imperialism, 1865–1900.* Chapel Hill: University of North Carolina Press.

Luce, H. R. 1941. "The American Century." *Life.* February 17, 1941.

Mann, M. 2004. "The First Failed Empire of the 21st century." *Review of International Studies* 30 (4): 631–653.

Merry, S. 2000. *Colonizing Hawai'i: The Cultural Power of Law*. Princeton: Princeton University Press.

Nasser, N. 2021. "American Imperialism in Hawai'i: How the United States Illegally Usurped a Sovereign Nation and Got Away With It." *Hastings Constitutional Law Quarterly* 48 (2): 319–353.

Nye, J. 1990. Soft Power. *Foreign Policy* (80): 153–171.

———. 2004. *Soft Power: The Means to Success in World Politics*. New York: Public Affairs.

Panitch, L. and S. Gindin. 2004. "American Imperialism and Eurocapitalism: The Making of Neoliberal Globalization." *Studies in Political Economy* 71(1): 7–38.

———. 2009. "Finance and American Empire." In *American Empire and the Political Economy of Global Finance*, edited by L. Panitch and M. Konings, 17–47. London: Palgrave Macmillan.

Sakellaropoulos, S. and P. Sotiris 2008. "American Foreign Policy as Modern Imperialism: From Armed Humanitarianism to Preemptive War." *Science and Society* 72(2): 208–235.

Schmitz, D. 1999. *Thank God They're on Our Side: The United States and Right-wing Dictatorships, 1921–1965*. Chapel Hill: University of North Carolina Press.

Sexton, J. 2011. *The Monroe Doctrine: Empire and Nation in Nineteenth Century America*. New York; Hill and Wang.

Stokes, D. 2005. The Heart of Empire? Theorising US Empire in an Era of Transnational Capitalism. *Third World Quarterly* 26(2): 217–236.

Turner, F. 1920. *The Frontier in American History*. New York: Henry Holt.

van Apeldoorn, B., and N. de Graaff. 2012. "Beyond Neoliberal Imperialism? The Crisis of American Empire." In *Neoliberalism in Crisis*, edited by H. Overbeek and B. van Apeldoorn, 207–228. London: Palgrave Macmillan.

16

ICONOGRAPHIC LANDSCAPES OF US BORDERS AND IMMIGRATION CONTROLS

Kenneth D. Madsen

Throughout US history, immigration has played an important but often contentious role and in recent decades there has been renewed pressure to restrict immigration (Nevins 2002). In turn this has increased demand for alternative means of entry at US borders and ultimately the construction of border barriers to stop such traffic. Despite the loosening of many closed borders after the Cold War, new barriers soon took their place and 9/11 in particular led to a further dramatic shifting of national priorities in favor of enhanced national security and immigration controls. First initiated on a large scale by the United States in the second half of George W. Bush's presidency, border-wide barrier construction was then re-started with less bipartisan support but even greater intensity and resolve under the Donald Trump Administration.

Border barriers represent political winds and national priorities. They reflect but also inform how societies think about international borders, national security, and their relationship with other countries (Bissonnette and Vallet 2021; Jones 2012; Vallet 2014). They are also dramatic means with which elected officials can make political statements to constituents and potential migrants. Fences and walls are the focus of this chapter, but policies, rhetoric, and even mundane cultural references reflect collective political anxiety over immigration as well. The ideals and meta-narratives represented by these structures are fought over in the halls of power and between everyday citizens. Ultimately, these interactions over the landscape and the images which represent it are a proxy for greater discussion of social change and national ideals.

This chapter assesses the ways in which political messaging is embedded in these contested landscapes. Border barriers have been viewed through the lens of sovereignty (Brown 2010) and recent concerns over globalization (Rosière and Jones 2012), but the power of these landscapes in shaping national identity (after Clark 2004) is also central. By embedding ideological concerns in the architecture of the border for all to see, distilling them into a few representative images, and broadcasting this imagery with political messaging, border walls have become iconographic landscapes. Specifically, what these landscapes mean varies by political persuasion, but despite specific interpretation they convey strong messages nonetheless: defense or division, security or cooperation, protection or compassion, action or status quo. Debates over the meaning of border barriers reveal divergent understandings of US society, what it is, and what it should be.

DOI: 10.4324/9781003121800-21

Even as meaning is in flux, the divide in understanding these landscapes is often stark. Landscapes in this case are most certainly a record of past politics, but they are also intensely contested because to create, manipulate, or embed meaning in, or even simply maintain such a landscape, is also a means of controlling one's surroundings and shaping the future. In border barriers we see this play out on dramatic national and international scales that make explicit some of the basic concepts about the study of landscapes and their critical interpretation.

Traditionally, landscapes offer a relatively straightforward way of understanding the built environment (see Meinig 1979). But even in a traditional landscape analysis there is always more than meets the eye: forgotten histories to uncover, layers of meaning to make sense of, and connections to make explicit. Critical understandings build on this to give greater attention to power dynamics and the making of meaning. How landscape is used as a tool of domination is important here, but also the way in which it reflects and reinforces ideology. In the case of border barriers there exists a fundamental divide in American society. While some citizens argue that immigration concerns do not justify investing billions of dollars in such a project, others counter that the security clearly outweighs the costs. Whatever action is taken, however, becomes a landscape record largely frozen in place physically—even as it continues to be fought over as a representation—for years to come (Madsen 2011).

The contentiousness over border barriers became even more highly politicized and recognized during the 2016 Presidential campaign when barriers became an ideology stamped with a clear Trump identity and quickly solidified into a cult of personality and signifier of a promise kept to his base. Being controversial made barrier construction all the more perfect for Trump's cause, pointing out the lack of action by others, burnishing his populist credentials, and emphasizing his ability to follow through despite political gridlock. But such an approach also opened the door for opposition to coalesce around this work and stopping or minimizing such landscapes became symbolic of the political opposition's success, defeat, or compromised values. Trump was wildly successful in building on nationalist sentiments to generate support for barriers, but only among his base. The Trump campaign in 2015–2016 was so successful that some supporters did not fully realize the extent to which barriers already existed along key stretches of border and that long-term immigration was actually in decline (Warren 2020).

Iconographic Landscapes

Ultimately, political landscapes reflect existing spatially based power dynamics and an attempt to control the trajectory of political expression. As controversial topics, borders and their representation have often been played to political advantage. To better appreciate this aspect of the built environment, this chapter utilizes the concept of iconographic landscapes. While drawing on the concept as articulated by Cosgrove and Daniels (1988), discussed here the idea is more ideological and intentional in nature. This chapter uses the term to refer to a manipulated symbol which has a public visuality as important as any actual impact in physical space. That symbolic representation, furthermore, is as much the point as the original construction. In the case of border barriers, in many ways on-the-ground presence itself is a means to communicate power more so than a direct expression of control. These landscapes are political in intent as much as practice, and this frequently outweighs their interpretation as practical or even cultural and historical phenomena. To understand them in that way, it is worth thinking of the study of landscape as a type of ethnography (after Cosgrove and Daniels 1988, 4) that seeks to make sense of US politics.

A prime example of such an iconographic landscape is the showcasing of border wall prototypes early on during the Trump Administration (Figure 16.1). Built to help identify new, more effective barrier styles, it was a follow-up to wall construction promised by President

Figure 16.1 Border wall prototypes in San Diego County. Constructed in early 2017, these iconographic structures were removed in February 2019. Photograph © Jill Marie Holslin, October 2017, used with permission.

Trump during the 2016 election campaign and illustrative of his plans. The call for proposals specifically invited the development of new and innovative design features and requested that models be "aesthetically pleasing" on the US facing side and "consistent with the surrounding environment" (US Customs and Border Protection 2017, 62). In other words, not just a landscape in an academic sense, but *landscaped* in the vernacular use of that term. In these phrases the Trump Administration recognized the importance of these structures as political and cultural landscapes which reflected on his administration not just in terms of immigration policy, but also his promise for the work to be "great" and "beautiful."

The prototypes were surrounded by a media frenzy and hit all the right buttons as an iconographic landscape. Despite only being constructed in short segments in isolation and therefore serving little purpose as a physical deterrent, the spectacle was prime political theater intended for rallying support, communicating deterrence, and laying out a future agenda (Gibson 2020; Holslin 2020). Unfortunately for the Trump agenda, durability tests produced less than stellar results and, more significantly, they were never funded. From a practical perspective these two setbacks were paramount as they were motivating factors in construction. But they were also ultimately less central to the prototypes' *raison d'être* in iconographic terms as the demonstration project energized supporters and inspired political compromise which led to the construction of other barrier types.

As part of a political compromise in March 2018, $1.375 billion was allocated for "fencing" but reference to Trump's favorite description "walls" was conspicuously absent. More substantially, funding was restricted to "operationally effective designs deployed as of the date of the Consolidated Appropriations Act, 2017," i.e. prior to the development of the Trump prototypes (US Congress 2018, 616–617). The design restriction had been settled on the previous budget year but with a quadrupling of funds for barriers now became a fixed benchmark restriction on barrier construction. Due to political compromise and inertia, every congressional budget passed throughout the Trump Presidency maintained this dollar amount, although the Administration would later also re-direct other funds to augment the amount. Theoretically the secondary fencing design in San Diego County funded in Fiscal Year 2018 was exempt from this restriction, but it had to allow for "cross-barrier visual situational awareness" and that condition eliminated six of the eight protypes which were solid and not see-through at ground level (US Congress 2018, 617). Ultimately the prototypes were demolished in February 2019—they had served their purpose, had been degraded through examination, and stood in the way of expansion of a second and more consistent layer of fencing.

The prototypes were designed as iconographic landscapes both in terms of what they represented and what they were prohibiting from representing. Opponents of the aggressive Trump agenda understood that by prohibiting use of the prototype designs and refusing to even use the word "wall" in the budget bill, they were denying advancement of his plan and full realization of a campaign promise. Even though funds were allocated and that itself was a win for the administration, by restricting the use and development of new designs, Congressional compromise reinforced the message that there were many miles of previous barrier already in place and there was therefore less need for anything new. This also signifies the power of iconographic landscapes. As a means of communicating with voters, barriers effectively demonstrate "doing something" about undocumented migration (Bhagwati 1986, 124). Alternatively, as a means of communicating to potential migrants, barriers convey deterrence. The compromise still did that but emphasized it would not be done entirely on Trump's terms.

Given that most people are likely to encounter imagery of border barriers than the physical structures in-person, walls are as useful in their virtual as their physical incarnations. This is especially true outside of urban border areas where direct visibility drops significantly. In many ways, the visual reproduction of these iconographic landscapes in the news, on social media, and as backdrops for politician photo ops is an equally if not more important reason for such construction in the first place.

Even the prototypes themselves were relatively remote. Constructed on the outer edges of San Diego where they stood as a bulwark between the overwhelming expansiveness of Tijuana and undeveloped land on the US side, the location emphasized their grandeur and paralleled rhetoric justifying border barriers as protecting the US from being overrun by immigration. The prototypes never served as actual deterrents to entry and even afterwards were not connected and employed as part of a real-life strategy of securing the border given compromised integrity due to testing and the cost of integrating them into more extensive secondary fencing. Most importantly, retaining the prototype landscapes would have served as a monument to Trump's attempt or, alternatively, emphasized his inability to follow-through. Yet the imagery of the protypes projected power and confidence for a brief time when rounding up support for Trump's border wall initiatives was crucial.

Like the study of rhetoric, one of the advantages of using landscapes to understand our relationship with borders and immigration is that the manifestation is public for everyone to see. Whereas clear motivations, back-room compromises, policy research, and cultural significance are sometimes challenging to unearth, landscapes are dramatically intentional. Even so, what we observe is not unfiltered but shaped by visual media and accompanying rhetoric (e.g., Huysseune 2010). As a result, imagery is sometimes prone to being used in strategic ways by advocates/opponents rather than being as open and democratizing as we may think.

Utilitarian Origins

Before the prototypes, there were other types of iconographic border barriers. These landscapes communicated the message of deterrence and taking action to address problems loudly and clearly as well, even if it did not saturate the American electorate to the same extent given the emphasis of later campaign promises and social media. Precursors to today's border walls include barbed wire, chain-link fencing, and metal mesh (Gilewitch 1992; Madsen 1999). In many ways every US presidential administration (and sometimes each collective US Congress) in recent decades has tried to out-securitize its predecessors, each claiming they are doing more than those before them. This can take different forms—more border agents, tougher immigration laws/regulations, stricter enforcement, more deportations, more miles, taller walls, less porous walls, etc.—but the visuality of border barriers is often key.

In the 1990s surplus military landing mats (Figure 16.2) communicated a solid, more durable approach to border security than predecessor structures and became a dominant visual way of conceptualizing border security. By today's measures they were relatively easy to climb over (especially when panels were installed horizontally, thereby providing climbing footholds) and they had the shortcoming of not allowing law enforcement to easily see potential crossing activity on the other side. These designs were readily installed by Border Patrol agents themselves and later members of the National Guard on temporary assignment to the border. The origins of the material provided an aesthetic that was as helpful optically as it was practically—sending a memo draped in military olive-green of being serious about border control. As with the prototypes there were two distinct audiences, again with similar messaging. First, and

Figure 16.2 Landing mat barrier, Calexico, California. Once dominant among US-Mexico border barrier designs, over time these types of walls were increasingly replaced by taller and more transparent construction styles. Photograph by the author.

arguably most important, was the landscape message to a US domestic audience that something was being done to seriously address migration (and to a lesser extent drug smuggling). Second was the landscape message to intending migrants: don't try to cross, we are serious about stopping you.

While electronic outlets for conveying border barrier imagery were less ubiquitous at this time, imagery of the walls made it into the popular consciousness in other ways that did not depend on large numbers of people experiencing and seeing the walls in-person themselves. Traditional print media conveyed the imagery and intended communication of the landing mat walls. Some people also experienced the walls first-hand, especially in large urban areas like San Diego and El Paso, but also in smaller border cities where tourists and shoppers crossed regularly and within visual distance of the border barriers. Walls and crossing points can even be a point of tourist attraction, a marker of difference and badge of exploration (Timothy 1995).

None of this is to say that on-the-ground impacts are not important and some argue that barriers were particularly useful when strategically deployed in urban areas. But these impacts were largely local shifts to a national issue. Crossings moved from California to Arizona, from bigger cities to smaller cities, and from smaller cities to rural areas. Some traffic moved to tunnels, was smuggled through ports-of-entry, or simply found ways to breach weak spots in the barriers themselves. But as iconographic messages the structures were still landscapes that delivered political messages.

Vehicle Barriers

Vehicle barriers are largely utilized to control shifting traffic from more urban areas. Given their lower cost and lighter environmental footprint, vehicle barriers were once the design of choice in many rural areas of the US southern border. With long distances to areas with food, water, and shelter, pedestrian activity without vehicular support is less likely in these areas. Ironically, expanding construction to more remote areas also meant increased road construction, which was used by migrants and smugglers as well as law enforcement.

The two most common types of vehicle barrier are post-on-rail and Normandy (Figure 16.3). Both styles were initially largely made from re-used railroad tracks, although new I-beams and other steel components became more common over time as demand outstripped supply and funding budgets expanded dramatically. The Normandy barrier was particularly popular during the pre-Trump era in sensitive habitat such as rivers and steep hillsides. While sometimes displaced during flash flooding, it was easily reset or replaced with new materials as needed.

Architecturally, the Normandy structures draw on memories of war-time barriers and specifically the overwhelming strength of Allied Forces in World War II and the invasion of Omaha Beach. The name "Normandy barrier" itself communicates deterrence and force almost as much as the structure itself. Once the name–structure connection is established the visual impact of this landscape alone is sufficient to draw on war-time imagery. Ironically, however, it was the Germans rather than the Allies that used Czech Hedgehogs—three metal beams connected as a giant jack game piece—and they were only loosely the architectural predecessor of today's Normandy border barriers. Furthermore, these structures were designed for use against tanks more than other vehicles and, despite all this, the structures were still breached by Allies on D-Day.

While the name is significant here, Normandy barriers became symbolic of porosity for some and many were replaced by more impenetrable pedestrian barriers in recent years. Functionally the structures served their purpose, but perception is key in understanding an iconographic landscape and vehicle barriers eventually lost their allure. During the Trump era it was no

Figure 16.3 Normandy vehicle barrier, Antelope Wells, New Mexico. The barbed wire and mesh livestock fence at left approximates the official border. Note improved access road for construction, maintenance, and monitoring. Photograph by the author.

longer about a cost–benefit analysis, but a construction-at-all-costs mentality that insisted on creating and then meeting public demand for a more clearly impenetrable style of barrier.

"Operationally Effective Designs"

Just as the landing mat fences replaced chain-link and other less enduring barriers, other styles of pedestrian barriers slowly replaced the utilitarian designs of the 1990s and early 2000s even as new vehicle barriers continued to be added in rural areas. As with their predecessors, these new styles appeared first in urban areas and gradually extended outward—an ongoing manifestation that politicians were still doing something about migration and smuggling. Perhaps the most common format was tall, relatively thin (4–6"), and narrowly spaced bollards (Figure 16.4). These structures are often topped with a Mexico-facing anti-climbing plate. Some have wider spacing or other accommodations for the passage of wildlife or floodwaters in desert washes—or even gates to be opened during flood season and for the removal of accumulating debris upstream.

This fencing style became widespread during construction following the Secure Fence Act of 2006. Although initiated under George W. Bush, construction in this era was completed under the Obama Administration. Construction then largely went on hiatus as the Obama Administration increasingly focused on other enforcement capabilities, especially developing a reputation for dramatic increases in deportation.

A major advantage of the popular bollard design was the ability of law enforcement to see potential staging activity on the other side. Architecturally, these barriers also evoke a prison

Figure 16.4 Bollard pedestrian barrier east of Nogales, Arizona. Note flat anti–climb plate facing the Mexican side of the barrier. Photograph by the author.

setting and the increasing enforcement and deterrent-based focus of borders in contemporary politics. While the Border Patrol and barrier advocates welcome that landscape message, it is also central to the concerns raised by borderlands residents of an increasing level of militarization.

When the Consolidated Appropriations Act of 2018 took a fatal swipe at the Trump Prototypes, the tall bollard design eventually became adopted by Trump as the most ideal of the allowed pre-2017 "operationally effective designs." It is not clear that construction funded through redirected funds would have been subject to the same restrictions, but the bollard design was also largely used for construction funded through declaration of a national emergency on the southern border in February 2019. The design was relatively intimidating and the incarceration imagery perfect for appropriation as a campaign promise made and kept. Although it never happened, stylistically Trump occasionally made efforts to have it painted black to heat the metal and therefore increase the deterrent factor, but also to fulfill another campaign promise of making the wall beautiful.

The biggest Trump contribution to this design, however, was its ever greater and more extensive expansion. Today the bollard wall has expanded so far that pedestrian barriers are more common than vehicle barriers even in rural areas, including high mountainous peaks where one would be hard-pressed to justify the environmental damage undertaken to install border barriers. In one particularly dramatic case in southeastern Arizona, access roads tore deep into otherwise inaccessible areas to further construction (see imagery at Kapoor and Brocius 2020). In another situation, blasting in Organ Pipe Cactus National Park was set off as part of a demonstration for visiting reporters amid concerns it was damaging to important

Indigenous archeological sites. At the same moment as this event, the Tohono O'odham Nation chairman was testifying before Congress about the threat of border wall construction to such sites (see Arizona 360 2020; Devereaux 2020). In dramatic images and moving video clips that were repeated in the news and on social media, these landscapes and landscapes-in-the-making had another intended impact—communicating to supporters and opposition alike that forging ahead with construction plans in even the toughest terrain and despite criticism is reflective of the Administration's resolve on this topic.

Given the importance of border walls to his presidency and his emphasis on aesthetics, President Trump liked to commemorate his efforts by visiting important locations. These visits were memorialized with wall signing ceremonies and plaques to solidify the significance of the construction—the first section in Calexico, 100 miles in Yuma, 200 miles in San Luis, and the wrap-up of his Presidency in Alamo, Texas. Such actions were further confirmation of the important symbolism these landscapes had to his presidential legacy—iconographic representations of restrictive immigration policies.

The Challenges of River Borders

The flow of water creates special challenges for borders. Although Normandy barriers were utilized early on when the border intersected small washes that are normally dry, in the second half of the Trump Administration these structures were increasingly replaced by tall bollard fencing with flood-release gates. When a border is defined by an active river, however, it is a very different problem and consideration of the floodplain is paramount.

The most common approach is to situate barriers some distance from the political dividing line itself which is defined as the *thalweg* or deepest channel of the river. Styles commonly included the pedestrian bollard walls seen in other areas, but also a new design which was the result of a political compromise and disguised funding for border barriers in a unique layout more popular locally—the levee wall (Figures 16.5a and b). Representative Enrique Robert Cuellar of Texas House District 28 is closely tied to securing funding for such structures, allowing him to go on record as bringing home money for additional flood protection even as he opposed the type of border barriers constructed elsewhere and which were less popular among his constituents. This type of barrier replaced the river-facing slope of a levee with a vertical concrete wall and allowed for much-needed improvements to flood protection levees while by-passing eminent domain proceedings otherwise needed to secure placement of a border wall in South Texas (Collier et al. 2017).

Levee walls combined with border barriers on top. This raised the border wall function to a minimum security height without raising the height of the levees themselves which is regulated by the International Boundary and Water Commission to avoid displacing floodwaters into Mexico. It also reflected the border mission even as the add-on came across more as a pedestrian safety feature. While bollard designs of various heights were most in-keeping with barrier architecture elsewhere, in some stretches vehicle guardrails were installed to run alongside gravel access roads that also topped the levees. Chain-link fencing generally parallels the guardrails as a pedestrian-safety feature where needed given the steep drop-off on the river-facing side of the retrofitted levees.

Made in large part of concrete, the resulting structures are reminiscent of the Berlin Wall even more so than other barrier styles. Given structural elements (tall, imposing, solid, grey, in-person and electronic monitoring) it is an easily invoked comparison given the iconographic importance of the Berlin Wall. But while occasionally and intentionally deployed to make a negative statement about US barriers, the levee walls are also largely hidden from public view in the US. From a design perspective it is almost as if the intent is to *not* be seen. While on the

(a)   (b)

Figures 16.5 a and b   Levee walls in western Hidalgo County, Texas. Figure 16.5a: gravel access roads are visible above and below the levee wall. 16.5b: a security camera box designed as a bollard tip (source of shadow on the wall) is visible. Photographs by the author.

Mexican-facing side these walls function as a largely impenetrable pedestrian barrier for border security purposes and the anti-immigration message is clear, for most US passers-by these are anti-iconographic landscapes that hide themselves as gentle sloping levees. This is no small advantage given many area residents oppose border walls. In this context, the structures' low US-facing profile also minimizes opposition.

This issue is more acute in South Texas where barriers and large population centers overlap in relatively close proximity. In addition to their law-enforcement function, levee walls do impact local lives here—most notably limiting access to personal property and recreation sites—but the visual reminders are made less prominent with the use of levee walls in the area. In many areas, gaps were also initially left open as funding, time, and the drive for barrier construction fizzled out in the wake of the Secure Fence Act. Under Trump, construction re-started in an even more brazen manner—greater publicity, wide swaths of vegetation clearance, taller barriers—amplifying what were previously subtle landscape nods to securitization and giving it a greater significance. This is not just an effort to build more barriers, but also an attempt to raise the structures' national visibility in support of a signature effort of the president and despite opposition.

Conclusion

In studying the landscapes of US–Mexico border barriers as immigration controls, their iconographic importance—both at the border and in their reproduced images—can be understood as a political stand-in for restrictive immigration policies. These symbolic representations help us make sense of the built landscape at a new level (Cosgrove and Daniels 1988). Built to communicate deterrence as much as to deter and built to communicate to the US electorate (or, in the case or levee walls, soften such communication), border barriers are highly ideological and intentional not only in their placement and physicality, but in their public presentation and representation. In turn, these political landscapes inform how we think about borders and ultimately serve as proxies by which diverse actors understand and discuss the political ideals attached to such construction.

References

Arizona 360 2020. "Tohono O'odham Chair Decries Ongoing Controlled Blasts for Border Wall Project." *Arizona Public Media*, February 28, 2020. news.azpm.org/p/news-splash/2020/2/28/166878-tohono-oodham-chair-decries-ongoing-controlled-blasts-for-border-wall-project/.

Bhagwati, J. N. 1986. "U.S. immigration policy: What next?" In *Essays on Legal and Illegal Immigration*, edited by S. Pozo, 111–128. Kalamazoo: W.E. Upjohn Institute for Employment Research.

Bissonnette, A., and É. Vallet, eds. 2021. *Borders and Border Walls: In-security, Symbolism, Vulnerabilities.* New York: Routledge.

Brown, W. 2010. *Walled States, Waning Sovereignty.* Cambridge: MIT Press.

Clark, G. 2004. *Rhetorical Landscapes in America: Variations on a Theme from Kenneth Burke.* Columbia: University of South Carolina Press.

Collier, K., T. C. Miller, J. Aguilar. 2017. "How a South Texas Bureaucrat Became a Multimillionaire Amid the Rush to Build a Border Fence." *The Texas Tribune*, December 29, 2017. www.texastribune.org/2017/12/29/how-south-texas-bureaucrat-became-multimillionaire-amid-rush-build-bor/.

Cosgrove, D., and S. Daniels. 1988. *The Iconography of Landscape: Essays on the Symbolic Representation, Design, and Use of Past Environments.* New York: Cambridge University Press.

Devereaux, R.. 2020. "The Border Patrol Invited the Press to Watch it Blow up a National Monument." *The Intercept*, February 27, 2020. https://theintercept.com/2020/02/27/border-wall-construction-organ-pipe-explosion/.

Gibson, D. W. 2020 *14 Miles: Building the Border Wall.* New York: Simon & Schuster.

Gilewitch, D. A. 1992. *The Border Fence at El Paso, Texas: Symbolism, Perceptions and Effectiveness of the "Tortilla Curtain".* Unpublished M.A. thesis, Geography, University of Kansas.

Holslin, J. M. 2020. "Hackeo de la Frontera: Un Desafío a la Cultura del Espctáculo" (Border Hack: Challenging Spectator Culture) *Revista Espiral Tijuana*, September 25, 2020. https://revistaespiraltijuana.org/2020/09/25/70-13/. English translation available at www.jillmarieholslin.com/writing/border-hack-challenging-trump-s-spectator-culture.

Huysseune, M. 2010. "Landscapes as a Symbol of Nationhood: The Alps in the Rhetoric of the Lega Nord." *Nations and Nationalism* 16 (2): 354–373. doi:10.1111/j.1469-8129.2010.00436.x.

Jones, R.. 2012. *Border Walls: Security and the War on Terror in the United States, India, and Israel.* New York: Zed Books.

Kapoor, M. L., and A. Brocius. 2020. "In Arizona, building a wall — and destroying a canyon. *High Country News*, October 30, 2020. www.hcn.org/articles/borderlands-border-wall-construction-and-environmental-destruction-in-a-remote-arizona-canyon.

Madsen, K. D. 1999. *The U.S.-Mexico Border Fencescape Along the Arizona-Sonora Boundary.* Unpublished M.A. thesis, Geography, Arizona State University.

Madsen, K. D. 2011. "Barriers of the US–Mexico Border as Landscapes of Domestic Political Compromise." *Cultural Geographies* 18 (4): 547–556. doi:10.1177/1474474011418575.

Meinig, D.W., ed. 1979. *The Interpretation of Ordinary Landscapes: Geographical Essays.* New York: Oxford University Press.

Nevins, J. 2002. *Operation Gatekeeper: The Rise of the "Illegal Alien" and the Making of the U.S.-Mexico Boundary.* New York: Routledge.

Rosière, S., and R. Jones. 2012. "Teichopolitics: Re-considering Globalisation Through the Role of Walls and Fences." *Geopolitics* 17 (1): 217–234. doi:10.1080/14650045.2011.574653.

Timothy, D. J. 1995. "Political Boundaries and Tourism: Borders as Tourist Attractions." *Tourism Management* 16 (7): 525–532. doi:10.1016/0261-5177(95)00070-5.

US Congress. 2018. "H.R. 1625. Consolidated Appropriations Act, 2018." Public Law 115–141 (132 Stat.). March 23, 2018. www.congress.gov/bill/115th-congress/house-bill/1625.

US Customs and Border Protection. 2017. Other Border Wall RFP sam.gov/opp/893ca637e 9c19e4be3845dcd4567a1a9/view#attachments-links (HSBP1017R0023_Other_Border_Wall_IDIQ_ RFP_-_SF_1442_(003).pdf).

Vallet, É., ed. 2014. *Borders, Fences and Walls: State of Insecurity?* Burlington: Ashgate.

Warren, R. 2020. "Reverse Migration to Mexico Led to US Undocumented Population Decline: 2010 to 2018." *Journal on Migration and Human Security* 8 (1): 32–41. doi:10.1177/2331502420906125.

17

(DE)CONSTRUCTING THE GRID

The Public Land Survey System and the Production of Abstract Space in the Early Republic

Timothy G. Anderson

"If you want to get to the west quarter, drive three miles west and then one mile south; the entrance to the field is halfway down the section line." In my adolescence in rural Oklahoma, this is the language everyone used to communicate directions to a given location; the use of the cardinal directions and terms like "quarter" and "section" as instruments of spatial orientation was commonplace. I did not fully comprehend it then, but such terminology dates from the late eighteenth century and is rooted in the lexicon of what came to be known as the Public Land Survey System (PLSS), a distinctive method of land survey and subdivision crafted by the federal government in 1785 and subsequently implemented in most regions of the United States west of the Appalachians. My local milieu was bounded and delimited within a *regular*, gridded landscape of straight lines with roads and property boundaries intersecting at right angles, oriented north–south–east–west (Figure 17.1). For those who spend much of their lives in an environment organized by means of such systematic boundaries, to encounter or live in a place where comparable rigid spatial organization is largely or even fully absent can lead to a very real sense of spatial disorientation, of being "out of place."

Anyone who has flown over the American Midwest or Great Plains can observe the most conspicuous, outward feature of the PLSS, often informally referred to as the "Township/ Range" system: a vast pattern of roads and property boundaries intersecting at right angles, oriented to the compass directions, fashioning a continental-scale checkerboard geometry upon the landscape. The federal government implemented the PLSS nationally by means of a series of land surveys built around the grid of latitude and longitude lines as the federal government acquired and added new tracts of territory. The division of space into such rectilinear grids as a method of land survey and subdivision is most certainly not an "American" innovation. The most commonly cited Old-World antecedents are the Roman orthogonal system of centuriation and the gridded urban plans of some Roman (Palet and Orengo 2011) and Greek (Mazza 2009) cities. But scholars have documented gridded urban arrangements dating from antiquity in numerous Old- and New-World pre-colonial settings as far afield as the Indus

DOI: 10.4324/9781003121800-22

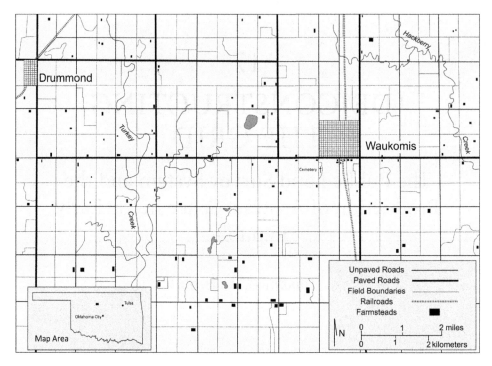

Figure 17.1 Road and cadastral patterns in a portion of Garfield County, Oklahoma. Google; Landsat/Copernicus. Map by Author.

Valley (Coningham and Young 2015), Mesopotamia (Algaze 2009), China (Knapp 2000), Japan (Johnson 1976), West Africa (Bigon and Hart 2018), and Mesoamerica (Gasparini 1993).

In an oft-cited early essay, geographer Dan Stanislawski (1946) identified the ancient city of Mohenjo-Daro in present-day Pakistan as likely the earliest (c. 3,000 BCE) example of a town deliberately planned on a grid pattern of intersecting streets and maintained that the idea of designing cities using an orthogonal plan subsequently diffused to India, Greece, and Rome. It should be noted, however, that Stanislawski was writing during an era in which many American geographers were preoccupied with pinpointing the original, singular source of cultural phenomena and with documenting the diffusion of innovations from an original source through time and across space. More recent scholarship calls into question the reasoning that underpins such quests to identify penultimate origins. James Blaut (1987), for example, argues that such thinking essentializes, privileges, and mythologizes the "original" at the expense of a more nuanced understanding of why innovations are adopted and how they are disseminated. Reuben Rose-Redwood refers to this as "doctrinaire diffusionism," and insists that even if the original source of an innovation can be determined, we should not assume that all subsequent adaptations are "imperfect" versions of the original form (Rose-Redwood 2008, 43). Following this line of thought, this essay forgoes any attempt to identify a specific antecedent of the PLSS grid. Instead, the focus here is on identifying how its implementation affected such tangible landscape features as road patterns and cadastral and administrative boundaries. And following in the vein of contemporary scholarship that scrutinizes cultural landscapes within the context of post-structural formulations, there is no attempt to isolate any "essential" meaning encoded within the PLSS. Rather, it is assumed that a multiplicity of meanings and discourses are imbedded in its use as a method of land survey and subdivision in the United States; several possible meanings are forwarded here for consideration. The essay begins with an overview of the genesis and implementation of the system set against the backdrop of federal land policies during the Early National period. Next, examples of visible impacts on settlement patterns and landscapes are presented. Finally, ideas of the production of space are employed as frameworks for identifying and understanding specific historical-political discourses encoded in the PLSS's rectilinear survey landscape, and as context for a more nuanced interpretation of the PLSS as a material manifestation of Enlightenment idea(l)s and as a striking example of spatial abstraction.

Origins and Implementation of the PLSS
Context and Earlier Survey Systems

The commercialization of land—the idea that land was a commodity to be bought and sold—was the underlying basis of European attitudes, traditions, and legal frameworks regarding land tenure and subdivision and was a central tenet of European overseas ventures operating within the framework of merchant capitalism during the colonial era; it arrived in the Americas with the very first Europeans. Vestiges of the ancient feudal system of land tenure, characterized by a highly uneven distribution of land and wealth concentrated in the hands of the aristocracy and the Church, remained codified in law and practice at the time of the initial colonization of North America. As such, many scholars maintain that Europeans equated land ownership with "independence," and in the case of the landless peasantry this meant independence from strict subject bonds as tenants or laborers within feudal social relations of production rather than political independence from king or state (Hofstra 2004). In commodifying the land, Europeans introduced a variety of land tenure and land subdivision methods and traditions to North America during the colonial era, producing a complex pattern of contrasting systems of land survey in different regions of the continent. Although now somewhat dated, Edward Price's (1995) detailed historical and regional account remains the single best source for insight into the assortment of survey systems employed in North America since colonial times, but Johnson's (1976) classic treatise and Hubbard's (2009) more recent analysis both provide good overviews of these systems as well. This mosaic of different survey systems in use prior to the Revolution likely played a key role in the federal government's desire to establish a standardized national system of land survey (Webster and Leib 2011).

In what is now the American Southwest, the Spanish outreach from the base of operations in the central highlands of Mexico that began in 1598 with colonization efforts in the upper Rio Grande valley of New Mexico was organized within the context of a highly formalized system of conquest, with significant "top-down" control over settlement processes and land tenure practices involving large tracts of land granted to settler families. Here, the Spanish (and later Mexican) authorities instituted survey systems that balanced local semiarid environmental conditions with the need to equitably partition irrigable land among colonists, including the division of riverine agricultural holdings into long, narrow strips (known as "long-lots") with frontage on rivers and irrigation ditches (Carlson 1975). In Texas, a unique and complex settlement history that took place under the auspices of several polities involving a variety of disparate population groups, produced a complex admixture of survey and land subdivision systems, including long-lot surveys and both "regular" (rectilinear) and "irregular" (non-rectilinear) systems (Jordan 1974). As such, Texas exhibits without a doubt the greatest variety of land survey systems of any state.

By contrast, the settlement of the American East by northwest Europeans during the colonial period, led by individuals, families, companies, and collectives operating within the paradigm of merchant capitalism, was much more experimental, diverse, and ambiguous in nature (Meinig

1986). In Tidewater Virginia and Maryland, for example, the practice of issuing patents (grants) of land to private individuals in the form of headrights emerged as the dominant method for alienating relinquished Crown lands. A headright was not a grant for a specific piece of land, but rather the right to claim any piece of land that had not been previously claimed. In the absence of any regular, standardized system of land survey, colonial surveyors employed irregular "metes and bounds" survey methods to delimit headright and plantation boundaries. In this system of land subdivision, surveyors described property boundaries using natural geographic features such as water courses or topographical features ("metes") as demarcation points that were then connected together by straight lines ("bounds"). This resulted in a rather chaotic settlement process without much oversight and often led to overlapping surveys that triggered legal disputes. Today, the use of this survey method remains evident in regions of the country settled prior to the adoption of the federal PLSS system (Price 1995). In colonial New England, while the legal process for establishing new towns was rather uniform, a standard system of land subdivision was never instituted. Likely based on English antecedents, the "town" (a large tract of land containing private agricultural holdings and common fields), together with the nucleated "village" at its center, emerged as the dominant settlement pattern. Although the "town and village" nucleated settlement pattern emerged as a highly romanticized New England ideal, local economic, social, and environmental conditions often hindered its full implementation (Wood and Steinitz 1992). Similarly, in colonial Pennsylvania, local conditions and historical circumstance thwarted William Penn's plan for a systematic and orderly arrangement of settlements bound together by a regular and methodical cadastral pattern that would be surveyed prior to settlement. Many of the earliest settlers "squatted" on tracts of land before obtaining legal warrants for them and arranged for surveys whose boundaries were often described using metes and bounds, resulting in a highly irregular cadastral pattern of dispersed farmsteads that can still be observed in the landscape of southeastern Pennsylvania today (Lemon 2002).

Iconic land tenure and subdivision patterns also emerged in New France owing to the implementation of the seigneurial system, a vestige of the French feudal system of land tenure. In this system, the Crown granted large blocks of land to *seigneurs* (manorial lords under the feudal system) in return for the promise to pay for the immigration of landless tenant farmers (known as *habitants*) to whom they would grant portions of their holdings for agricultural purposes. With respect to land subdivision, the most idiosyncratic feature of the system was the practice of subdividing seigneuries into individual farms in long-lot fashion in which long, thin strips of land were cleared from the forest, each with frontage on a river or stream. This system was first implemented in the St. Lawrence valley, but with time expanded into the interior via the Great Lakes, the Illinois Country along the Ohio and its tributaries, and eventually down the Mississippi to Louisiana as French colonists established trading posts, forts, missions, and small agricultural settlements in claimed territories in the North American interior. In locales where such settlements were established, the practice produced permanent impacts on local cadastral patterns and road networks as they were incorporated later into the PLSS. Examples of this can be observed at Detroit and the River Raisin region in Michigan (Figure 17.2), Kaskaskia, Illinois, Vincennes, Indiana, and, most extensively, in the lower Mississippi river valley north and west of New Orleans (Gentilcore 1957; Harris 1984).

The Land Ordinance of 1785 and the US Public Land Survey System

In addition to ending the Revolutionary War and recognizing the United States as a sovereign, independent state, the Treaty of Paris, signed in late 1783, also established the political boundaries of the new republic, transferring hundreds of millions of acres between the Appalachians

Figure 17.2 Road and cadastral patterns reflecting the influence of French long-lot surveys in a portion of Monroe County, Michigan. Google; NOAA. Map by author.

and the Mississippi River from Great Britain to the United States. From the very beginning, the federal government faced an array of highly complex issues related to the administration, management, and control of such a vast territory, compounded by the subsequent addition of even more territory as a consequence of the Louisiana Purchase (1803) and the Mexican–American War (1846–1848). First, although many Americans at the time considered these lands to be largely primeval, they in fact comprised the ethnic homelands of hundreds of disparate Indigenous population groups who had occupied the continent for thousands of years. As such, the federal government needed to address the issue of territorial claims on the part of these tribal groups. Second, Congress had to manage conflicting claims to territory west of the Appalachians by several eastern states resulting from some of the earliest "sea-to-sea" colonial charters that recognized the Pacific Ocean as their western boundary. Finally, because a central goal was to "alienate" these public lands (collectively referred to as "Congress lands" or the "public domain") into the hands of private individuals, the first Congress grappled with the problem of how to achieve this in an orderly and "practical" way (Anderson 2014).

The distressing, violent history of the forced Anglo-European/American expulsion of Native American populations from their ancestral lands is well documented. Regarding Indigenous territorial claims, the American government essentially continued the former British policy of "excluding" Indians from their territorial homelands through a series of military ventures and quasi-legal treaties that eventually resulted in the cession of virtually all claimed ancestral territories (Meinig 1986). The relatively recent paradigm of "settler colonialism" seeks to reframe this history by situating the settlers themselves as central actors in the record of Indigenous dispossession. Walter Hixson provides a comprehensive summary of this argument, contextualizing settler colonialism as "a history in which settlers drove indigenous [sic] populations from the land in order to construct their own ethnic and religious national communities" and viewing Anglo-American settlers as colonizers whose aim was to "occupy the land permanently." This permanent occupation of territory and the creation of structural legal and cultural frameworks for achieving it distinguished settler colonialism from "conventional" colonialism. Because "Indians participated at every level of the colonial encounter and, contrary to settler fantasies, the indigenes did not 'vanish'," settler colonialism assigns agency to Indigenous peoples as central actors—rather than passive participants—in the triangular relationship between settlers, the state, and indigenes that the paradigm articulates (Hixson 2013, 4–5, 197). Regarding land claims by eastern states in the areas north and west of the Ohio River (collectively referred to as the "Northwest Territory"), Congress eventually won the concession of nearly all these claims after a complex and sometimes contentious series of negotiations, in return for federal assumption of the states' war debts. These territories subsequently became part of the "public domain" of the federal government (Thrower 1966; Figure 17.3).

Finally, in negotiating a solution to the challenge of the orderly dispossession of public domain lands into the hands of private individuals, the federal government eventually adopted a systematic, standardized system of rectilinear survey and subdivision tied to the geographic grid of latitude and longitude. The process of negotiating this solution was lengthy, complex, and at times contentious. From the viewpoint of the federal government, the value and utility of its public domain lands shifted over time, but in the beginning, it understood them chiefly as a source of revenue to be sold to individual settlers and investors with the aim of replenishing the treasury. There were experimentations with selling large tracts of land to private land speculation companies, especially in western New York and Ohio, but few of these ventures were economically successful and did not raise the amount of revenue officials anticipated. Congress eventually decided on a system under which the federal government would sell tracts of the public domain directly to private individuals. A committee headed by Thomas Jefferson in

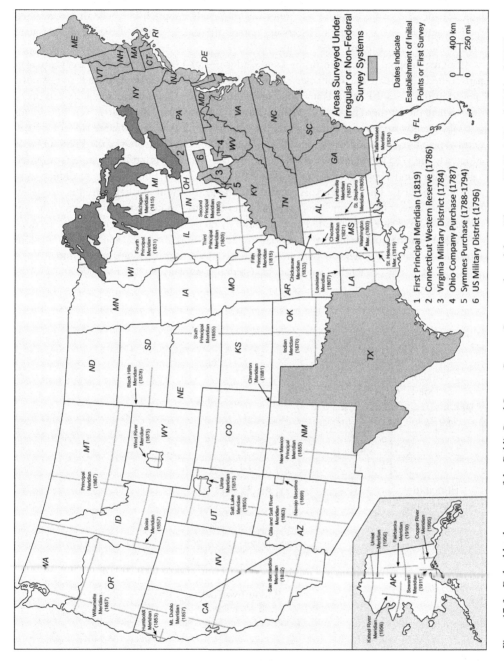

Figure 17.3 Federal land surveys of the Public Land Survey System. Sources: adapted from Thrower (1966, 6–7) and Hubbard (2009, 309). Map by author.

1784 drafted a plan for the survey of public domain lands into one-hundred-square-mile tracts, further subdivided into one-square-mile lots that would be sold at public auction for $1 per acre. Disagreement among members of Congress regarding this system, especially the ideal size of lots, however, eventually resulted in a system that was not based on decimal measures of land (a basic grid of thirty-six square miles was ultimately adopted) but nevertheless retained the fundamental characteristics of Jefferson's system, that is a rectilinear survey grid and square subdivision of land (Pattison 1957; White 1983).

Congress adopted this system in the Land Ordinance of 1785, and its basic precepts remain part of the PLSS today. The ordinance called for the survey of public domain lands and was first applied in the so-called Seven Ranges in present-day eastern Ohio, beginning at an imaginary point where the southern boundary of Pennsylvania intersected with the Ohio River. The system was subsequently implemented throughout the country in thirty-seven different surveys, undertaken as new territories and states were added (Figure 17.3). The basic unit of the PLSS survey as adopted in the 1785 Ordinance was the thirty-six-square-mile "township." Each survey comprised a grid built around a beginning north–south line (the "principal meridian") and a beginning east–west line (the "base line"). Surveyors constructed the grid around north–south lines (township lines) and east–west lines (range lines) every six miles from the "initial point" where the principal meridian and base line intersected. The basic unit of land became the one-square-mile "section" (640 acres), which could be further subdivided into half sections (320 acres) and "quarter" sections (160 acres). The government revised the system for numbering townships several times, but a 1796 act of Congress established a permanent system that remains in place today (Figure 17.4).

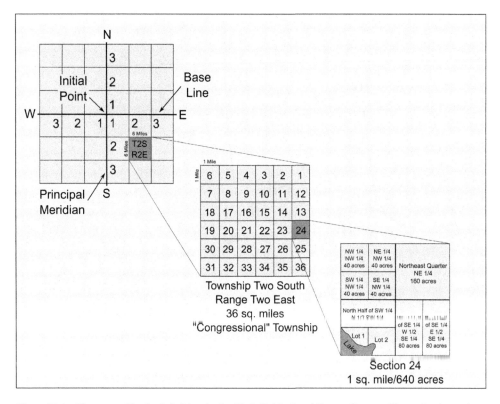

Figure 17.4 Elements of land subdivision in the U. S. Public Land Survey System. Illustration by author.

The Public Land Survey System became the legal standardized method of land survey and subdivision throughout the country, with the exception of areas settled prior to 1785 under previous survey systems (discussed above): the original thirteen states; Kentucky, Tennessee, and West Virginia; Hawaii; parts of New Mexico and Texas initially settled under Spanish jurisdiction; parts of Louisiana, Wisconsin, and Michigan originally settled under French rule; and the Virginia Military District in Ohio (Figure 17.3). Non-federal rectilinear surveys that are not part of the PLSS system also appear in a handful of locales as a result of unique, local settlement circumstances. Examples include the Holland Land Company purchase in western New York (Wyckoff 1986), Texas' own rectangular survey system instituted in much of the western and northwestern part of that state, and the Connecticut Western Reserve in northeast Ohio. Canada also adopted the US PLSS for the survey of its "dominion lands" in the prairie provinces. The Canadian survey also employs the basic thirty-six-square-mile township, but instead of multiple surveys the system begins at a principal meridian and then "restarts" at meridians spaced four degrees east and west of that meridian (Hubbard 2009).

American geographers have devoted considerable attention to assessing the outward, tangible effects of the PLSS survey in the American cultural landscape, documenting its impact on such features as field patterns (Hart 1968), road networks and administrative boundaries (Thrower 1966), cadastral patterns, and the size and shape of agricultural holdings (Johnson 1976). In their concise summary of this long line of empirical research on the subject, Webster and Leib contend that the implementation of the PLSS has outwardly impacted the American landscape in a number of ways, including the nature of the country's transportation network, the subdivision of the rural landscape into square parcels, the size and shape of county boundaries, street and road naming, and the grid's apparent disregard for topography and local environmental features. Perhaps most profoundly, they argue that the implementation of the PLSS on a national scale, with the square as its most basic geometric form, profoundly affected the nation's "sense of symmetry" (Webster and Leib 2011, 2129).

Re-Reading the PLSS

Materialized Discourses and the Production of Space

While there can be little doubt that the application of the PLSS grid on a continental scale produced concrete, palpable stamps on the American landscape, a preoccupation with documenting the material manifestations of "culture" by an earlier generation of geographers produced a tendency to overlook potential meaning that might be encoded within such features and within institutions such as the PLSS. Thinking about the cultural landscape in this way, how might we "read" or interpret the PLSS within the framework of more recent geographical scholarship and ideas of spatialization? A common theme in cultural geography scholarship that has grown more explicit in recent years concerns the "authorship" of cultural landscapes, especially with regard to how and by whom they are reproduced, and the meanings embedded within them. Richard Schein (1997), for example, draws upon James Duncan's (1990) concept of discourses encoded in space and place in offering an alternative "framework" for evaluating any given landscape. Duncan argues that decisions people make about the use of space on an everyday basis reflect attitudes, beliefs, and perceptions, which he refers to as "discourses," defined as "shared meaning[s] [and] ideologies which are socially constituted." Schein has shown how such discourses are embedded within cultural landscapes, demonstrating that decisions made by individuals and groups of individuals reflect a variety of discourses and serve to create and modify landscapes, becoming expressions of such discourses in the form of tangible landscape elements. In this way any cultural landscape "can itself capture different, even competing, sets of meaning, or independent, thematic networks of knowledge ... [E]ach seemingly individual decision behind any ... landscape is embedded within a discourse. When the action results in a tangible landscape element, or total ensemble, the cultural landscape becomes the *discourse materialized*." In this way, any given cultural landscape element can be contextualized as a "material component of a particular discourse or set of intersecting discourses" (Schein 1997, 663).

Henri Lefebvre's scholarship regarding the production of space offers another perspective on the meanings encoded in landscapes that when combined with Schein's (and Duncan's) concept of materialized discourses allows for an even deeper understanding of social and political discourses embedded within the PLSS grid. The key contribution of Lefebvre's *The Production of Space* ([1974] 1991) is the idea that space is socially produced through the human fabrication of social relations of production and the assignment of use-value to goods and services. For Lefebvre, the "social" spaces within which social relations of production work are defined by a set of fundamental dialectics and dualities. Social space is "itself the outcome of past actions," but is at the same time "what permits fresh actions to occur, while suggesting others and prohibiting yet others." It is at once "a *field of action* ... and a *basis of action* ... a collection of *materials* ... and an ensemble of *matériel*" (emphases in the original) (Lefebvre [1974] 1991, 73, 191). Lefebvre theorizes a tripartite production of space predicated on the dialectical relationships between three classes of space: "spatial practice," the cohesive patterns and spaces of routine social activity; "representations of space," the spaces conceived by engineers, architects, cartographers and the like that configure a system of signs and codes used to organize and direct spatial relations; and "representational space," the "underground" spaces of the imagination envisioned by artists and others who seek to portray alternative spaces (Lefebvre [1974] 1991, 36–46). Further, Lefebvre devotes many pages to how spatial experience has shifted over time, arguing that in the West prior to the Early Modern era space and time were "absolute" in nature, experienced and manifested in large part through "local, lived conditions." The Renaissance, however, engendered the development of complex mathematical systems that "allowed space to be broken into fixed units which could be mapped ... establishing a system of abstraction allowing for exact measurement and location" and led to the development of what Lefebvre labels "abstract" space (Lefebvre [1974] 1991, 229–291). David Harvey extrapolates on this concept, noting that the revolutionary changes in the conceptualization of space and time during and after the Renaissance produced the "conceptual foundations" for the Enlightenment and its thinkers, for whom the "conquest" and "rational ordering" of space was a strategic instrument of power and authority. The rational and empirical measurement of time and space employing chronometers and maps as "abstract and strictly functional systems for the factual ordering of phenomena in space" became key to the Enlightenment model of state organization, administration, and social control (Harvey 1989, 249; Anderson 2020).

A Multiplicity of Meanings and Discourses

When viewed within the framework of contemporary theories of spatialization, the social and political significance encoded within the PLSS grid becomes clearer, and reveals a multiplicity of meanings, discourses, and implications associated with its implementation, three such possible meanings are offered here by way of conclusion. First, Lefebvre's ideas on the production of space help us understand that the PLSS represents an abstraction of space—and as such as a materialization of Renaissance and Enlightenment idea(ls) regarding the (re)ordering of space—on a grand scale. For Anglo-Americans, the grid was a means of imposing order upon the disordered spaces that lay beyond a frontier that denoted a spatial demarcation between order and chaos, security and insecurity. Such notions of the frontier and of the "West" as a place of "savagery" (occupied by "savage" peoples) are deeply ingrained in the country's imagination and in the tropes associated with our frontier past. In a very real sense, then, the grid imposed occidental notions of "law and order" upon the landscape and was (and still is) the template upon which such notions continue to be reproduced (Blomley 2003). Contextualized in this way, it can be argued that the PLSS grid represented a reorganization of "the framework through which social power is expressed" (Harvey 1989, 255).

Second, the PLSS grid was an instrument of governmentality that imposed both a literal and figurative framework for the territorial control of the public domain and for the social control of the people living within it. In the process, western notions regarding the commodification of land and the legal apparatuses for its appropriation and control became codified as normative ideals. As Matthew Hannah has argued, territorial organization is key to social control, and the ultimate goal of governmentality is the control of territory in order to exact social control over the state's population. Seen in this way, the PLSS grid was both a means for, and an end product of, "territorial integration" on the part of the federal government (Hannah 2000, 39). Pier Aureli contends that the grid "was an all-encompassing system that coalesced geometrical order, surveying, and financial value into one efficient apparatus" (Aureli 2018, 156). In this way, the PLSS can be interpreted as a reconfiguration of space in order to impose a legal framework for supporting capitalist production and exchange. That is, the grid provided a systematic, "geo-coded" legal structure that facilitated private ownership of parcels and capitalist agricultural enterprises, both of which were central goals of the state (Rose-Redwood 2006, 479).

Finally, in a "winner-takes-all proposition that demanded the removal of indigenous [sic] peoples and the destruction of their cultures," the PLSS was an instrument of dispossession and settler colonialism that facilitated the systematic, state-sanctioned removal of Indigenous peoples through the design and implementation of a survey system designed to replace them methodically and systematically by redistributing the public domain into the hands of Anglo/Euro-American settlers (Hixson 2013, 197). Ironically, because "citizenship" was from the beginning of the American experiment predicated on the ownership of land, the removal of Indigenous peoples and the opening of their ancestral lands to private ownership came to be seen as the "precondition for stable and open democracy" (Cosgrove 1996, 8). Hidden within the rigid, abstract geometry of the PLSS grid is a violent history of "appropriation and the consolidation of that appropriation through subdivision" by which "civilization ... [was brought to] a land imagined as populated by people incapable of such spatial mastery." As an instrument of American governmentality that facilitated settler colonialism, the ubiquity of the grid in the American cultural landscape thus masks its central role in the commodification of the land (Aureli 2018, 157–158).

References

Algaze, G. 2009. *Ancient Mesopotamia at the Dawn of Civilization: The Evolution of an Urban Landscape.* Chicago: University of Chicago Press.

Anderson, T. G. 2014. "Dividing the Land." In *North American Odyssey: Historical Geographies for the Twenty-first Century*, edited by C. E. Colten and G. L. Buckley, 209–226. Lanham, MD: Rowman & Littlefield.

Anderson, T. G. 2020. "Cameralism and the Production of Space in the Eighteenth-century Romanian Banat: The Grid Villages of the Danube Swabians." *Journal of Historical Geography* 69: 55–67.

Aureli, P. V. 2018. "Appropriation, Subdivision, Abstraction." *Log* 44: 139–167.

Bigon, L. and T. Hart. 2018. "Beneath the City's Grid: Vernacular and (Post-)colonial Planning Interactions in Dakar, Senegal." *Journal of Historical Geography* 59: 52–67.

Blaut, J. M. 1987. "Diffusionism: A Uniformitarian Critique." *Annals of the Association of American Geographers* 77 (1): 30–47.

Blomley, N. 2003. "Law, Property, and the Geography of Violence: The Frontier, the Survey, and the Grid." *Annals of the Association of American Geographers* 93 (1): 121–141.

Carlson, A. W. 1975. "Long-Lots in the Rio Arriba." *Annals of the Association of American Geographers* 65 (1): 48–57.

Coningham, R. and R. Young. 2015. *The Archaeology of South Asia: From the Indus to Asoka, c. 6500 BCE–200 CE.* Cambridge: Cambridge University Press.

Cosgrove, D. 1996. "The Measure of America." In *Taking Measures Across the AmericanLandscape,* edited by J. Corner and A. S. MacLean, 3–13. New Haven: Yale UniversityPress.

Duncan, J. S. 1990. *The City as Text: The Politics of Landscape Interpretation in the Kandyan Kingdom.* New York: Cambridge University Press.

Gasparini, G. 1993. "The Pre-Hispanic Grid System: The Urban Shape of Conquest and Territorial Organization." In *Settlements in the Americas: Cross-Cultural Perspectives,*edited by R. Bennett, 78–109. Newark: University of Delaware Press.

Gentilcore, R. L. 1957. "Vincennes and French Settlement in the Old Northwest." *Annals of the Association of American Geographers* 47 (3): 285–297.

Hannah, M. 2000. *Governmentality and the Mastery of Territory in Nineteenth-Century America.* New York: Cambridge University Press.

Harris, R. C. 1984. *The Seigneurial System in Early Canada: A Geographical Study.* Kingston: McGill-Queen's University Press.

Hart, J. F. 1968. "Field Patterns in Indiana." *Geographical Review* 58 (3): 450–471.

Harvey, D. 1989. *The Condition of Postmodernity: An Enquiry into the Origins of Cultural Change.* Oxford: Oxford University Press.

Hixson, W. L. 2013. *American Settler Colonialism: A History.* New York: PalgraveMacmillan.

Hofstra, W. R. 2004. *The Planting of New Virginia: Settlement and Landscape in the Shenandoah Valley.* Baltimore: Johns Hopkins University Press.

Hubbard, B., Jr. 2009. *American Boundaries: The Nation, the States, and the Rectangular Survey.* Chicago: University of Chicago Press.

Jordan, T. G. 1974. "Antecedents of the Long-Lot in Texas." *Annals of the Association of American Geographers* 64 (1): 70–86.

Johnson, H. B. 1976. *Order Upon the Land: The US Rectangular Land Survey and the Upper Mississippi Country.* New York: Oxford University Press.

Knapp, R. G. 2000. *China's Walled Cities.* Oxford: Oxford University Press.

Lefebvre, H. [1974] 1991. *The Production of Space.* Translated by D. Nicholson-Smith. Oxford: Blackwell Publishing.

Lemon, J. T. 2002. *The Best Poor Man's Country: Early Southeastern Pennsylvania.* Rev. ed. Baltimore: Johns Hopkins University Press.

Mazza, L. 2009. "Plan and Constitution – Aristotle's Hippodamus: Towards an 'Ostensive' Definition of Spatial Planning." *The Town Planning Review* 80 (2): 113–141.

Meinig, D. W. 1986. *The Shaping of America: A Geographical Perspective on 500 Years of History.* Vol. 1: *Atlantic America, 1492–1800.* New Haven: Yale University Press.

Pattison, W. 1957. *Beginnings of the American Rectangular Land Survey System, 1784–1800.* Chicago: University of Chicago, Department of Geography, Research Paper 50.

Palet, J. M. and H. A. Orengo. 2011. "The Roman Centuriated Landscape: Conception, Genesis, and Development as Inferred from the Ager Tarraconensis Case." *American Journal of Archaeology* 115 (3): 383–402.

Price, E. T. 1995. *Dividing the Land: Early American Beginnings of Our Private Property Mosaic.* Chicago: University of Chicago Press.

Rose-Redwood, R. S. 2006. "Governmentality, Geography, and the Geo-Coded World." *Progress in Human Geography* 30 (4): 469–486.

———. 2008. "Genealogies of the Grid: Revisiting Stanislawski's Search for the Origin of the Grid-Pattern Town." *Geographical Review* 98 (1): 42–58.

Schein, R. H. 1997. "The Place of Landscape: A Conceptual Framework for Interpreting an American Scene." *Annals of the Association of American Geographers* 87 (4): 660–680.

Stanislawski, D. 1946. "The Origin and Spread of the Grid-Pattern Town." *Geographical Review* 36: 105–120.

Thrower, N. J.W. 1966. *Original Survey and Land Subdivision: A Comparative Study of The Form and Effect of Contrasting Cadastral Surveys.* Chicago: Association of American Geographers.

Webster, G. R. and J. Leib. 2011. "Living on the Grid: The US Rectangular Public Land Survey System and the Engineering of the American Landscape." In *Engineering Earth*, edited by S. D. Brunn. doi:10.1007/978-90-481-9920-4_117.

White, C. A. 1983. *A History of the Rectangular Survey System.* Washington: United States Bureau of Land Management.

Wood, J. S. and M. P. Steinitz. 1992. "A World We Have Gained: House, Common, and Village in New England." *Journal of Historical Geography* 18 (1): 105–120.

Wyckoff, W. W. 1986. "Land Subdivision on the Holland Purchase in Western New York State, 1797–1820." *Journal of Historical Geography* 12 (2): 142–161.

18

LANDSCAPE AND POLITICS

The 2020 Presidential Election in the United States

Fred M. Shelley and Heather Hollen

Since ancient times, humans have manipulated natural environments to create landscapes. These landscapes often reflect the values and cultures of those who create them, and they have symbolic meaning to people who interact with them. But the process of landscape creation also reflects power differentials within a society. Those in power can design landscapes that benefit themselves, often at the expense of the less powerful and the downtrodden. Hence, landscapes are inherently political.

Landscapes are often created by those holding power, but the meanings of the landscapes vary between those in power and the powerless. Over the course of history, such power was held by royalty and the nobility in a society. Although the United States does not recognize titles of royalty and nobility, such power differentials have affected American politics since colonial times. At the time of American independence, only white male property owners could vote. Not until the late twentieth century did all adults have the right to vote, although in many places lower-income individuals and non-whites have less access to the ballot box. Often, the political process is linked to issues associated with the meaning and control of landscapes. This chapter addresses these relationships through analysis of the presidential election of 2020.

The 2020 Presidential Election

The 2020 presidential election, in which former Vice President Joe Biden defeated incumbent President Donald Trump, was one of the most polarizing in American history. Increasing polarization across the country contributed to the record voter turnout, as voters from each political party viewed the election outcome as pivotal to the future of the country.

Tension between the Democratic and Republican parties was amplified further by tensions within the two parties themselves. The Democratic Party experienced tension between its moderate and progressive wings. Divisions between these wings of the Democratic Party emerged in 2016 with Hillary Clinton's narrow victory in the party's nominating contest over Senator Bernie Sanders, her loss to Republican Trump in the general election of 2016 (for which Clinton and Sanders supporters each blamed the other), and the Democrats picking up 41 House seats in the 2018 midterm elections.

DOI: 10.4324/9781003121800-23

Throughout his term, Trump was immensely popular among members of his Republican base. However, his popularity ratings among Democrats and independent voters were consistently very low. As the 2020 campaign began, Trump's overall unpopularity led many Democrats to predict that he could be beaten in the upcoming general election. Accordingly, more than 20 Democrats competed for the party's presidential nomination.

The most significant candidates included moderates such as Biden, former Mayor Pete Buttigieg of South Bend, Indiana, and Senator Amy Klobuchar of Minnesota. Others—including Sanders and Senator Elizabeth Warren of Massachusetts—were regarded as progressives. Beginning in the summer of 2019 and continuing into 2020, these and other candidates waged a spirited contest for the nomination. By early March, however, Biden had clinched enough support among delegates chosen by primary voters to be recognized as the party's presumptive nominee (Shelley and Hollen 2021). Before the Democratic Party's virtual national convention in August, Biden selected Senator Kamala Harris of California, one of the original candidates for the 2020 Democratic nomination, as his running mate. On the Republican side, Trump and Vice President Mike Pence were renominated without serious opposition. For the most part, Trump mitigated ongoing tension between the moderate and conservative wings of the Republican Party. However, some prominent Republicans refused to support Trump, and some endorsed Biden publicly.

The question of ballot access became a significant issue during the campaign. Many people voted by absentee or by mail, and some of these voters cast their ballots in this fashion for the first time. Some states and localities reduced the number of polling places in which voters could cast their ballots in person. In many places, these shutdowns were concentrated in areas inhabited primarily by racial minorities and/or low-income individuals, some of whom lacked transportation to get to the polls. In parts of Atlanta, for example, people waited in line for five hours or more to cast their ballots in the June primary election (Fowler 2020). And in the general election, many voters also stood in line for several hours before they could cast their ballots. Significantly, many of these reductions in polling places occurred in the Atlanta metropolitan area, which Biden carried by a margin large enough to offset Trump's strength in other parts of the state (Cohn et al. 2020). The divergence between Democratic strength in Atlanta and Republican strength in rural areas paralleled this divergence in many other parts of the country, illustrating increasing polarization between urban and rural places in recent years.

Limited voting access in low-income and minority-dominated areas across the United States likely contributed to the large number of Democrats using mail-in or absentee ballots. In contrast, Trump dissuaded Republicans from using mail-in and absentee voting methods. (This was ironic since Trump himself voted via mail-in/absentee ballot in Florida.) Thus, a majority of the mail-in and absentee votes in 2020 were cast for Biden.

The high number of mail-in and absentee ballots meant that the 2020 winner was unlikely to be declared on election night. Many states had provisions allowing mail-in and absentee ballots that were postmarked by Election Day to arrive later and still be counted. In addition, counting mail-in and absentee ballots required ensuring that the voter's signature on the ballot matched the signature on file in the voter record. If the signatures were not a close enough match, the individual had to be contacted to resolve the issue. This process delayed the vote count for several days.

Because the majority of mail-in and absentee votes were cast for Biden, on election night Trump appeared to have a lead in the Electoral College nationwide. However, when mail-in and absentee votes were finally counted, Trump's lead dwindled. The media did not declare

Biden to have won the election until Saturday, November 7, four days after Election Day. Biden ended up with approximately seven million more popular votes nationwide than did Trump. He won by a margin of 306 to 232 votes in the Electoral College, winning every state that Clinton had carried in 2016 and capturing five states that Trump had won in 2016—Arizona, Georgia, Michigan, Pennsylvania, and Wisconsin.

The Global Pandemic and the 2020 Election

Political tensions in 2020 were compounded by the global spread of the novel coronavirus disease, named COVID-19 for the year it was identified. The virus first emerged in China in late 2019. It spread rapidly in China and then throughout the rest of the world in early 2020. The first confirmed COVID-19 case in the United States was reported on January 20, with the first American fatality reported on February 6. By March, the virus was recognized as a pandemic. The number of infections and deaths in the United States increased throughout the year, resulting in more than 20 million infections and nearly 400,000 deaths by the end of December 2020.

Given the ease of transmission and severity of infection, the Centers for Disease Control and Prevention (CDC) discouraged large gatherings and crowds. The CDC encouraged Americans to wear face masks in public places and to social distance by keeping themselves at least six feet apart from one another. As the disease spread, Trump and his supporters tended to downplay the severity of the pandemic. Some Republicans regarded the CDC's efforts as unnecessary and alarmist. Concern about how the Trump administration had handled the pandemic would prove to be an important issue in the fall campaign.

The CDC's advice was received very differently by Democratic and Republican leaders and voters. Some states, counties, and cities mandated mask use, while others refused. Leaders of conservative states—such as South Dakota and Wyoming—claimed that mask usage was a matter of personal freedom and that mask laws infringed on civil liberties. In keeping with traditional Republican emphasis on decision-making at the local level, the Republican governor of Oklahoma refused to order a statewide mask mandate but did not object to mask mandates imposed by individual cities and towns. Alternatively, leaders in more liberal states such as California and Vermont claimed that mask mandates were necessary to protect public health and the wellbeing of all citizens. The act of simply wearing a mask became politicized to the point that wearing a mask in public places (even without a mandate) was synonymous with liberals, and refusing to wear a mask in public was linked to conservatives. Some conservatives who did wear masks chose masks with images and slogans supporting the Trump campaign. This practice has continued since the election.

Throughout the pandemic, Trump referred to COVID-19 repeatedly as the "Chinese virus" and the "kung flu," implying that China was to blame for the pandemic. Such rhetoric led to sharp increases in levels of harassment and violence against Asian-Americans. For example, two Asian-American women in San Francisco were stabbed in broad daylight at a bus stop, another in New York was hit over the head by a hammer, and an 84-year-old Thai immigrant in San Francisco died after being violently shoved to the ground (BBC News 2021). According to one estimate, more than 3,800 hate incidents against Asian-Americans were reported between March 2020 and February 2021, an increase of nearly 150% over the previous year. A majority of these incidents were directed against Asian-American women (Yam 2021). These actions were especially evident in Chinatowns, which are the historic centers of Chinese culture in many cities and whose landscapes reflect this Chinese-American culture. In some communities, residents were assaulted and injured by non-Asians, and businesses owned by Chinese–Americans were looted or robbed. In response, police erected barricades around these communities. Civilian foot patrols were organized by local community leaders to protect residents (particularly the elderly), while community residents organized rallies in support of their rights as American citizens.

These differing reactions to the pandemic affected the campaign strategies of both candidates and their parties. During the summer, Biden held small rallies and required attendees to social distance and wear masks. Some rallies took place in parking lots in which his supporters could remain in their cars while watching him speak. However, Trump continued to hold large rallies. His first rally following the outbreak of COVID-19 took place in Tulsa, Oklahoma on June 20. Thousands of attendees crowded into a large arena that is used normally for basketball games and concerts, and many chose not to wear masks. The city of Tulsa and other areas of Oklahoma observed an increase in COVID-19 cases following the rally and attributed some of this increase to the rally. Another of Trump's early rallies took place on July 4 near the sculpture of George Washington, Thomas Jefferson, Abraham Lincoln, and Theodore Roosevelt at Mount Rushmore, South Dakota amid reports that Trump had pushed to have his image added to the monument (Martin and Haberman 2020).

Landscapes of Inclusion, Diversity, and Racial Justice

Issues associated with racial diversity and justice also played an important role in the campaign, and these issues were reflected in landscapes. Concern about the shooting deaths of several Blacks in cities across the United States led to the establishment of the Black Lives Matter (BLM) movement, which was organized to protest racially motivated violence and police brutality. Although BLM was first organized on an informal and decentralized basis in 2013, it achieved national attention in 2020 following the death of an unarmed Black man, George Floyd, in Minneapolis. On May 25, Floyd suffocated and died after a white police officer knelt on his neck for nearly ten minutes.

Floyd's death generated widespread protests across the United States and around the world. Murals showing images of Floyd and support for BLM were painted and added to landscapes throughout the country (Steele and Almond 2020; Figure 18.1). Washington, DC's Mayor Muriel Bowser ordered that the words "Black Lives Matter" be painted on a street in front of the White House (Figure 18.2), and she had the street renamed Black Lives Matter Plaza (Nirappil et al. 2020).

In early June, protestors occupied part of the Capitol Hill district in Seattle, Washington. The protestors referred to the area as the Capitol Hill Autonomous Zone, and barricades were erected to keep authorities out of the area (Baker 2020). Signs reading "You are now entering the Capitol Hill Autonomous Zone" and similar messages were placed at the perimeter of the area. A large Black Lives Matter mural was painted by local artists, and a community garden was established in a local park. In addition to protesting police brutality and the murder of George Floyd, those occupying the Autonomous Zone demanded free access to health care, rent control, and the reallocation of public funds from policing to community development projects. Similar efforts to establish such autonomous zones took place in Minneapolis, where George Floyd had lived and was murdered, and in Portland.

These activities around the country—in Minneapolis, DC, Seattle—and their manifestations vis-à-vis the landscape reinforced what many saw, and many others now a witnessed a need for change in the country's recognition of various forms of racism, particularly institutional, and other injustices.

Figure 18.1   George Floyd Mural, Scholes St. and Graham Ave., Brooklyn, March 30, 2021. Library of Congress, Prints and Photographs Division, photograph by Camilo J. Vergara [reproduction number, LC–DIG–vrg–15595]. Public domain image at https://www.loc.gov/pictures/item/2021636975/

Figure 18.2   People walking on the Black Lives Matter large mural painted on Black Lives Matter Plaza, June 19, 2020. Library of Congress, Prints and Photographs Division, photograph by Carol M. Highsmith [reproduction number, LC–DIG–highsm–63670]. Public domain image available at https://www.loc.gov/pictures/item/2020720125/

The Party Conventions

The ways in which each party responded to COVID-19 and to issues of racial diversity were also reflected in their national conventions. The Republican convention was originally scheduled to take place in Charlotte, North Carolina at the end of August. When North Carolina's Democratic governor announced limitations on public gatherings in the state (including a limit on the number of people allowed to attend public gatherings), Trump planned to have the convention moved to Jacksonville, Florida. Eventually, however, Trump abandoned this plan. Some of the convention events were held in Charlotte, but without the large crowds that Trump had wanted. Trump's acceptance speech and other convention events took place remotely. (In keeping with the tradition that the party out of power holds its convention prior to the other party's convention, the Republican convention was held after the Democratic convention had concluded.)

Prior to the pandemic, Democrats had selected Milwaukee, Wisconsin, as their convention site. After the continuing spread of COVID-19, however, party leaders decided to make the convention almost entirely virtual. Traditionally, at in-person conventions delegates from each state announce the number of votes cast for each candidate from the convention floor. Under the virtual format, however, each state's party leaders prepared a video in which its votes were announced. This format gave party leaders in each state opportunities to use landscape images to emphasize political issues in their states as they impacted the upcoming general election campaign. Some of these landscape images contained within the states' videos showed places with historical and/or contemporary political importance. For example, Alabama's video showed the Edmund Pettis Bridge that spans the Alabama River in Selma. On March 7, 1965, hundreds of persons crossed the bridge as they began to march to the state capital of Montgomery to demand the right to vote. After crossing the bridge, some of the marchers were attacked and beaten by white police officers, and some of the marchers were hurt seriously. More than 65 demonstrators including John Lewis, who later served as a member of the US House of Representatives for 34 years before his death in 2020, were hospitalized or treated for injuries in what became known as Bloody Sunday. This event led to the passage of the Voting Rights Act of 1965. Thus, the Edmund Pettis Bridge symbolized, and continues to symbolize, the struggle for voting rights. Civil rights were also emphasized in Georgia's video that featured a mural in Atlanta depicting Dr. Martin Luther King, Jr.

New Mexico's video highlighted the state's racial diversity with a photograph of a desert landscape in the background. The speaker in the video, State Representative Derrick Lente, is a member of the Sandia Pueblo Nation and gave part of his speech in the Pueblo language. The Alaska video called attention to climate change by showing coastal communities threatened by sea-level rise associated with global warming, which has been especially evident throughout that state. Collectively, the videos used landscape images to call attention to themes to be emphasized by the Democratic Party in the fall campaign against Trump, including racial diversity, poverty, and environmental quality. (The videos from each state can be accessed at www. youtube.com/watch?v=AEDAGeGC5yg.)

Landscape and the Fall Campaign

Political campaigns affect landscapes. Some of these landscape effects are temporary. For example, many people and some businesses place signs and flags supporting candidates of their choice on their own private properties and sometimes in public spaces. Most communities require election signs posted in public places to be removed within a specified number of days after the election, although some persons have kept their signs and flags on their own properties on display for weeks or even months after Election Day.

However, some landscape changes in public places associated with elections are more permanent. For example, murals such as those memorializing George Floyd and protesting anti-Asian hate crimes remain on the urban landscape long after they are painted. Statues with political implications are constructed or removed in association with political campaigns. In June 2020, the US House of Representatives approved a resolution to remove statues of five historical figures who had been associated with the Confederacy from the United States Capitol building. Statues of other prominent figures who have been associated with racism, sexism, and genocide against Indigenous peoples have also been removed.

Place names also affect landscapes and their meanings, as has been the case of Black Lives Matter Plaza in Washington, DC. Throughout the world, public spaces have been renamed to commemorate prominent historical figures, while names associated with slavery, inequality, and other past injustices have been removed (Rose-Redwood et al. 2018). Schools, other buildings, and streets whose previous names were associated with past injustice have also been renamed, while the Washington Redskins of the National Football League and the Edmonton Eskimos of the Canadian Football League have been renamed the Washington Commanders and the Edmonton Elks respectively. In July, 2021, the Cleveland major league baseball team announced that it would change its nickname from the Indians to the Guardians.

The issue of immigration was also an important issue in the campaign. During his presidency, Trump took a hard line against immigration from Mexico and Central America. This was embodied by the continuation of a border wall of several hundred miles in length that extended along parts of the border between the US and Mexico. This wall, of course, has a profound and possibly permanent impact on landscapes on both sides of the international boundary. Supporters of maintaining and expanding the border wall regarded doing so as enhancing border security. However, opponents pointed to what they considered inhumane treatment of immigrants who crossed the border illegally, including separating small children from their parents.

After the Election

For several weeks after the election, Trump and some of his supporters claimed that the election results were fraudulent and that he had actually won. The argument was based on a supposition that some of Biden's votes were illegitimate, and that Trump had received more votes in the closely contested states than he had received according to the official counts. Trump claimed that the mail-in voting process was fraudulent and that it gave Biden an unfair and illegitimate advantage. He and his supporters filed numerous lawsuits to try to overturn the results of the election, but these lawsuits were thrown out of court by federal and state judges in each of the closely contested states.

Although Trump's efforts failed in the courts, his supporters continued to claim victory and continued to contest the outcome. Matters came to a head on January 6, 2021, when Congress assembled in the US Capitol to formally count the Electoral College votes, thereby certifying Biden's election. That day, thousands of Trump supporters stormed the Capitol and entered it illegally, vandalizing and looting the building and occupying the House and Senate chambers. The building was placed under lockdown over concerns for the safety of Vice President Pence, members of Congress, and their staffs. Before the insurrection was subdued, five persons were killed, and more than 140 others were injured. Extensive property damage, including defacing of statues of prominent Americans in the Rotunda of the Capitol, took place. It became evident that Trump had provided at least tacit support for the protestors, and he was blamed heavily for the actions of the rioters. The following week, the House impeached Trump for a second time, although he was acquitted by the Senate after he left office. Eventually, hundreds of Trump supporters were arrested on various charges associated with the insurrection.

Biden and Harris were inaugurated formally two weeks later, on January 20. To prevent further rioting and disruption, unprecedented security measures were taken. Thousands of National Guard troops from around the country were sent to Washington, DC, to maintain law and order. Barricades were placed around the Capitol and the National Mall. The inaugurations of Biden and Harris took place without incident.

The barricades represent a significant change in the meaning of the landscape of Washington, DC. Historically, the Capitol has been regarded as "the people's house." Any person could enter the building, express his or her views on issues of public concern to members of Congress and their staffs, and observe debates on the House and Senate floors. But the barriers have resulted in a changed meaning, symbolizing restricted rather than unrestricted access. Today, the barricades and other security measures illustrate the degree to which access to the Capitol has become restricted.

Concluding Thoughts: Looking Toward the Future

Although the 2020 election is now over, the impacts of politics on landscape and the impacts of landscape on politics will continue into the foreseeable future. After the election, some states considered or enacted legislation addressing questions of ballot access. Republican governors and state legislators supported bills that would limit ballot access. In March 2021, Georgia's legislature passed a particularly restrictive ballot access law that was signed into law by the state's governor. Among other provisions, the law declared it illegal to provide water or food to persons standing in line at the polls waiting to vote, despite the long lines and waiting times at polling places. Implementation of this law has resulted in several efforts to boycott Georgia-based corporations and to keep conferences and other public events out of the state. Major League Baseball pulled its 2021 All-Star game out of Atlanta and moved it to Denver.

Environmental issues remain important to the political process, as evidenced by the Keystone Pipeline controversy concerning the construction of a new oil pipeline from the Canadian province of Alberta through parts of Montana, South Dakota, and Nebraska. The project, if approved, would impact rural landscapes throughout the region, and it has been criticized because the proposed route traverses the environmentally sensitive Sand Hills of western and central Nebraska. The proposed Keystone Pipeline extension not only addresses questions of environmental impact directly, but it also symbolizes questions of American dependence on fossil fuels. Such dependence has been linked by scientists to ongoing global warming and climate change, which is seen as catastrophic to the planet's environment in the years ahead. This concern has been reinforced by increasing numbers of extreme weather events, wildfires, and droughts as well as by continued increases in surface temperatures worldwide.

Addressing these concerns, some prominent progressives have proposed what has been termed the Green New Deal. The Green New Deal consists of a series of policy proposals aimed at reducing American dependence on fossil fuels, reducing carbon emissions that are recognized generally as contributing to global warming, and facilitating a transition to reliance on clean energy. Such transition to clean energy will have ongoing effects on material landscapes. Already, thousands of turbines used to create wind energy are in place throughout the rural Great Plains. Other rural landscapes may remain more pristine, for example with fewer disruptions associated with pipeline construction and maintenance. Opponents of the

Green New Deal concept claim that its implementation will impose extraordinary costs on American taxpayers, but supporters respond that this tax increase will be offset by job creation and reduced expenditures needed to respond to extreme weather events and associated natural disasters.

In summary, the 2020 election represented an excellent illustration of relationships between the political process, election campaigns and outcomes, and the material landscape. As societal evolution continues and as new issues of public importance come to the forefront, these relationships will no doubt evolve as well.

References

Baker, M. 2020 "Free Food, Free Speech, and Free of Police: Inside Seattle's 'Autonomous Zone'." *New York Times*, June 11, 2020. www.nytimes.com/2020/06/11/us/seattle-autonomous-zone.html.

BBC News. 2021. "COVID 'Hate Crimes' Against Asian Americans on Rise." *BBC*. May 19, 2021. www.bbc.com/news/world-us-canada-56218684.

Cohn, N., M. Conlon, and C. Smart. 2020. "Detailed Turnout Data Shows [sic] how Georgia Turned Blue." *New York Times,* November 17, 2020. www.nytimes.com/interactive/2020/11/17/upshot/georgia-precinct-shift-suburbs.html.

Fowler, S. 2020. "Why Do Nonwhite Georgia Voters Have to Wait in Line for Hours? Too Few Polling Places." *NPR*, October 17, 2020. www.npr.org/2020/10/17/924527679/why-do-nonwhite-georgia-voters-have-to-wait-in-line-for-hours-too-few-polling-pl.

Kaur, H. 2021. "As Attacks Against Asian Americans Spike, Advocates Call for Action to Protect Communities." *cnn.com*, February 13, 2021. www.cnn.com/2021/02/13/us/asian-american-attacks-covid-19-hate-trnd/index.html.

Martin, J., and M. Haberman. 2020. "How Kristi Noem, Mt. Rushmore and Trump Fueled Speculation About Pence's Job." *New York Times*, August 8, 2020. www.nytimes.com/2020/08/08/us/politics/kristi-noem-pence-trump.html#click=https://t.co/pTPfdX6b2o.

Nirappil, F., J. Zauzmer, and R. Chason. 2020. "'Black Lives Matter: In Giant Yellow Letters, DC Mayor Sends Message to Trump. *Washington Post*, June 5, 2020. www.washingtonpost.com/local/dc-politics/bowser-black-lives-matter-street/2020/06/05/eb44ff4a-a733-11ea-bb20-ebf0921f3bbd_story.html.

Rose-Redwood, R., D. Alderman, and M. Azaryahu, eds. 2018. *The Political Life of Urban Streetscapes: Naming, Politics, and Place*. London and New York: Routledge.

Shelley, F. M., and H. Hollen. 2021. "The Primary Elections," in *Atlas of the 2020 Elections*, edited by R. Watrel, R. Weichelt, F. Davidson, J. Heppen, and E. H. Fouberg. Boulder, Colorado: Rowman and Littlefield, forthcoming.

Steele, A., and K. Almond. 2020. "George Floyd Murals are Popping Up All Over the World." *CNN*, June 26, 2020. www.cnn.com/2020/06/06/world/gallery/george-floyd-murals-trnd/index.html.

Yam, K. 2021. "There Were More Than 3,800 Anti-Asian Racist Incidents, Mostly Against Women, in Past Year." *NBC News*, March 16, 2021. www.nbcnews.com/news/asian-america/there-were-3-800-anti-asian-racist-incidents-mostly-against-n1261257.

19

AMERICA'S CONTEMPORARY COMMEMORATIVE LANDSCAPES

Themes and Practices in Memorial Mania

Erika Doss

America's commemorative landscape has changed dramatically in recent decades. Whereas earlier generations erected bronze and marble statues of famous heroes in public parks, commemoration today is often a more historically complex undertaking informed by memory work (a constant process of negotiation between us and our past), felt experience (physical and emotional levels of participation), and civil rights. Many contemporary memorials are transactional spaces where audiences are encouraged to challenge notions of a progressive and united national history. In 2003, for example, the Indian Memorial at the Little Bighorn Battlefield National Monument in Montana was dedicated to the Indian victors of the 1876 battle famously lost by George Custer. In 2018, the National Memorial for Peace and Justice in Montgomery, Alabama was dedicated to the African American victims of lynching and racial terrorism.

Scholarship on contemporary commemorative landscapes is deeply interdisciplinary. Traditional surveys of place and space are enhanced by reflection on the ideas, beliefs, values, and actions that shape and define memorials. Many scholars study commemorative landscapes that address the disruptions and contradictions of American history, including racial discrimination (Alderman 2013; Linenthal 2008; Upton 2015), violence, trauma, and terrorism (Bednar 2020; Foote 2003; Linenthal 2003), and wars and epidemics (Sturken 1997). Others are attentive to institutional practices and hierarchies that determine commemorative spaces (Bogart 2018; Bruggeman 2017; Savage 2009;). Contemporary commemorative landscapes are often examined through the theoretical framework of "memory studies" (Halbwachs 1992; Nora 1989), which considers how individual and collective memories of the past shape contemporary understandings of identity and belonging. They are further informed by studies that highlight how public feelings and felt experiences shape contemporary understandings of self and nation (Doss 2010).

This chapter focuses on *Danzas Indigenas*, a memorial designed by Los Angeles artist Judy Baca for a commuter rail station in Baldwin Park, California (Figure 19.1). Like other commemorative landscapes that focus on remembering those who have been forgotten or marginalized in American history (Aden 2014; Young 2000), this memorial links the past with the present through historical symbolism and emotional response.

DOI: 10.4324/9781003121800-24

Figure 19.1 Judy Baca, entrance to Danzas Indigenas, 1993. Baldwin Park Metrolink Railway Station, Baldwin Park, California. Courtesy of Judy Baca.

Memorial Mania

Expansive and wide-ranging interests today in memory and history, accompanied by urgent needs to claim both in visible, public landscapes, characterize "memorial mania" (Doss 2010). This differs from the "statue mania" that gripped America from the 1870s to the 1920s, when thousands of "great men" monuments were erected to promote American exceptionalism, white supremacy, and masculine dominance, among other national norms. Contemporary memorial mania is less convinced of common national narratives, or unified social and political beliefs.

Many memorials today focus on conflict, rupture, and loss. Many are designed in a Minimalist modern art style and seek certain emotional, or affective, responses from audiences. Many are specifically commissioned to address, rather than erase, difficult and complex historical subjects.

The Vietnam Veterans Memorial, for example, designed by Maya Lin and dedicated in Washington, DC in 1982, features two intersecting walls of black granite inscribed with the names of the more than 58,000 US soldiers who died in that divisive war. Its highly polished surfaces and horizontal, earth-hugging forms encourage audience response: visitors see the reflection of their own bodies, touch the names of the dead, and often leave notes and flowers (Figure 19.2). Stylistically and emotionally, the Vietnam Veterans Memorial challenges more

Figure 19.2 Section of the Vietnam Veterans Memorial, designed by Maya Lin, dedicated 1982, in Washington, D.C. Photo by Chris W. Post.

traditional forms of commemoration on the National Mall, such as the Washington Monument and Lincoln Memorial. Interviewed by filmmaker Freida Mock in 2004, Lin related that she wanted to design a deeply moving place, remarking "I really did mean for people to cry" (Mock 2004). Today, the Vietnam Veterans Memorial remains one of Washington's most popular memorials: in 2019, more than 4.58 million visitors were counted at the site.

Contemporary Commemorative Landscapes and Public Feeling

Memorial mania is contextualized by an "experiential turn" in how Americans understand history and memory (LaCapra 2004, 3). Americans today often seek embodied and emotional encounters as sources of knowledge and identity formation. Many want to visit the places where "history" happened, and touch and photograph things they encounter. By extension, the meaning making of contemporary commemorative landscapes is informed by expectations of felt experience, or perceiving history in terms of particular places, things, and feelings. As Lin observed in 1995, "I don't make objects; I make places. I think that is very important—the places set a stage for experience and for understanding experience" (Lin 1995, 13).

Memorials are places of public affect. They are "repositories of feelings and emotions" encoded in their material forms, narrative content, and the "practices that surround their production and reception" (Cvektovich 2003, 7). Commemorative landscapes fabricate understandings of history and memory through different affective states, including feelings of fear, gratitude, grief, shame, and anger (Doss 2010). Memorials can induce anger, for example, when audiences see them as offensive historical symbols and narratives. In recent years, civil disobedience against symbols of white supremacy led to the toppling of many Confederate memorials.

Anger can also induce social and political activism: "My response to racism is anger," American author Audre Lorde observed:

> Women responding to racism means women responding to anger; the anger of exclusion, of unquestioned privilege, of racial distortions, of silence, ill-use, stereotyping, defensiveness, misnaming, betrayal, and co-optation … Anger expressed and translated into action in the service of our vision and our future is a liberating and strengthening act.
>
> *(Lorde 1984, 124, 127)*

Much of the controversy surrounding Baldwin Park's memorial involved feelings of anger.

Box 19.1 More-Than-Representation

Jacque Micieli-Voutsinas

Traditionally, heritage landscapes have been understood in representational terms—their histories are described to site visitors in expository texts that dictate their continued social relevance to visiting publics. Combined with size and scale, their historical significance is communicated through visual prominence and spatial dominance in the landscape. What we learn from these traditional representational practices is that their interpretive power is limited to heritage sites themselves. Here, historical meaning is quite literally set in stone: because it exists, it is important. Breaking the rules of traditional memorial design, including figuration, iconography, and doctrinal elements, anti-monumental aesthetics emerged in favor of the abstract, spatial, and experiential elements of memorial architecture. Moving beyond representational conventions, anti-monumentalism prioritized evocative experiences of heritage space to dictate historical relevance. Here, the *more-than*-representational, or sensory and emotional experiences of place, are given prominence.

More-than-representational understandings of places of memory have gained prominence in shaping visitors' experiences of the past, over the last 15 years. As this scholarship acknowledges, heritage spaces "are all multi-sensual sites, alive with intense and often lingering sounds, smells, and sights" (Waterton 2014, 830). This trend toward *more-than*-representation "prioritizes spatiality and the affective dynamics of memorial design," with the ultimate goal of instilling visitors with more evocative connections to the past, directly acquired through embodied, experiential encounters within and within heritage environments (Micieli-Voutsinas and Person 2020, 1). This is especially true at sites commemorating or preserving difficult, or traumatic pasts. The Vietnam Veterans Memorial in Washington, DC, for example, is a pre-eminent example of anti-monumental aesthetics. The mnemonic site invites visitors to experientially interpret the site's meaning as they move alongside the memorial's dark granite walls, etched with the names of the dead. As a site of contentious, or unresolved social history, the Vietnam Veterans Memorial does not overtly tell you why this history matters, or what its continued social relevance is. Rather, visitors are asked to *feel* the power of the heritage site as a mnemonic presence and interpret the evocative *absence* of the nearly 60,000 American lives lost, instinctively. Here, embodied, affect-oriented experiences of heritage design are mobilized to direct cultural memory-making through emotional and experiential modes of communication that both elicit and reinforce narratives of place sensorially, rather than stating them outright. Although conventional memorial design has never been fully abandoned in western practices of remembrance, this shift towards "affective heritage" (Micieli-Voutsinas 2017) has become commonplace in contemporary heritage architecture following World War II.

Affective heritage has shifted our "thinking about the spaces of heritage... from the static 'site' or 'artefact' to questions of engagement, experience, and performance" (Waterton 2014, 824). Here, traditional museal techniques of scientific inquiry and cognitive learning have been abandoned in favor of more affective modes of story-telling, geared towards the visitor's emotional learning, or moral inquiry into the past (Micieli-Voutsinas 2021). This shift has also produced its own set of challenges and questions: For example, how *exactly* do heritage environments inform human behavior and shape visitors' emotional experiences? Is something akin to "user choice" built into, or negotiated within these evocative spaces? And if so, how do visitors exert agency over their physiological and psychological responses to such affect-oriented designs?

References

Micieli-Voutsinas, J. 2017. "An Absent Presence: Affective Heritage at the National September 11th Memorial & Museum." *Emotion, Space and Society* 24 (1): 93–104.
———, J. 2021. *Affective Heritage and the Politics of Memory After 9/11: Curating Trauma at the Memorial Museum.* London: Routledge.
Micieli-Voutsinas, J. and Person, A. 2020. *Affective Architectures: More-than-Representational Approaches to Geographies of Heritage.* London: Routledge.
Waterton, Emma. 2014. "A More-Than-Representational Understanding of Heritage? The 'Past' and the Politics of Affect." *Geography Compass* 8 (11): 823–833.

Danzas Indigenas: *History and Design*

In 1993, the Los Angeles County Metropolitan Transportation Authority commissioned Baca to create a commemorative landscape for the Baldwin Park Metrolink rail station. A public artist and director of SPARC (Social and Public Art Resource Center, Venice, California), Baca is internationally recognized for *The Great Wall of Los Angeles* (1976–1983), a 2,740-foot mural painted on the walls of the Tujunga Wash Flood Control Channel in Van Nuys (Doss 1995). Her memorial in Baldwin Park, designed in collaboration with Siegel Diamond Architects of Los Angeles, consists of a 20-foot archway, 100-foot plaza, and 400-foot train platform with kiosk shelters, benches, and planters. Named *Danzas Indigenas*, or "Indigenous dances," the memorial features historical symbols specific to Baldwin Park, a city in the San Gabriel Valley, about twenty miles east of Los Angeles. "We believe it's important to make the train system not only efficient but attractive and user-friendly," explained Jessica Cusick, the director of the county's Art for Rail Transit Program. "It's important," she emphasized, to "provide communities with a sense of ownership in the stations" (Dubin 1993). Commemorating the histories and hopes of Baldwin Park residents, *Danzas Indigenas* is a democratizing model of community-based and site-specific public art.

Floor patterns on the train platform, for instance, replicate the site plans of the four Franciscan missions built near Baldwin Park (San Gabriel, San Fernando, Santa Barbara, and San Juan Capistrano) in the eighteenth century, when the area consisted of ranches, small farms,

Figure 19.3 Judy Baca, detail of *Danzas Indigenas* showing train platform tiles. Courtesy of Judy Baca.

and vineyards (Hudson 1993, J-4). In 1906, the town was named after Elias "Lucky" Baldwin, a wealthy landowner whose vast estate was noted in Helen Hunt Jackson's classic California novel *Ramona* (1884). Interwoven with historic details of the missions are abstracted designs representing the steps of tribal dances from the Gabrielinos (Tongva) and Chumash people who lived in the area before the Europeans. Phrases in five different languages—Gabrielino, Chumash, Luiseno, Spanish, and English—reference the region's multiple inhabitants and histories. Letters in one section of the platform, for instance, read, in Gabrielino and English, "Ne Hiin Taxatem=My people" (Figure 19.3). Statistics on cattle raised at nearby missions, and acres of grape vines, are juxtaposed with the numbers of Indigenous peoples who were forced to labor at the missions, and the numbers who died.

The roofs of the kiosk shelters and the gateway arch, which is made of stucco and exposed adobe brick, allude to the Spanish Colonial architecture of the San Gabriel Mission, established in 1771 about one mile from the Baldwin Park rail station. A stone prayer mound in front of the arch, placed where the altar would be located in the San Gabriel Mission, is dedicated to Toypurina, a 23-year old Tongva medicine woman who led a revolt against the mission and Spanish colonization in 1785 (Hackel 2003; Figure 19.4). "I was interested in why the San Gabriel Mission looked like a fortress," Baca recalls, noting: "As I did research I discovered its dark history as a place where indigenous people were literally worked to death" (Baca 2006). Inscribed in tile underneath the arch are the words, "When the Indians died, the villages ended," made by a Tongva elder who was a consultant on the memorial.

The two sides of the gateway further historicize the region. Challenging a sanitized Spanish Colonial—or "Taco Bell"—aesthetic, Baca designed the arch as a partial ruin to show the impact of Spanish conquest and colonization in California (Hudson 1993, J-4). In addition to their layers of history, Baca's commemorative landscape projects are typically informed, or "personalized," by the concerns and feelings of their audience. The side of the arch that

Figure 19.4 Judy Baca, detail of *Danza Indigenas* showing prayer mound dedicated to Toypurina.
Courtesy of Judy Baca.

commuters pass through when they exit the train features the words "Baldwin Park" and comments Baca collected from local residents about their city, including "A small town feeling," "Not just adults leading but youth leading too," "Use your brain before you make up your mind," and "The kind of community that people dream of, rich and poor, brown, yellow, red, white, all living together." Another statement, tucked into a corner of the arch, was made by a local white politician discussing Mexican immigration after World War II: "It was better before *they* came." Baca left the source of these words unexplained and ambiguous, inviting viewers to imagine for themselves who "they" might be.

The other side of the arch, the entrance to the train platform, is marked "Sunigna," the original Tongva name for this region of Southern California. It features drawings based on indigenous pictographs and a fragment of a poem by Gloria Anzaldúa, which reads: "This land was Mexican once, was Indian always and is. And will be again." The poem, which opens Anzaldúa's 1987 collection *Borderlands/La Frontera*, reflects on the precarious circumstances of bordered cultures and contested lands, and speaks to the complex historical mosaic that is the Southwest, and all America (Anzaldúa 1987, 3). In 1770, before the Franciscans arrived, the Tongva numbered about 10,000 people and covered a territory of some 1,500 square miles in the greater Los Angeles Basin. Until 1848, Baldwin Park, like all of Southern California, was owned by Mexico. Today, only about 300 Tongva people remain and only a handful speak the language. "I wanted to put memory into a piece of land once owned by the American Indian cultures," Baca remarks. "Memory and willpower are what any culture, the ones living then and those living now, has to have to preserve itself" (Baca 2006).

Imagining Baldwin Park as a layered site of memory, *Danzas Indigenas* recollects how Spanish conquest and colonization aimed to erase the history of Indigenous California. It simultaneously remembers the history of Indigenous resistance that refused that erasure by including the figure of Toypurina. "The missions were the first sweat shops of California," Baca observes. "My hope was to bring Toypurina's spirit back to the site so she would be an inspiration to young women today, to show them they can organize and lead" (Baca 2006). Baldwin Park's memorial is scattered with other subtly subversive historical referents, from the ruined façade of the gateway arch to bronze seals on the benches—so that when commuters sit down they are "branded" like the Indians and animals that were enslaved in nearby missions. Ironically, these defiant historical and aesthetic details were completely overlooked in the angry protests that erupted over the memorial in 2005, more than a decade after it was built.

Commemorative Landscapes as Immigration Battlegrounds

Controversy began when a group called SOS (Save Our State) claimed that Anzaldúa's words were "anti-American," and demanded their removal (Pierson and Biederman 2005). SOS accused Baca of making a memorial to "reconquista," a Chicano nationalist concept from the 1970s advocating the return of the US Southwest to Mexico in order to restore the Aztec homeland of Aztlán. Baca recalls: "I was never engaged in 'reconquista'—someone had to explain to me what it is all about. I couldn't believe that SOS was hung up on this—and didn't seem to notice the monument's historical details" (Baca 2006). In a specious lawsuit, SOS charged the city with promoting racism, separatism, and "treasonous sentiments" in a "tax-supported monument."

SOS targeted *Danzas Indigenas* to promote its anti-immigration agenda. Linked to the Minuteman Project (armed vigilantes who, starting in 2005, policed the US–Mexico border) and neo-Nazi groups like National Vanguard and Stormfront, SOS was founded in 2004 by Joseph Turner, a white supremacist living in Ventura, 85 miles northwest of Baldwin Park. Informed by Proposition 187, a 1994 California ballot initiative that sought to deny public and social services such as schools and health care to undocumented immigrants (and was declared unconstitutional by a federal judge in 1997), SOS denounced undocumented immigrants as threats to national security and as financial burdens to "citizen taxpayers." In the early 2000s, Turner organized SOS rallies encouraging violence against people of color. Two decades later, "reinvigorated" by the election of Donald Trump, he marched with the Proud Boys and formed American Children First, a group dedicated to denying K-12 education to the children of undocumented workers (Landa 2017). Turner's aim, he told one reporter, was to "send up a flag, to tell people, 'I know you're angry and I know you want to get involved, and here's a way to do it'" (Alvarez 2005). The Southern Poverty Law Center (SPLC) designated SOS a hate group in 2005 (Kelley 2018).

On its website, SOS honed its race baiting with a "Hall of Shame" vilifying Latino leaders, and stating: "Californians are tired of radical revisionist history and militant separatist rhetoric … Californians are tired of watching their communities turn into Third World cesspools as a result of a massive invasion of illegal aliens." SOS "actions" included boycotting day labor hiring centers and pressuring a Los Angeles TV station to remove billboards referencing Mexican culture. "We have reached the point where we can no longer sit back and allow our government to aid and abet the illegal alien invasion," Turner declared in 2005, "We must respond as our founding fathers would have responded. We must refresh the tree of liberty." Arrested several times for assault and battery during SOS rallies, Turner said his group's goal was the "transference of pain." He explained: "If you do not make it painful enough for an organization or entity to continue doing what they're doing, then they're simply going to keep doing it" (Doss 2010, 371–372). SOS's goal in Baldwin Park was to cause financial pain, to force the city to support its commemorative landscape to the point of bankruptcy.

In 2005, Turner organized two protests at *Danzas Indigenas*, each held with permits obtained from the city and each met by strong opposition. Promoted on the *John & Ken Show*, a local talk radio show whose alt-right hosts supported the Minuteman Project and the deportation of "no-good illegal immigrants," "Baldwin Park 1" took place on May 14, 2005 and featured 60 SOS supporters, 500 counter-protesters, and 70 specially hired riot police. A police helicopter circled above the memorial during the entire rally. SOS supporters held signs stating: "Stop Illegal Immigration" and "You Have the *Right* to Be Deported!" Counter-protesters held placards stating: "Stop Embarrassing White People" and "Just Say No to Racism."

Both sides were angry; each group inflamed the other. After an SOS member was hit by a water bottle, the group sought $1 million in damages from the City of Baldwin Park and demanded that it "eliminate the racist comments set forth in the challenged monument." (The lawsuit was thrown out of court.) Sneering "bring it" to the angry crowd of counter-protesters, Turner boasted "They're communicating my message better than I could" to reporters covering the rally (Pierson and Biederman 2005). Discussing SOS's small numbers, he added: "I don't measure success or victory by the size of the crowd turnout. I'm looking to inflict economic damage and I'm also looking to use the size of the crowd that opposes us to our propaganda advantage, with respect to showing the public all of the Mexican flags waving, the American flags they're kicking, their violence against us, the epithets they hurl" (Doss 2010, 372).

Commemorative Landscapes and Creative Activism

"Baldwin Park 2," held on June 25, 2005, was even more emotionally charged. SOS intended to simply restage its violent anti-immigration platform. *Danzas Indigenas*'s supporters, however, reconsidered the form and nature of their protest. "I was surprised at the absence of nonviolent protest in the first rally, the style of protest that was used during the civil rights and farmworkers' movements," Baca later remarked. "And along with others I decided we needed to rethink our whole approach to activism."

Rather than fueling SOS's "dualistic, binary" anger, Baca and SPARC encouraged the development of counter-demonstrations to give "a third space for a plurality of voices" and to articulate issues of freedom of expression, creative integrity, and anti-racism. "The groups who oppose us," Baca explained on SPARC's website,

> welcome confrontation so that they can broadcast their message of fear to others through the media. We will not succumb to their tactics … They will offer cynicism and we will offer ceremony. They will raise criticism and we will offer culture. They will condemn art and we will simply make more of it. They will paint a picture of weakness and we will celebrate our strength, for in our eyes, the law protects us, our creativity dignifies us.

Promoting the second rally as "The Reconquest of Justice, Peace, Liberty and Love," Baca and SPARC designed posters advertising the protest as a "celebratory festival to confront this latest incarnation of terrorism by disarming its violence with love, humor, dignity, compassion, understanding, Indigenous spirituality, inclusivity, and resistance" (Doss 2010, 372–374).

At the second rally, SOS was represented by fewer than 50 people. *Danzas Indigenas*, in contrast, was defended by more than 1,000 Baldwin Park residents, including high school and college students, local politicians, musicians, poets, dancers, performance artists, guerrilla theater groups, and peace organization activists. Employing the "innovative tactical repertoire" of labor-organizing drives and anti-racism rallies held in Southern California since the 1990s, the supporters of the commemorative landscape constituted a highly creative multi-ethnic coalition (Davis 2000, 147).

Baca and SPARC made a 90-foot mobile memorial titled *You Are My Other Me* that featured placards with images and quotes on both sides (Figure 19.5). Held aloft by about 30 people and flipped back and forth during the three-hour rally, the signs read "Good art confuses racists," "The land does not belong to us, we belong to the land," and "America turns its back on hate groups." Earlier on the day of the second rally, the Baldwin Park City Council issued a resolution honoring Baca and *Danzas Indigenas* and stating: "The strong sentiments expressed by people who make various interpretations of its meaning after 12 years, is a testament to its value as an artwork." Baldwin Park's mayor, Manuel Lozano, declared: "This monument represents our city, our people, and all America. And we support it" (Doss 2010, 373). "Baldwin Park 2" saw fewer angry outbursts from counter-demonstrators, although Turner did his best to provoke them.

Defending *Danzas Indigenas* was neither cheap nor effortless. Baldwin Park spent more than $40,000 to hire security guards and riot police during the two rallies. Baca, SPARC, and other groups spent hundreds of hours strategizing ways to defeat SOS's hateful anti-immigrant actions. There were other damages, too. Baldwin Park city officials and Baca were inundated by racist emails and even death threats during the two-month time span of the two rallies.

Figure 19.5 You *Are My Other Me*, mobile mural displayed at "Baldwin Park 2" rally, June 25, 2005, Baldwin Park, California. Courtesy of Judy Baca.

SOS's targeting of *Danzas Indigenas*, which amounted to the racial profiling of an entire community, stimulated many to rethink the point and purpose of commemoration—and to determine the depth of their commitment to the issues and histories that memorials embody in their landscapes. "Hate groups want to waste our time," Baca observes. "Public culture is forced to be defensive, always defending itself against hate and intolerance" (Doss 2010, 374).

Border Metaphors and Landscapes

Controversy over *Danzas Indigenas* was grounded in the contemporary dynamics of American citizenship and national memory: of who "counts" as an American today, and what America itself means and represents. These dynamics are especially charged by the policies and politics of the southwestern border. In the early 2000s, anti-immigration politicians such as Russell Pearce (R-AZ) and Tom Tancredo (R-CO) called the "crisis" of the border "the No. 1 issue facing America" (Lelyveld 2006, 40).

The US–Mexico border has long played a symbolic role in constituting concepts of US territoriality and nationalism. As a physical place and symbolic object, it is deeply unstable. First, shifting, sandy soils make a solid, physical border with a fixed, permanent fence or wall practically impossible. The border is an unrealizable and divisive political metaphor. Second, the US has historically depended on immigration for "national growth and national greatness." Immigration fuels the nation's economic progress and cultural and social diversity (Yglesias 2019). The "big, beautiful wall" that was the signature promise of Trump's 2016 election campaign, like the surge of militarized nation-state border walls erected in recent years around the world (India–Bangladesh, Israel–West Bank, Hungary–Serbia, Botswana–Zimbabwe), was intended to define and project US territorial power and control (Rogers and Bailey 2020). Paradoxically, Wendy Brown argues, such walls symbolize the diminishing authority and deep anxieties of the nations they are supposed to contain and keep safe, especially in the economic, social, and cultural shift to the neoliberal terms of globalization (Brown 2010).

The border between the US and Mexico is, of course, very real on legal and military terms: a 2,000-mile line that delimits the land claims of two different nations; a barrier to the human rights claims of immigrants hoping for refuge and opportunity in America. More than just a line separating Mexico and the United States on a map, the border looms throughout the US, producing a "wall mindset" that seeks to divide Americans on multiple fronts (Yang 2017). Racist stereotypes about the US–Mexico border, Mike Davis explains, police Latino populations wherever they live, imposing rules on public access, mobility, and visibility (Davis 2000, 60–61). Whether Nebraska, Georgia, Maine, or California, US border politics are perpetuated each time ICE (US Immigration and Customs Enforcement) raids an American workplace or community.

Porous but rigid, unstable yet constant, the border is perhaps the most obvious metaphor for America today. Throughout its history, the US has been fundamentally shaped by immigration, diversity, and multiculturalism, and has repeatedly pursued its own re-invention. It is also, however, inhabited by people who angrily contest the variable dimensions and fluctuating terms of American history and identity. Immigration—because it represents a threat to what some see as a white homeland and history—is their primary target; the "big beautiful" Border Wall is their final solution. No wonder a commemorative landscape like *Danzas Indigenas*—not a wall, but an inviting doorway—generated so much anger from SOS.

In June 2005, comments on SOS's website included "We are angry! We are seething with anger and boiling with rage," followed by:

> The monument in Baldwin Park is not just a rock, it is a disgusting testament to how pathetically apathetic Americans have grown in response to the hostile takeover attempt by the Mechistas [Chicano separatists] and the massive illegal alien invasion. It is a slap in the face to all Americans.
>
> *(Doss 2010, 375)*

Such rage was, of course, utterly misdirected. Rather than targeting *Danzas Indigenas* to argue that Mexico and Mexicans threaten America and Americans, SOS might have surveyed the aspirational goals of neoliberal capitalism and globalization, which center on "the interests of the unimpeded operations" of "free" markets (Duggan 2003, xii). But that would require a post-nationalist political alignment across racial and ethnic lines that SOS—whose angry anti-immigration platform is fixated on white supremacy, racist nativism, and zealous isolationism—could not imagine.

The Affective Possibilities of Commemorative Landscapes

Anger's transformative potential, its capacity to raise consciousness rather than fuel further conflict, was evident at "Baldwin Park 2" in the mobile memorial *You Are My Other Me*. Striving for creative cultural discourse, the counter-demonstrations of the second rally were thoughtful, productive, and politically progessive. In an online forum organized by SPARC, artist Suzanne Lacy wrote, "Maybe this is what art is supposed to do. Maybe this is how art becomes something more than concrete and steel, paint on a canvas" (Doss 2010, 376).

Similarly reflecting on the themes and purpose of the Baldwin Park memorial gateway, Baca observed,

> Our capacity as a democracy to disagree and to coexist is precisely the point of this work. No single statement can be seen without the whole, nor can it be removed without destroying the diversity of Baldwin Park's voices. Silencing every voice with which we disagree, especially while taking quotes out of context, either through ignorance or malice, is profoundly un-American.
>
> *(Baca 2006)*

A year later, in May 2006, more than two million people took to the streets in Los Angeles to support immigrant rights, as did millions more throughout the US.

The affective possibilities of anger lie in ameliorating its negative effects: in practicing self-restraint in the interest of democratic politics and social reform; in channeling irrational fury into something passionate and progressive. The struggle over *Danzas Indigenas* shows that America's commemorative landscapes have the capacity to unleash anger's productivity by providing the spaces and subjects that permit cultural and political creativity, and prompt acts of "good" citizenship. From the Baldwin Park rallies in 2005 to the rallies for Black Lives Matter in recent years, twenty-first century Americans recognize the links between public culture and emotionally productive politics.

References

Aden, R. C. 2014. *Upon the Ruins of Slavery: The President's House at Independence National Historical Park and Public Memory*. Philadelphia: Temple University Press.

Alvarez, F. 2005. "A Street-Fighter Mentality on Illegal Immigration." *Los Angeles Times*, June 27, 2005: B-2.

Alderman, D. H. 2013. "Street Naming and the Politics of Belonging: Spatial Injustices in the Toponymic Commemoration of Martin Luther King, Jr." *Social and Cultural Geography* 14 (2): 211–233.

Anzaldúa, G. 1987. *Borderlands/La Frontera: The New Mestiza*. San Francisco: Spinsters/Aunt Lute Book Company.

Baca, J. 2006, interviewed by Erika Doss, Milwaukee, Americans for the Arts Conference, June 4.

Bednar, R. M. 2020. *Road Scars: Place, Automobility, and Road Trauma*. Lanham, MD: Rowman & Littlefield.

Bogart, M. H. 2018. *Sculpture in Gotham: Art and Urban Renewal in New York City*. London: Reaktion Books.

Brown, W. 2010. *Walled States, Waning Sovereignty*. New York: Zone Books.

Bruggeman, S. C., ed. 2017. *Commemoration: The American Association for State and Local History Guide*. Lanham, MD: Rowman & Littlefield.

Doss, E.1995. *Spirit Poles and Flying Pigs: Public Art and Cultural Democracy in American Communities*. Washington, DC: Smithsonian Institution Press.

———. 2010. *Memorial Mania: Public Feeling in America*. Chicago: University of Chicago Press.

Davis, M. 2000. *Magical Urbanism: Latinos Reinvent the US City*. New York: Verso, 2000.

Dubin, Z. 1993. "Journey Back in Time: At the Baldwin Park Rail Station, UCI Art Professor Judy Baca Pays Homage to the Area's Indian Heritage." *Los Angeles Times*, July 27, 1993, www.latimes.com/archives/la-xpm-1993-07-27-ca-17395-story.html

Duggan, L. 2003. *The Twilight of Equality? Neoliberalism, Cultural Politics, and the Attack on Democracy*. Boston: Beacon Press.

Foote, K. E. 2003. *Shadowed Ground: America's Landscapes of Violence* Austin: University of Texas Press, 2003.

Hackel, S. W. 2003. "Sources of Rebellion: Indian Testimony and the Mission San Gabriel Uprising of 1785." *Ethnohistory* 50 (4): 643–669.

Halbwachs, M. 1992. *On Collective Memory*. Edited and translated by L. A. Coser. Chicago: University of Chicago Press.

Hudson, B. 1993. "Riding the Rails of History." *Los Angeles Times*, July 25, 1993: SG J-1, J-4.

Kelly, B. J. 2018. "California Nativist Joseph Turner Re-Emerges to Push Anti-Immigrant Legislation." "Hatewatch," Southern Poverty Law Center, January 9, 2018, www.splcenter.org/hatewatch/2018/01/09/california-nativist-joseph-turner-re-emerges-push-anti-immigrant-legislation

LaCapra, D. 2004. *History in Transit: Experience, Identity, Critical Theory*. Ithaca: Cornell University Press.

Landa, J. 2017. "This Group Wants to Bar Children in the U.S. From Attending Some Schools." *Los Angeles Times*, April 5, 2017: B-3.

Lelyveld, J. 2006. "The Border Dividing Arizona." *New York Times Magazine*, October 15, 2006), www.nytimes.com/2006/10/15/magazine/15immigration.html.

Lin, M. 1995. "Maya Lin." In *Grounds for Remembering: Monuments, Memorials, Texts*, 8–14. Berkeley, CA: Doreen B. Townsend Center for the Humanities.

Linenthal, E. T. 2008. "Remembrance, Contestation, Excavation: The Work of Memory in Oklahoma City, the Washita Battlefield, and the Tulsa Race Riot." In *Public Culture: Diversity, Democracy, and Community in the United States*, edited by M. S. Shaffer, 52–66. Philadelphia: University of Pennsylvania Press.

Lorde, A.1984. "The Uses of Anger: Women Responding to Racism." *Sister Outsider: Essays and Speeches*. Trumansburg, NY: Crossing Press.

———. 2003. *The Unfinished Bombing: Oklahoma City in American Memory*. New York: Oxford University Press.

Mock, F. L., dir. 1994. *Maya Lin: A Strong Clear Vision*. Santa Monica, CA: American Film Foundation, 2003. DVD.

Nora, P. 1989. "Between Memory and History: Les Lieux de Mémoire." *Representations* 26: 7–24.

Pierson, D. and P. W. Biederman. 2005. "Protest Over Art Forces Police to Draw the Line." *Los Angeles Times*, May 15, 2005: B-1.

Rodgers, L. and D. Bailey. 2020. "Trump Wall: How Much Has He Actually Built?" *BBC News*, October 31, 2020, www.bbc.com/news/world-us-canada-46824649.

Savage, K. 2009. *Monument Wars: Washington, DC, the National Mall, and the Transformation of the Memorial Landscape* Berkeley: University of California Press.

Sturken, M. 1997. *Tangled Memories: The Vietnam War, the AIDS Epidemic, and the Politics of Remembering.* Berkeley: University of California Press.

Upton, D. 2015. *What Can and Can't Be Said: Race, Uplift, and Monument Building in the Contemporary South.* New Haven: Yale University Press.

Yang, M. 2017. "The Trump Wall: A Cultural Wall and a Cultural War." *Lateral: Journal of the Cultural Studies Association* 6 (2), csalateral.org/issue/6-2/trump-wall-cultural-war-yang/.

Yglesias, M. 2019. "Immigration Makes America Great." *Vox*, August 12, 2019, www.vox.com/policy-and-politics/2017/4/3/14624918/the-case-for-immigration.

Young, J. 2000. *At Memory's Edge: After-Images of the Holocaust in Contemporary Art and Architecture.* New Haven: Yale University Press.

20

PLACE (RE)NAMING

Jordan P. Brasher

The world, spurred in part by events in the United States, sits in the midst of a renaming moment. Neighborhoods, parks, schools, streets, bridges, and other landmarks and features of the human-built landscape are being rechristened and undergoing dramatic symbolic change at an unprecedented rate. From the symbolic renaming of major urban thoroughfares to "Black Lives Matter Way" to the renaming of primary schools and college campus buildings that honored Confederates, Ku Klux Klansmen, settlers, enslavers, and other white supremacists, the American place name landscape is caught up in a new wave of feminist and anti-racist protest sweeping the globe. Activists and advocates are challenging long-standing place names that valorize historical figures associated with slavery, racial segregation, settler colonialism, and imperialism, and confronting the lack of racial, ethnic, and gender diversity in the place name (i.e., toponymic) landscape.

Although commemorative place (re)naming is a highly charged public issue that occupies the interests of journalists, citizen groups, elected officials, laypeople, and scholars, there remains a lack of appreciation of the power and importance of place (re)naming in the American cultural landscape. Often derided as an endeavor in "political correctness" or in somehow "erasing" or "destroying" history, commemorative place (re)naming is an affective process important to people's lived material experiences, political-emotional wellbeing, and broader place-making practices as they inhabit, claim, and create places (Alderman, forthcoming). In this sense, place names are at least as much about the present as they are about the past. After all, rewriting the toponymic landscape is not necessarily the same as rewriting history; the place name landscape has never been a *tabula rasa*, but is better conceptualized as rugged territory over which the right to narrate the past and belong in the present is actively worked out—and indeed it is work.

As the burgeoning field of critical place name studies shows, commemorative place names are subjective; they tell stories from a particular point of view, selectively remembering parts of and persons from the past while delineating the contours of sociospatial order, spatial reference and wayfinding, and an affective sense of belonging in the present. Commemorative place names can work to uphold the prevailing social order or work to challenge it. Place names are likewise vulnerable to usurpation and appropriation by various groups with sometimes divergent, convergent, or outright competing political interests. Some in power even use

DOI: 10.4324/9781003121800-25

the toponymic landscape to placate without meeting the demands of the oppressed, like at the University of North Carolina-Chapel Hill, when in 2015 the university renamed a building to "Carolina Hall" instead of Hurston Hall (Brasher et al. 2017; Menefee 2018). UNC students, faculty, and other advocates for place name change demanded the building be renamed for Zora Neale Hurston, prolific Black author who attended the university informally prior to integration.

This chapter traces the history and development of place (re)naming studies, emphasizing major theoretical trends and shifts in the ways geographers have written about the American place name landscape. Given the political urgency of the global struggle for Black lives, a new wave of which has been spurred by the public police execution of George Floyd in Minneapolis, Minnesota on May 25, 2020, place names and place (re)naming are contextualized within the racialized politics of commemoration and commemorative (re)naming in the United States. Recent struggles of activists on American university campuses to un-name and rename buildings that commemorate(d) white supremacists are highlighted. In doing so, place (re) naming is situated within the framework of "memory-work" (Till 2012; Till and Kussito-Arponen 2015) to emphasize the importance of not just un-naming but also renaming problematically christened places. Finally, the chapter concludes by reflecting on the reparative possibilities and limits of place (re)naming to heal deep-seated and entrenched structural wounds in the United States. Caution is offered against renaming problematic places without any accompanying material or structural policy change. Ultimately, the chapter argues that despite the limitations of place (re)naming in carrying out memory-work and doing full justice to the struggles and contributions of marginalized peoples, scholars, activists, and others interested in place (re)naming should take seriously the symbolic and sociocultural capacity that place names hold for repairing entrenched racialized and gendered structural harm and expanding the senses of belonging for marginalized peoples.

From Catalog to Symbolic Capital

While the study of the place name landscape has a long history in the United States, American place name scholars initially neglected the role of power and politics behind the place (re) naming process. Initial studies of the place name landscape were often more interested in compiling catalogs that contained information about the etymology or linguistic origins of place names, rather than analyzing how sociopolitical power and ideology influence the naming process. Questions such as who gets to control the naming process under what circumstances and for what (and whose) political goals remained unanswered. This started to change in the 1990s and early 2000s.

In response to wider trends in the study of cultural geography in the 1990s and early 2000s that foregrounded critical social theory and re-conceptualized the meaning of culture to include the collision and working out of conflicting social interests, a "critical turn" in place name studies emerged (Alderman 2008; Azaryahu 1996, 2011; Berg and Vuolteenaho 2009; Rose-Redwood et. al 2010; Rose-Redwood and Alderman 2011; Yeoh 1992, 1996). Scholars began to re-conceptualize place names, and indeed the American cultural landscape more broadly, as sites of power and politics over which social groups with different interests exert uneven control and fight for cultural recognition and legitimacy, as well as lay claim to territory and the right to belong to place. Rather than mere passive artifacts of the past, place names and the naming process started to be conceptualized as "symbolic capital" used to associate places with selective, consumable, and exclusionary visions of the past and the right to belong to them in the present (Bourdieu 1984, 1991; Alderman 2008; Post and Alderman 2014).

The concept of symbolic capital, coined by cultural critic Pierre Bourdieu, connotes the various forms of distinction and status acquired through cultural recognition (Bourdieu 1991; Rose-Redwood 2008). In other words, symbolic capital is the sociocultural currency employed by various social actors in the symbolic struggle over the production of (geographic) common sense. With the critical turn in place name studies, the (re)naming of streets became implicated, explored, and analyzed as a part of this process of meaning-making (Azaryahu 1996; Rose-Redwood 2008; Alderman and Inwood 2013). According to Maoz Azaryahu (1996), for example, commemorative street naming is a practice that aims to "introduce an authorized version of history into ordinary settings of everyday life" (p. 312). Numerous social groups have attempted to legally rename places as an important strategy for acquiring legitimacy, prestige, and cultural recognition in the form of symbolic capital (Azaryahu 1996; Alderman 2008; Rose-Redwood 2008; Duminy 2014; Brasher et al. 2020). Increasingly, marginalized groups reject legally authorized channels for renaming and rely instead on informal and unauthorized forms of renaming as ways to reclaim places named after white supremacists and/or by colonial governments (Cumbe 2016; Brasher et al. 2017; Bigon 2020).

Yet, struggles over the toponymic landscape and its symbolic capital are not always exclusively about commemoration and the reputational politics of historical figures. The development of the nation-state and the modern global order has been accompanied by a concerted effort on the part of national and international authorities to standardize place names and place naming practices, reflected via authorized naming boards like the US Board on Geographic Names and the United Nations Group of Experts on Geographical Names (Monmonier 2006; Berg and Vuolteenaho 2009). The computerization of cartography and mapmaking and its accompanying standardizing practices has forced such naming boards to (in some cases) legally adjudicate public concern over racist place names that disparage women and Black, Indigenous, and other people of color (Monmonier 2006). Within the context of US settler colonialism, this process has historically often involved erasure of Indigenous place names, what geographer and map historian Brian Harley called "toponymic silencing" (2001, 99). Authorities' place name standardization as part of the creation of the "grammar of place" (Goeman 2014) mediated by linguistic politics and selective choices that serve the geographic interests of the state in naming, claiming, referring to—and ultimately, controlling—place(s). Yet, as shown in the next section, marginalized people find ways to disrupt prevailing hegemonic grammars of place (Goeman 2014) and the oppressed often consider place names as essential to "grammars of reckoning" (Safransky 2021) that take seriously their role in creating, upending, redressing, repairing, and re-organizing linguistic and sociospatial orders.

Rose-Redwood (2008) shows how renaming numbered streets in New York City links to an elite project of symbolic erasure on the one hand and the cultural recognition of marginalized African Americans on the other. His study highlights the fact that while place names are imbricated in symbolic struggles between the elite and the marginalized, the privileged and the subaltern, the symbolic struggle over street (re)naming in New York City—and in the American cultural landscape more broadly—cannot be reduced to such a binary. After all, fissures exist within both elite and marginalized social groups and there are multiple, intersecting axes of exclusion at stake in the production and (re)naming of commemorative places. Additionally, the symbolic capital associated with place names can be co-opted and re-appropriated to serve different interests.

The American university campus landscape provides a useful setting for exploring the politics of place (re)naming. Though the campus landscape has long witnessed the struggle over the right to belong and to equal access to education and upward economic mobility (Rodriguez 2012; Wilder 2013; Combes et al. 2016; Mustaffa 2017), recent waves of antiracist protest and uprisings associated especially with the Black Lives Matter movement since 2013 have targeted campus landscapes, especially place names commemorating Confederates, Ku Klux Klansmen, enslavers, and other white supremacists, as legitimate political arenas. To emphasize the importance of place (re)naming to these demands, the next section surveys battles over place names on several American campus landscapes through the lens of memory-work. It illustrates how universities have authorized place renaming differently in different social, political, and geographic contexts, drawing lessons from the ways in which campus officials have stymied and limited the work of memory and attempts to recover a Black sense of place and the past on campus.

Place (Re)naming and Memory-Work: A View from American University Campuses

American university campus landscapes have drawn increasing public attention and scrutiny in recent years for their roles in maintaining and perpetuating the white male supremacist settler social order. American campus landscapes historically in many cases were built with the labor of enslaved people on the unceded land of Indigenous people whose descendants were then barred from accessing the benefits of an education or for whom the education process served as a form of colonial annihilation (Lomawaima et al. 2021). Early American universities were heavily invested in the purchase and sale of land to be worked with enslaved labor and turned for a profit to fund the school (Wilder 2013). The use of enslaved labor influenced college campuses so much that according to Wilder (2013), it determined the difference between financial success or failure for many early colonial schools, suggesting that these institutions' reliance and dependence on the violence of enslavement extends well beyond the ending of the Civil War. For this reason, university campuses in particular can be considered "wounded places" given how they emerged historically and geographically from racist, patriarchal settler economic, political, and cultural institutional structures (Wilder 2013; Bonds and Inwood 2016; Combes et al. 2016; Inwood and Bonds 2016; Brasher et al. 2017).

Attempts to shed light on the difficult histories enshrined in American campus landscapes can be considered what geographer Karen Till and her colleagues call "memory-work" (Till 2012; Till and Kuusito-Arponen 2015). Memory-work, among other things, creates new forms of public memory, commits to building social capacity, and opens up opportunities for marginalized people to lay claim to a sense of belonging on the memorial landscape. American campus landscape reform efforts such as renaming problematic buildings on campus, led especially by students, faculty, and other activists of color, have emerged alongside wider sets of demands to challenge and change the university on an institutional scale out of recognition that the American university continues to be a site for the (re)production of racial and gendered inequality and settler colonial domination (Luke and Heynen 2021). For example, admissions policies continue a tradition of providing preferential treatment to white applicants through "legacy admissions"; universities remain disproportionately white and male in terms of student and faculty rolls; inequalities in the tenure and promotion process for faculty of color remain an issue (Lamb 1993; Lee and Leonard 2001; Massey and Mooney 2007; Durodoye Jr. et al. 2020). The university is also implicated in the creation of the social order of the "frontier" that evolved out of the variable histories of the "settler-native-slave triad" of plantation agriculture and Indigenous dispossession (Luke and Heynen 2021). Thus, consideration of the demands of marginalized and colonized peoples for toponymic landscape reform should be considered alongside their wider material demands for repairing the intersecting violences through which universities were forged.

One university that has in recent years had a particularly difficult time responding to the demands of students and activists for place name change is Middle Tennessee State University (MTSU). MTSU's campus maintains a building known as Forrest Hall, named after Nathan Bedford Forrest, the first grand wizard of the Ku Klux Klan and a controversial (Carney 2001) Confederate general. The 2015 Charleston Massacre (Kurtz 2016) spurred a new wave of demands for landscape change on the campus, building on a long history of activism to remove white supremacist landscape iconography from MTSU's landscape (Calise 2018; Allen and Brasher 2019). Students and activists from campus organizations like the Talented Tenth (a Black student organization on campus) submitted a list of demands to the university president that included—alongside demands for the un-naming and renaming of Forrest Hall—a plan to increase the Black student retention rate, hire more and improve the tenure rate for Black faculty, and commission a study to analyze the racism that Black students experience(d) on MTSU's campus (Forrest Hall Protest Collection 2020). MTSU student-activists clearly connected the moniker commemorating Forrest with the wider material realities and lived experiences of students of color on campus.

Although MTSU student-activists ultimately succeeded in petitioning the university to un-name and rename the building, their demand has not yet come to full fruition. When the university president approved the recommendation to rename the building as the Army ROTC Building and took it before the Tennessee Historical Commission (THC)—which oversees the changing of the commemorative landscape on public lands—the THC denied the request, citing the state's Heritage Preservation Act, which prohibits the renaming, removal, or relocating of any military monument or item honoring a military unit or person. This decision by the commission highlights the difficulty of accomplishing toponymic landscape change through authorized channels, common in other places especially in the southeastern US with similar heritage preservation laws (see Wahlers 2016; Plott 2020), given the multiple scales at which place name change must be authorized in order to take place through legal means.

Out of understandable frustration at the difficulty of accomplishing toponymic landscape reform through authorized means, a number of activist groups have resorted to the informal, unauthorized renaming of places to reclaim them for their students and their communities. The University of North Carolina's (UNC) 2015–2016 attempt to rename Saunders Hall is one such example. William Saunders—the building's original namesake—was an 1854 graduate of UNC and a former North Carolina Secretary of State widely known for leading the expansion of the Ku Klux Klan in the area (Brasher et al. 2017; Menefee 2018; McDaniel 2020). UNC students again in response to the 2015 Charleston Massacre organized an effort to rename Saunders Hall for Zora Neale Hurston—a prolific Black woman writer and anthropologist who unofficially attended UNC prior to integration. The university ignored students' demands and chose instead to rename the building as "Carolina Hall," an ode to the school's mascot and an attempt to "move on from valorizing the overt white supremacy of past generations without dealing with the structural wounds that supposedly race-neutral policies produce in the neoliberal university" (Brasher et al. 2017, 297; Figure 20.1). Rather than engaging with memory-work, the university chose instead to do "public relations work" or "PR work" (Allen and Brasher 2019)—an increasingly popular strategy not just of university authorities but also mayors and city planners—that gets the place name issue off the administrator's desk and out of the headlines without creating space to meaningfully grapple with the legacies of racist violence that the place name represents (see Brasher et al. 2020). UNC officials concluded their deliberations by instituting a 16-year moratorium on place name change in the campus landscape.

Figure 20.1 A display inside of Carolina Hall shows visitors the changing of the building's name from its original Saunders title. Photo courtesy of Stephen Birdsall.

In response, UNC students and faculty informally renamed Saunders/Carolina Hall to "Hurston Hall" in an effort to create an oppositional politics of belonging that reclaims a Black sense of place and the past and honors the legacy and influence of a famous Black woman intellectual and informal UNC alumna. What became apparent from the UNC case is that creating an oppositional politics of belonging is not simply about removing symbols to white male supremacy but also resolving the invisibility of people of color, women, and other marginalized groups within the university's social memory, suggesting that un-naming alone is not enough but counter-naming or what might be described as "reparative naming" (Alderman and Rose-Redwood 2020; Brasher et al. 2020) is also required. Reparative naming involves the careful selection of a surrogate name to replace the old problematic name by working in consultation with—and taking seriously—the toponymic landscape change demands of Black, Indigenous, and other people of color, as well as white women. Reparative naming is about repairing the harm inflicted on the sense of belonging of marginalized peoples by conferring a sense of legitimacy to and valorization of the memories, experiences, contributions, and struggles of marginalized groups in the campus' or city's landscape. Yet, the selection of a suitable surrogate name in and of itself does not fully close the reparative loop and in fact can be a symbolic landscape change meant to divert attention from larger material demands of student-activists (Tichavakunda 2020). Thus, by way of conclusion, it is worth reflecting briefly on the possibilities and limits of place (re)naming in advancing reparative memory-work.

The Possibilities and Limits of Reparative Place (Re)naming

A number of (student-)activist groups affiliated with Black Lives Matter and other civil rights advocacy organizations are increasingly making toponymic landscape change a core part of their demands for repairing the entrenched structural harm caused by the legacies and ongoing impacts of white male settler colonial supremacy in the cultural landscape. At the same time, they are also demanding that these symbolic changes be accompanied by material changes to their wellbeing, not just stylized aesthetic symbols gesturing toward Blackness designed to placate and divert attention away from (meeting) their demands for justice and structural change (see Summers 2019, 2020).

One lingering question, from a decolonial perspective, is what are the limits of reparative renaming on university campuses that remain on stolen Indigenous land? If every building, road, park, and landscape feature that commemorates a white supremacist male is successfully renamed, but the relations of land, property, and capital that buttress the settler state remain intact, what has really changed? In a May 2021 article in *The Guardian* assessing how George Floyd's public police execution has "changed the world" one year after its occurrence, a Minneapolis activist described her ambivalence about American place name landscape change this way:

> I want to see more of my people and my ancestors reflected in the world that I live in … But how about we just build that world instead of just changing the street signs in this one?
>
> *(Douglas, Chrisafis, and Mohdin 2021)*

In this sense, there are important limitations to symbolic place renaming not accompanied by wider material change. First, any optimism about the power of place (re)naming to engage in meaningful reparative memory-work should be tempered by an awareness of those naming authorities who would rather do "PR work" than memory-work and who prize Black aesthetic emplacement over the demands of Black people for justice and (toponymic landscape) change. Additionally, although place (re)naming holds the potential for rendering visible the long silenced and forgotten or ignored contributions, experiences, memories, and struggles of marginalized peoples, done without connection to wider material demands for reparative social, political, and geographic change, place (re)naming can also be rendered ineffective for advancing social justice.

Ultimately, place names, although certainly part of wider networks of symbolic and material landscapes and resources that represent opportunities to honor and recognize with dignity and distinction the lives and struggles of the marginalized, are not by themselves fully capable of bringing about social, economic, and political change to repair the deeply entrenched wounds of structural white male supremacy. Yet, when carried out appropriately, the removal of racist monuments can be more than hollow symbolism, but part of the project of decolonizing higher education (Kearns 2020) and of the place-based, institutional work of listening to and caring for marginalized students, faculty, and staff.

References

Alderman, D. 2008. "Place, Naming, and the Interpretation of Cultural Landscapes." In *The Ashgate Research Companion to Heritage and Identity*, edited by B. Graham and P. Howard, 195–213. Abingdon: Routledge.

Alderman, D. Forthcoming. "Naming the World: The Politics of Place Naming." In *Naming Places*, edited by F. Giraut and M. Houssay-Holzschuch. ISTE-Wiley. Published in French.

Alderman, D. and J. Inwood. 2013. "Street Naming and the Politics of Belonging: Spatial Injustices in the Toponymic Commemoration of Martin Luther King Jr." *Social & Cultural Geography* 14 (2): 211–233. doi:10.1080/14649365.2012.754488.

Alderman, D. and R. Rose-Redwood. 2020. "The Classroom as 'Toponymic Workspace': Towards a Critical Pedagogy of Campus Place Renaming." *Journal of Geography in Higher Education* 44 (21): 124–141. doi:10.1080/03098265.2019.1695108.

Allen, D. and J. Brasher. 2019. "#ChangetheDamnName: MTSU Student Activism Against Forrest Hall, White Supremacy, and a Reparative Approach to Heritage Preservation." *The Activist History Review.* November 22, 2019. activisthistory.com/2019/11/22/changethedamnname-mtsu-student-activism-against-forrest-hall-white-supremacy-and-a-reparative-approach-to-heritage-preservation/.

Azaryahu, M. 1996. "The Power of Commemorative Street Names." *Environment and Planning D: Society and Space* 14 (3): 311–330. doi.org/10.1068%2Fd140311.

———. 2011. "The Critical Turn and Beyond: the Case of Commemorative Street Naming." *ACME: An International E-Journal for Critical Geographies* 10 (1): 28–33. acme-journal.org/index.php/acme/article/view/883.

Berg, L. and J. Vuolteenaho eds. 2009. *Critical Toponymies: Contested Politics of Place Naming.* Aldershot: Ashgate.

Bigon, L. 2020. Towards Creating A Global Urban Toponymy – A Comment. *Urban Science* 4 (4): 75. www.mdpi.com/2413-8851/4/4/75.

Bonds, A. and J. Inwood. 2016. "Beyond White Privilege: Geographies of White Supremacy and Settler Colonialism." *Progress in Human Geography* 40 (6): 715–733. doi:10.1177%2F0309132515613166.

Bordieu, P. 1984. *Distinction: A Social Critique of the Judgement of Taste.* Translated by R. Nice. Cambridge, MA: Harvard University Press.

Bordieu, P. 1991. *Language and Symbolic Power.* Translated by G. Raymong and M. Adamson. Cambridge, MA: Harvard University Press.

Brasher, J., D. Alderman, and J. Inwood. 2017. "Applying Critical Race and Memory Studies to University Place Naming Controversies: Toward A Responsible Landscape Policy." *Papers in Applied Geography* 3–4: 292–307. doi:10.1080/23754931.2017.1369892.

Brasher, J., D. Alderman, and A. Subanthore. 2020. "Was Tulsa's Brady Street Really Renamed? Racial (In) Justice, Memory-work, and Neoliberalism's Mandate of Least Disruption." *Social & Cultural Geography* 21 (9): 1223–1244. doi.org/10.1080/14649365.2018.1550580.

Calise, S. 2018. "Protesting the Confederacy on Campus." *The Activist History Review,* July 25, 2018. activisthistory.com/2018/07/25/protesting-the-confederacy-on-campus/.

Carney, C. 2001. "The Contested Image of Nathan Bedford Forrest." *The Journal of Southern History* 67 (3): 601–630. doi:10.2307/3070019.

Combes, B. H., K. Dellinger, J. T. Jackson, K. A. Johnson, W. M. Johnson, J. Skipper, J. Sonnett, J. M. Thomas, and the Critical Race Studies Group, University of Mississippi. 2016. "The Symbolic Lynching of James Meredith: A Visual Analysis and Collective Counter Narrative to Racial Domination." *Sociology of Race and Ethnicity* 2 (3): 338–353. doi:10.1177%2F2332649215626937.

Cumbe, C. 2016. "Formal and Informal Toponymic Inscriptions in Maputo: Towards Socio-linguistics and Anthropology of Street Naming." in *Place names in Africa,* edited by L. Bigon, 195–205. Springer: Cham. doi:10.1007/978-3-319-32485-2_13.

Douglas, D., A. Chrisafis, and A. Mohdin. 2021. "One Year On, How George Floyd's Murder Has Changed the World. *The Guardian,* May 22, 2021. www.theguardian.com/us-news/2021/may/22/george-floyd-murder-change-across-world-blm?fbclid=IwAR1nDVLbbM1RhnJJSsPUbb8CNKOK0WBsDOCPCMgp0jx7TsHZmm5DtN9VMw0.

Durodoye Jr., R., M. Gumpertz, A. Wilson, E. Griffith, and S. Ahmad. 2020. "Tenure and Promotion Outcomes at Four Large Land Grant Universities: Examining the Role of Gender, Race, and Academic Discipline." *Research in Higher Education* 61(3): 628–651. doi:10.1007/s11162-019-09573-9.

Duminy. J. 2014. "Street Renaming, Symbolic Capital, and Resistance in Durban, South Africa." *Environment and Planning D: Society and Space* 32 (2): 310–328. doi:10.1068%2Fd2112.

Forrest Hall Protest Collection. 2020. "MTSU Al Gore Research Center Archives." Last accessed on December 23, 2020. digital.mtsu.edu/digital/collection/p15838coll11.

Goeman, M. 2011. "Disrupting Settler-Colonial Grammar of Place. The Visual Memoir of Hullcah Tsinhnahjinnie." In A. Simpson and A. Smith eds. *Theorizing Native Studies.* doi:10.1515/9780822376613.

Harley, J. B. 2001. *The New Nature of Maps.* Baltimore, MD: Johns Hopkins University Press.

Inwood, J., and A. Bonds. 2016. "Confronting White Supremacy and a Militaristic Pedagogy in the U.S. Settler Colonial State." *Annals of the Association of American Geographers* 106 (3): 521–29. doi:10.1080/24694452.2016.1145510.

Kearns, G. 2020. Topple the Racists 1: Decolonising the Space and Institutional Memory of the University. *Geography* 105: 116–125.

Kurtz, H. 2016. "Introduction to the Special Forum on the Charleston Massacre of 2015." *Southeastern Geographer* 56 (1): 6–8. www.jstor.org/stable/26233765?seq=1.

Lamb, J. D. 1993. "Real Affirmative Action Babies: Legacy Preference at Harvard and Yale." *Columbia Journal of Law & Social Problems* 26 (3): 491–521.

Lee, L. J. and C. A. Leonard. 2001. "Violence in Predominately White Institutions of Higher Rducation." *Journal of Human Behavior in the Social Environment* 4 (2–3): 167–86. doi:10.1300/J137v04n02_09.

Lomawaima, K. T., K. McDonough, J. M. O'Brien, and R. Warrior. 2021. "Editor's Introduction: Reflections on the *Land-Grab Universities* Project." *Native American and Indigenous Studies* 8 (1): 89–91.

Luke, N. and N. Heynen. 2021. "Abolishing the Frontier: (De)Colonizing 'Public' Education." *Social & Cultural Geography* 22 (3): 403–424.

Massey, D. and M. Mooney. 2007. "The Effects of America's Three Affirmative Action Programs on Academic Performance." *Social Problems* 54 (1), 99–117. doi:10.1525/sp.2007.54.1.99.

McDaniel, B. 2020. "Silent Sam and Saunders Hall: Protests and Reactions at the University of North Carolina-Chapel Hill." PhD diss., University of South Carolina. scholarcommons.sc.edu/etd/5769/.

Menefee, H. 2018. Black Activist Geographies: Teaching Whiteness as Territoriality on Campus. *South: A Scholarly Journal* 50 (2): 167–186.

Monmonier, M. 2006. *From Squaw Tit to Whorehouse Meadow: How Maps Name, Claim, and Inflame.* Chicago: University of Chicago Press.

Mustaffa, J. B. 2017. "Mapping Violence, Naming Life: A History of Anti-Black Oppression in the Higher Education System." *International Journal of Qualitative Studies in Education* 30: 711–727. doi.org/10.1080/09518398.2017.1350299.

Plott, E. 2020. "For a Civil Rights Hero, 90, A New Battle Unfolds on His Childhood Street." *New York Times*, December 25, 2020. www.nytimes.com/2020/12/25/us/politics/fred-gray-rosa-parks-montgomery.html.

Post, C. and D. H. Alderman. 2014. "'Wiping New Berlin Off the Map': Political Economy and the De-Germanisation of the Toponymic Landscape in First World War USA." *Area* 46 (1): 83–91. doi:10.1111/area.12075.

Rodríguez, D. 2012. "Racial/Colonial Genocide and the 'Neoliberal Academy': In Excess of a Problematic." *American Quarterly* 64(4): 809–813. doi:10.1353/aq.2012.0054.

Rose-Redwood, R. 2008. "From Number to Name: Symbolic Capital, Places of Memory and the Politics of Street Renaming in New York City." *Social & Cultural Geography* 9 (4): 431–452. doi:10.1080/14649360802032702.

Rose-Redwood, R., D. H. Alderman, and M. Azaryahu. 2010. "Geographies of Toponymic Inscription: New Directions in Critical Place-name Studies." *Progress in Human Geography* 34 (4): 453–470. doi:10.1177%2F0309132509351042.

Rose-Redwood, R. and D. H. Alderman. 2011. "Critical Interventions in Political Toponymy." *ACME: An International E-Journal for Critical Geographies* 10 (1): 1–6. acme-journal.org/index.php/acme/article/view/879.

Safranksy, S. 2021. "Grammars of Reckoning: Redressing Racial Regimes of Property." *Environment and Planning D: Society and Space.* doi:10.1177%2F0263775821999664.

Summers, B. 2019. *Black in Place: The Spatial Aesthetics of Race in a Post-Chocolate City.* Chapel Hill: The University of North Carolina Press.

———. 2020. "We Need Action to Accompany Art." *Boston Globe*, June 11, 2020. www.bostonglobe.com/2020/06/11/opinion/we-need-action-accompany-art/.

Tichavakunda, A. 2020. "A Critical Race Analysis of University Acts of Racial 'Redress': The Limited Potential of Racial Symbols." *Educational Policy*, published online first. doi:10.1177%2F0895904820983031.

Till, K. 2012. "Wounded Cities: Memory-work and a Place-based Ethics of Care." *Political Geography* 31: 3–14. doi:10.1016/j.polgeo.2011.10.008.

Till, K. and A. Kuusito-Arponen. 2015. "Towards Responsible Geographies of Memory: Complexities of Place and the Ethics of Remembering." *Erdkunde* 69 (4): 291–306. www.jstor.org/stable/24585777.

Wahlers, K. E. 2016. "North Carolina's Heritage Protection Act: Cementing Confederate Monuments in North Carolina's Landscape." *North Carolina Law Review* 94 (6): 2176–2200. doi: scholarship.law.unc.edu/nclr?utm_source=scholarship.law.unc.edu%2Fnclr%2Fvol94%2Fiss6%2F8&utm_medium=PDF&utm_campaign=PDFCoverPages.

Wilder, C. 2013. *Ebony and Ivy: Race, Slavery, and the Troubled History of America's Universities.* New York: Bloomsbury.

Yeoh, B. 1992. Street Names in Colonial Singapore. *Geographical Review* 82 (3): 313–322.

———. 1996. Street-naming and Nation-building: Toponymic Inscriptions of Nationhood in Singapore. *Area* 28 (3): 298–307.

21

THE LANDSCAPE OF INCARCERATION IN THE UNITED STATES

Matthew L. Mitchelson

Landscape is important because it really is *everything* we see when we go outside. But it also is everything that we do not see.

Don Mitchell (2008, 47)

The United States operates the largest archipelago of jails and penitentiaries in the world. And yet it can be hard to find the prison in today's landscape.

Brett Story (2019, 181)

The purpose of this chapter is to welcome landscape scholars to the geographic study of the prison and the landscape of incarceration in the United States. My general argument is that the study of "carceral geography" (cf. Moran 2015) is a worthwhile subject for landscape scholars, and that landscape studies have much to offer to the subfield. My specific argument is that the prison must be seen; yet it is seldom what (or where) it appears to be, because the prison obscures as much as it reveals. The paper is divided in two sections, which mirror one another. The first reviews the geographic literature on incarceration over the past 25 years and the emergence of carceral geography. The second presents a case study from prison research in southwestern Virginia (cf. Mitchelson 2011). Brett Story's observation (above) is an important theme in both sections. And, as we will explore in what follows, the stakes are quite high.

Both sections in this chapter are oriented towards Mitchell's "new axioms for reading the landscape" (2008). That essay reflects (also from roughly 25 years) on a seminal piece by Peirce Lewis (1979). The Mitchell essay does not engage directly with prisons, yet the axioms serve as a meaningful heuristic for considering the landscape of incarceration and a powerful method for its analysis. At its core, this chapter is premised on Mitchell's assertion that "landscape *obscures*" and similarly seeks to provide a foundation for what incarceration's landscape "*is* and *does*, and for what a more just landscape might *be*" (Mitchell 2008, 33 [italics in original]).

Perhaps the most substantive point of departure from the Lewis essay is the prison population, which stood at roughly 500,000 people in 1979. As of this writing, that figure has more than quadrupled. The numbers can distract from the *always human* geographies on which they are based (cf. McKittrick 2012). No society has ever caged as many people in as many places as are currently held across the US landscape of incarceration. Today, "the prison" in the United

DOI: 10.4324/9781003121800-26

States confines 2.3 million people within thousands of jails, prisons, migrant detention facilities, and such (Sawyer and Wagner 2020). The country is 50 years into the largest-scale confinement of human lives in history. We turn now to its landscape.

Part I: The Emergence of "Carceral Geography" over the Past 25 Years

This chapter would have been unlikely 25 years ago. Although the prison population was already growing at an unprecedented scale and pace, there had not been much room in academic geography for the prison. There were notable exceptions (e.g., Valentine and Longstaff 1998). However, *the Dictionary of Human Geography*'s entry for "prisons" states that "the geographical literature on prisons is scant" (Herbert 2009). Yet, geographers over the past 25 years have created and contributed to a remarkable body of scholarship—and the subdiscipline of carceral geography is now growing at a remarkable pace. The study area—"carceral space"—is understood as, "the sites and relations of power that enable and incentivize the systemic capture, control, and confinement of human beings through structures of immobility and dispossession" (Story 2019, 14). Although this chapter centers on the landscape of incarceration in the United States, the field of carceral geography is increasingly global, with a particularly strong presence in the United Kingdom.

Carceral geography seldom starts or ends within the prison. Instead, the roots of US-based geographic approaches to the prison trace over the political-economic context of imprisonment and to the *life* and *families* of the imprisoned (cf. Gilmore 2008). In *Golden Gulag*, Gilmore champions the understanding "that prison is not a building 'over there' but a set of relationships that undermine rather than stabilize everyday lives everywhere" (Gilmore 2007, 242; cf. Bonds 2012). The resonance with critical landscape studies here is substantive. For example, landscape approaches appear a dozen times in *Golden Gulag*, as in: the "extraordinary lens" of "how the urban social experience fits into the rural landscape" (p. 17); or, how "California's deep political-economic restructuring reconfigured the social reproductive landscape, as well as the world of work" (p. 184); or, how "California's expanding criminal justice system overlaid the state's restructuring landscape with new prisons" (p. 222); and, with *always humanizing* stakes held and driven "in the landscape of home" (p. 177).

In turn, US geographers concerned with the landscape of incarceration have most often produced scholarship that resonates strongly with Gilmore's work as well as Mitchell's axioms, and his call to pay attention to political economy and social justice, specifically. The sections mirror one another, working through the axioms in descending and then ascending order, so as to start and end with the dire stakes of imprisonment (Figure 21.1).

Axiom 6: Landscape is the Spatial Form that Social Justice Takes

Carceral geography is understood and analyzed at the *social* scale. This is an important point of departure from the highly individualized, neoliberal epistemologies at the root of contemporary criminalization, policing, and popular opinion (cf. Peck 2003; Story 2019). Within those frameworks, atomistic individuals make deviant choices and enter a punitive system, which processes offenders throughout a carceral landscape that includes the highly abstract "bedspace" of the prison (Mitchelson 2014). There is little room for one's humanity or social justice within such a place. Yet, that is the state of social justice across the prison "in its material form—as it really is, not as we wish it to be" (Mitchell 2008, 47).

The landscape of incarceration is riddled with the tragic consequences of structural racism, xenophobia, economic inequality, sexism, and untold social ills that haunt the prison's broader

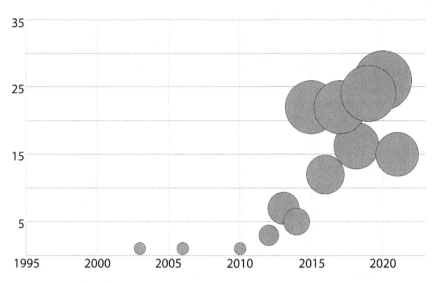

Figure 21.1 Frequency of Publications with the Topic "Carceral Geography" (as of December 2021). Web of Science.

context. The landscape of incarceration is then "made to be functional within (to align with)" those very social problems (Mitchell 2008, 46). Geographic studies that consider these sociospatial processes are composing a brutal chronicle of "actually existing" social justice in its spatial form (Bonds 2019; Catte 2018; Loyd, Mitchelson, and Burridge 2012; McKittrick 2012; Morin 2013, 2016, 2018; Shabazz 2015; Sexton and Lee 2006) across what amounts to a landscape of trauma (Mountz 2017). For example, Morin's analysis identifies, "spaces of unprecedented fear, terror, violence, and death" (2013, 382). For Mitchell, critical landscape analysis has a revelatory potential, "even as one of the functions of landscape is precisely to make those social relations obscure" (2008, 47). The state of social justice in the prison is a reflection of power relations.

Axiom 5: Landscape is Power

Carceral geographies do work on millions of people, and the types of work that the prison can do stand in stark contrast to the personal freedom(s) of their bodies. For example, the carceral landscape holds host to such punitive socio-spatial processes as banishment, exclusion, disposession, and disappearance (Herbert and Beckett 2009; Mountz et al. 2013; Anne Bonds 2012; Story 2019). In the United States, the prison is also the acute location of the state's monopoly on the right to execute its citizenry (cf. Inwood 2015). It seems reasonable to expect that these state powers would be limited and marginalized instead of persistent and expansive. They are. However, these state powers are only limited and marginalized *for some*. It takes a specific type of work to empower the proliferation of the prison and its social (in)justice. That work is done by *criminalization*.

Criminalization is the process at the root of carceral expansion. For Gilmore (quoted in Loyd 2012, 45), criminalization is "to move the line of what counts as criminal to encompass and engulf more and more people into the territory of prison eligibility." This explanation destabilizes the often taken-for-granted relationship between crime and incarceration (crime goes up, criminals go to prison, and crime goes down), which belies the empirical reality that rates of crime and incarceration are often unrelated (Gilmore 2007). In other words, the root of social (in)justice across the landscape of incarceration traces back to the *social* (in)justice of criminalization. Yet, it assumes the discursive and ideological appearance of *personal* responsibility. The disappearance of this relationship—the dialectical relationship between personal *and* social behaviors and responsibilities (i.e., structure, agency, and context)—discursively produces the prison as a *necessary* institution, because it obscures a nearly infinite number of creative social solutions to social problems beyond walls and cages (Loyd, Mitchelson, and Burridge 2012). With the prison—as with other landscapes—"the preeminent power that landscape might express is the power to erase history, signs of opposition, alternative readings, and so forth" (Mitchell 2008, 43). Yet, every prison has a history.

Axiom 4: History Does Matter

The prison is often cited as a textbook example of a Not In My BackYard (NIMBY) land use, one to be resisted strongly by local property owners. However, this has shifted radically in recent decades, particularly in regions characterized by primary commodity production or deindustrialization. As Bonds explains, "rural prison recruitment has become so ordinary that the notion of YIMBYism ["Yes In My BackYard"] seems more commonplace with respect to correctional facilities than NIMBYism (Bonds 2013, 1402). The construction of "rural prisons" was commonplace during the 1980s and 1990s (Beale 1996). How could such a profound reversal of attitudes come to be?

Boosters often boast that the prison is recession-proof, and prison construction is touted as a means of economic development. This refrain is found in vastly distant (and often vastly different) locales (Che 2005; Gilmore 2007; Bonds 2013; Story and Schept 2018), and despite substantive evidence to the contrary (Glasmeier and Farrigan 2007; Hooks et al. 2010). The message is politically salient where people struggle to make a living—where the mines and farms in prospective "prison towns" may still produce commodities at high levels, but jobs have often shifted elsewhere (or been "creatively destroyed"). Following Mitchell's logic (2008, 43), all of this *made* the landscape of incarceration, but it is very hard to find *in* the landscape. Thus, we must consider "the local" through its often distant social relations.

Axiom 3: No Landscape is Local

The third axiom argues that "to understand [a given place or landscape] requires looking not there" (Mitchell 2008, 40). Story similarly invites us, "to see the prison clearly, and therefore differently… we might do well to cast our gaze, at least for a period, everywhere but the actual penitentiary" (2019, 183). The prison is a tricky building, full of other places. *Golden Gulag* starts and ends with people on a bus, as does Story's fascinating documentary *The Prison in Twelve Landscapes* and the bus garners a full chapter treatment in *Prison Land* (Story 2019). Buses manifest the prison's connections to distant people and places as they travel to and from each prison. These teach us that however powerful the spatial processes addressed with Axiom 5 may be (e.g., banishment or exclusion), and however dehumanizing a punitive process may be intended to be, if a geography of the prison is *always human* then it is always produced by and through dynamic social relations. These relations can be productive of distance, fixity, and flow.

Cages may suggest a timelessness and fixity. But that fixity requires flow: flows of always human bodies, flows of innumerable objects, flows of resources, capacities, and capital (Gilmore and Gilmore 2008). Recent contributions acknowledge the "spatial fix" that confinement entails, but, also account for flow as a component of carceral *circuitry* (Gilmore 2007; Gill et al. 2018; Martin 2021. Very little of the prison is local. All of this landscape moves or holds host to circulation(s). In other words, because the landscape of incarceration is a reflection of the spatial form of social justice and power well beyond the prison's walls, the prison is much more than its architectural facts and its locational footprint. The point is not to neglect or forget the cage; the point is to see it as one of many positions in a broader carceral landscape. That landscape is always *put to work* (and *working*, in its own right) towards particular functions.

Axiom 2: Any Landscape is (or was) Functional

Mitchell is insistent that "landscapes have a role to play in social life: they exist for a reason" (2008, 35). So, what function(s) does the landscape of incarceration serve? The answer is simultaneously obvious and elusive, because the functions are contradictory. Both coercion and consent are in play. Criminalization is a foundational problem. However, even after the juridical facts of "crime" itself, criminal justice systems—and the popular criminology upon which they are founded (e.g., "broken windows" policing)—consistently exacerbate social differences to the political-economic ends of neoliberalism (Herbert and Brown 2006). Again, the historical-geographic record of carceral injustice is simply brutal. Why would any society permit such a system to proceed?

For Gilmore and Gilmore, "the state is remaking itself using the newly vast prison system's coercive powers on some parts of the population to produce consent, among others" (2008, 158). On the one hand, this empowers the state to manage "surplus populations." Within all of this is a tremendous disappearing, from the right to vote to the right to personhood (Varsanyi 2008, 2012), to the prisoner's body and life itself (Mitchelson 2014; Martin 2021). This is a specific configuration of social relations: a concentration of coercion within one "criminal class"; and, simultaneously, a concentration of consent within a second class of "law-and-order abiding citizens.". Any harms—including "collateral consequences"—disappear, in a sense. Out of sight (for some), out of mind (for some). As the prison "both expresses and naturalizes difference" (Mitchell 2008, 37), it does a massive, sprawling type of work and it produces a massive, sprawling landscape.

Axiom 1: The landscape is produced; it is actively made: it is a physical intervention in the world and thus is not so much our "unwitting autobiography" (as Lewis put it) as an act of will.

You can find a prisoner in more than 7,000 different prisons across the United States. There are nearly 1,943 state and federal prisons; 3,134 jails; 1,772 juvenile facilities; 218 detention facilities; and more than 100 institutions of another type (Sawyer and Wagner 2020). Not one exists by coincidence. Each physical intervention was actively made. Each consumes vast resources. Each reflects the social relations through which it was produced (cf. Mitchell 2008, 34). We will transition now to a case study, revisiting each axiom in reverse.

Part II: Case Study: Producing an "Alcatraz in the Sky" in Appalachia

"It must be here."

A geographer is lost, down by the side of the road, in southwest Virginia. It is a bright, cloudless day in the early spring. A paper map is unfolded across the trunk of his car, because his digital map is out of service. He is tired and talking to himself, muttering the same four words.

"It must be here."

In fairness, there's a lot to take in. The Powell Valley is a beautiful place, and a geological wonder in which the elevation drops two thousand feet over a short distance. Of all the sights to see here, however, he is looking for a prison.

"It must be here."
There is a large prison just ahead, up on the ridge. A prison on a mountaintop.
He would see it if he looked just right.

Finding 1: The Landscape of Incarceration is Produced; it is Actively Made

Virginia's Wallens Ridge State Prison in Wise County is an act of will, hailed as an "Alcatraz in the Sky" (cf. Mitchelson 2011). When the Secretary of Public Safety announced that a prison was coming to southwest Virginia the regional paper's headline read, "Hard work makes prison dream reality" (Mitchelson 2011, 1752). It was eventually paired and co-constructed with a twin facility—Red Onion State Prison—some 30 miles away. Things did not proceed as planned. Everything from cost and state politics, to construction site complexity and a nearby Koch Carbon Company lawsuit, seemed destined to prevent their construction. Yet, years later the same paper announced "Two prisons break ground," and declared that "1995 was the year of the prison project in southwest Virginia" (Mitchelson 2011, 1751).

Finding 2: The Landscape of Incarceration is Functional

While local officials were working to will the prison(s) into existence, the state reconfigured itself. New Governor George Allen championed "tough on crime" politics. He abolished parole and lengthened sentences statewide, and aggressively pursued prison construction in southwest Virginia. He replaced seven of the eight Board of Corrections members who had voted against a deed transfer from the Pittston Coal Group for the proposed Red Onion State Prison site. One of Allen's new appointees to the Board was Jack Rose, who had initially approached Pittston Coal about the land. The prison figured prominently in Allen's vision of the state.

But, there was a problem for Allen's vision. There were not enough in-state prisoners to fill the prisons. When the state initially kept Wallens Ridge empty, Allen's appointed Secretary of Public Safety said, "here we have a state-of-the-art facility... that is doing nothing. It is like having a new car and making payments on it but not driving it" (Mitchelson 2011, 1757). Like so much of what produced these prisons, prisoners themselves would have to come from elsewhere.

Finding 3: The Landscape of Incarceration is Not (only) Local

Thousands of prisoners have come to Wallens Ridge and Red Onion from Virginia. Thousands more have also come from remarkably distant places: Wyoming, New Mexico, Hawaii, and the Virgin Islands. Connecticut state prisoners were also held in Wallens Ridge until the deaths of several prisoners prompted protests, investigations, and a federal civil rights lawsuit. Their removal resulted in a multi-million dollar decline in revenues for Virginia. While prisoners are necessary in the production of modern prisons, they are only one part of their circuitry.

Public prisons such as these are "vetted by the power of commodities—right down to the last bolt and washer" (Mitchell 2008, 34), and the capital circuitry of prison construction is remarkably non-local. Prison sites were managed by firms hundreds of miles away (one in Richmond and one in Laurel, Maryland). The integrated engineering/architectural firm that oversaw both prisons is part of AECOM, headquartered in Los Angeles. There are more than

1,400 cages in Wallens Ridge, in which two prisoners share an area just over 7' x 13'. They were prefabricated in Petersburg, two cells to a module, and shipped 400 miles to their place in the coal country archipelago.

Finding 4: History Does Matter

Today, four large state prisons—with an average daily population of 3,632 between them—are located in the seven-county coalfields where there were once more than 120 coal company towns. Wallens Ridge and Red Onion in Wise County were preceded by Keen Mountain State Prison in adjacent Dickenson County, and they are proceeded by the Pocahontas State Correctional Center in Tazewell County, which opened in 2007. Red Onion was physically and symbolically built atop a reclaimed mine. Prison jobs pay a fraction of what miners once earned in the region (Catte 2018). Yet, today there are 30% more prisoners confined in the coalfields than there are miners throughout the entire state. In this economically restructured context, prison construction appealed to many.

The corrections director's position was clear: "prisons are an economic development program" (Mitchelson 2011, 1760). Years before, while promising the jobs required to build "everything you need to run a little city," one state official's remarks clearly echo the YIMBY phenomenon as he described "a community such as yours where there's a united voice saying 'come on down, be our neighbors'" (Mitchelson 2011, 1752). The executive director of a regional planning commission asserted "there ain't too many alternatives… if we don't take advantage of it from an economic development standpoint, I'm afraid we'll go to hell in a handbasket" (p. 1760). Just before the state signed the 150-acre land option atop Wallens Ridge, the mayor publicly wished that the Department of Corrections would "put something in the region's Christmas stocking" (p. 1754). Ground was soon broken in front of hundreds of locals as the Powell Valley High School band played.

Finding 5: The Landscape of Incarceration is Power

Wallens Ridge—like all landscapes—was produced through an (albeit amalgamated) act of will. The ridge was lowered (by the length of a football field) and flattened as 2.3 million cubic yards were excavated from the summit. Excavation expenses alone were $17 million. When Governor Allen detonated "The Last Blast," turning 17,000 cubic yards into rubble, he said, "I've never been able to blow up a mountain. I also never had this much fun with firecrackers" (Mitchelson 2011, 1756). Years before, at another public prison rally, those in attendance reportedly laughed as he said, "Some didn't want another prison in southwest Virginia because they thought it would be too far for visitors." Amidst the crowd's laughter he said, "That was kind of my reaction, too" (Mitchelson 2011, 1758).

As the first prisoners were bused up the mountain the mark of power on the landscape appeared to be absolute. The Governor, for example, seemed to have power; prisoners and their loved and loving ones seemed to be without. Yet, there were struggles against the prison. At a public forum one teacher said, "I just want to know why you can afford to build new prisons but can't afford to build better schools"; and, longtime resident Sister Beth Davies (Congregation of Notre Dame) argued forcefully that "building more prisons to fight crime is like building more graveyards to fight terminal disease" (Mitchelson 2011, 1761). Ultimately, the long roads to Wallens Ridge and Red Onion lead back to the criminalization at the root of incarceration's political economy. So little of that is visible on the landscape. Yet, it is there.

Finding 6: *The Landscape of Incarceration is the Spatial Form that Social Justice Takes*

There is so much invested in the prison, and the magnitude of investment can be approximated. Last year, Wallens Ridge held roughly 1,075 prisoners per day at a total cost of more than $36.4 million; Red Onion held an average of just under 750 prisoners each day for more than $35.1 million (Virginia Department of Corrections 2020). In Wise County, where more than one in five residents is reportedly living in poverty, the median household income is 15% higher than the expense per prisoner at Wallens Ridge; the cost per prisoner at Red Onion is actually 20% higher than median household income for the county (US Census Bureau 2019). These are political-economic choices. In order to try to make sense of such choices, we might consider another community, some 2,500 miles away.

Gilmore tells a powerful geographic story through the work of Juana Gutierrez against a prison siting in her Sacramento neighborhood, as "urban and rural households struggle from objectively similar but subjectively different positions across the prison landscape… A principled sense of mortal urgency gets grassroots activists to go to meetings, makes them board buses, and inspires hope" (2007, 250–251). Indeed, a principled sense of mortal urgency can guide us to the questions that might empower us to collectively respond to social problems through means less costly—in every meaning of the word—than walls and cages. Perhaps, in time, it will.

The same four words are still stuck in his head, but things have changed.

The geographer can see the prison now. He's come back, after dark, and he couldn't miss it if he tried. The 1,000-watt high-pressure sodium lamps are burning on their high mast towers. Up on the mountain, they look like signal fires. Unmistakable, even from 10 miles away. Burning energy in this form, out of a universe of possibility and opportunity for change.

The geographer speaks the same four words, remade as the question that the prison demands.
"Must it be here?"

References

Beale, C. 1996. "Rural Prisons: An Update." *Rural Development Perspectives* 11 (2): 25–27.

Bonds, A. 2012. "Building Prisons, Building Poverty: Prison Sitings, Dispossession, and Mass Incarceration." In *Beyond Walls and Cages: Prisons, Borders, and Global Crisis*, edited by J. Loyd, M. Mitchelson and A. Burridge, 129–142. Athens, GA: The University of Georgia Press.

———. 2013. "Economic Development, Racialization, and Privilege: "Yes in My Backyard" Prison Politics and the Reinvention of Madras, Oregon." *Annals of the Association of American Geographers* 103 (6): 1389–1405.

———. 2019. "Race and ethnicity I: Property, Race, and the Carceral State." *Progress in Human Geography* 43 (3): 574–583.

Catte, E. 2018. *What You Are Getting Wrong About Appalachia*. Cleveland, OH: Belt Publishing.

Che, D. 2005. "Constructing a Prison in the Forest: Conflicts Over Nature, Paradise, and Identity." *Annals of the Association of American Geographers* 95 (4): 809–831.

Gill, N., D. Conlon, D. Moran, and A. Burridge. 2018. "Carceral circuitry: New Directions in Carceral Geography." *Progress in Human Geography* 42 (2): 183–204.

Gilmore, R. W. 2007. *Golden Gulag: Prisons, Surplus, Crisis, and Opposition in Globalizing California*. Book American Crossroads. Berkeley: University of California Press.

———. 2008. "Forgotten Places and the Seeds of Grassroots Planning." In *Engaging Contradictions: Theory, Politics, and Methods of Activist Scholarship*, edited by C. R. Hale, 141–162. Berkeley, CA: University of California Press.

Gilmore, R. W., and C. Gilmore. 2008. "Restating the Obvious." In *Indefensible Space: The Architecture of the National Insecurity State*, edited by M. Sorkin, 141–162. New York: Routledge.

Glasmeier, A. K., and T. Farrigan. 2007. "The Economic Impacts of the Prison Development Boom on Persistently Poor Rural Places." *International Regional Science Review* 30 (3): 274–299.

Herbert, S. 2009. "Prisons." In *The Dictionary of Human Geography*, edited by D. Gregory, R. Johnston, G. Pratt and S. Whatmore, 582–583. Malden, MA: Wiley-Blackwell.

Herbert, S., and K. Beckett. 2009. *Banished: The New Social Control in Urban America.* Edited by M. Tonry and N. Morris. *Studies in Crime and Public Policy.* New York, NY: Oxford University Press.

Herbert, S., and E. Brown. 2006. "Conceptions of Space and Crime in the Punitive Neoliberal City." *Antipode* 38 (4): 755–777.

Hooks, G., C. Mosher, S. Genter, T. Rotolo, and L. Lobao. 2010. "Revisiting the Impact of Prison Building on Job Growth: Education, Incarceration, and County-Level Employment, 1976–2004." *Social Science Quarterly (Wiley-Blackwell)* 91 (1): 228–244.

Inwood, J. F. 2015. "Introduction: Geographies of Capital Punishment in the United States: The Execution of Troy Davis." *ACME: An International E-Journal for Critical Geographies* 15 (4): 1058–1065.

Lewis, P. 1979. "Axioms for Reading the Landscape." In *The Interpretation of Ordinary Landscapes: Geographical Essays*, edited by D. Meinig and J.B. Jackson, 11–32. New York, NY: Oxford University Press.

Loyd, J. M. 2012. "Race, Capitalist Crisis, and Abolitionist Organizing: An Interview with Ruth Wilson Gilmore." In *Beyond Walls and Cages: Prisons, Borders, and Global Crisis*, edited by J. Loyd, M. Mitchelson and A. Burridge, 42–54. Athens, GA: The University of Georgia Press.

Loyd, J. M., M. Mitchelson, and A. Burridge, eds. 2012. *Beyond Walls and Cages: Prisons, Borders, and Global Crisis.* Edited by D. Cowen, N. Heynen and M. W. Wright. Vol. 14, *Geographies of Justice and Social Transformation.* Athens, GA: The University of Georgia Press.

Martin, L. 2021. "Carceral Economies of Migration Control." *Progress in Human Geography* 45 (4): 740–757.

McKittrick, K. 2012. "On Plantations, Prisons, and a Black Sense of Place." *Social & Cultural Geography* 12 (8): 947–963.

Mitchell, D. 2008. "New Axioms for Reading the Landscape: Paying Attention to Political Economy and Social Justice." In *Political Economies of Landscape Change: Places of Integrative Power*, edited by J. Wescoat and D. Johnston, 29–50. Heidelberg: Springer Netherlands.

Mitchelson, M. L. 2011. "'Alcatraz in the Sky': Engineering Earth in a Virginia (USA) Prison." In *Engineering Earth*, edited by S. D. Brunn, 1749–1764. New York, NY: Springer.

———. 2014. "The Production of Bedspace: Prison Privatization and Abstract Space." *Geographica Helvetica* 69 (5): 325–333.

Moran, D. 2015. "Carceral Geography Spaces and Practices of Incarceration Introduction." In *Carceral Geography: Spaces and Practices of Incarceration.* Aldershot: Ashgate Publishing Ltd.

Morin, K. M. 2013. 'Security Here is Not Safe': Violence, Punishment, and Space in the Contemporary US Penitentiary." *Environment and Planning D: Society and Space* 31 (3): 381–399.

———. 2016. "Carceral Space: Prisoners and Animals." *Antipode* 48 (5): 1317.

———. 2018. *Carceral Space, Prisoners and Animals.* London: Routledge.

Mountz, A. 2017. "Island Detention: Affective Eruption as Trauma's Disruption." *Emotion Space and Society* 24: 74–82.

Mountz, A., K. Coddington, R. T. Catania, and J. M. Loyd. 2013. "Conceptualizing Detention: Mobility, Containment, Bordering, and Exclusion." *Progress in Human Geography* 37 (4): 522–541.

Peck, J. 2003. "Geography and Public Policy: Mapping the Penal State." *Progress in Human Geography* 27 (2): 222–232.

Sawyer, W., and P. Wagner. 2020. "Mass Incarceration: The Whole Pie 2020." The Prison Policy Initiative. Last Modified March 24, 2020. www.prisonpolicy.org/reports/pie2020.html.

Sexton, J., and E. Lee. 2006. "Figuring the Prison: Prerequisites of Torture at Abu Ghraib." *Antipode* 38 (5): 1005–1022.

Shabazz, R. 2015. *Spatializing Blackness: Architectures of Confinement and Black Masculinity in Chicago.* Urbana, IL: University of Illinois Press.

Story, B. 2019. *Prison Land: Mapping Carceral Power across Neoliberal America.* Minneapolis, MN: The University of Minnesota Press.

Story, B., and J. Schept. 2018. "Against Punishment: Centering Work, Wages, and Uneven Development in Mapping the Carceral State." *Social Justice* 45 (4 (154)): 7–34.

US Census Bureau. 2019. www.census.gov/quickfacts/wisecountyvirginia.

Valentine, G., and B. Longstaff. 1998. "Doing Porridge." *Journal of Material Culture* 3 (2): 131–152.

Varsanyi, Monica W. 2008. "Rescaling the "Alien," Rescaling Personhood: Neoliberalism, Immigration, and the State." *Annals of the Association of American Geographers* 98 (4): 877–896.

———. 2012. *Fighting for the Vote: The Struggle Against Felon and Immigrant Disenfranchisement.* University of Georgia Press.

Virginia Department of Corrections. 2020. *Management Information Summary Annual Report for the Fiscal Year Ending June 30, 2020.* Richmond, VA: The Budget Office Division of Administration.

PART IV

Urban and Economic Landscapes in the US

Chris W. Post

In a Venn diagram, the geographies of economics and urbanization share a lot of space. As US economic activities have developed from agricultural origins to industrial might and to service-oriented emphasis, people have clustered and created the nation's towns, cities, and suburbs. This section combines these two fundamental themes of human geographies and landscape.

The shifts in our economic focus have radically changed our cities. The Cuyahoga River near me runs cleaner and Chicago's air breathes easier. Simultaneously we've forged the inefficient (and unsustainable) development of suburbs, the agglomeration of retail businesses in strip and shopping malls, and abandonment of factories on our riverfronts and coastlines. Problems remain and new concerns have emerged. Segregation and environmental racism continue to evidence themselves daily. Dry day flooding plagues cities such as Miami and Norfolk, Virginia. Chicago's basements flood from a shifting Lake Michigan (Egan 2021). On the whole, we seem to have traded in some concerns for new ones as racism and a warming planet continue to manifest. In some cases, our cities morph in response, still sensitive to our decisions.

The contributors in this section signal several themes that anchor the relationship between urban and economic landscapes over the past quarter century. The first of these is change. The American economy's post-industrial shift has altered where and how we work, shop, eat, drive, and thrive. Part and parcel with this shift is the digitization of our lives—the development of e-commerce and the resulting "gig" labor through online services. Third, a new sense of justice has entered the American urban–economic landscape—not that justice has been entirely achieved. The arc of labor justice has never really ceased. While it perhaps has slowed in recent decades as CEO pay skyrockets in response to pleasing investors, there are signs of unionization in the gig economy and increased wages during our collective recovery from the COVID-19 pandemic that give hope. Our cities are becoming "smarter" and greener, and the tourism economy has also shown more equity in recent decades than it once expressed as both legal segregation of sites and their environmental impacts have waned.

With these shifts in the "how" and the "where" of our economic practices and resulting urban spaces, has come a new set of landscapes that situate these actions. Be they tourist spaces, urban farms, brownfield development, or Amazon lockers, our economic and urban landscapes continue to shift under our feet as we make new decisions about how we live.

DOI: 10.4324/9781003121800-27

Our first chapter investigates the environmental costs produced through our economic and urban development through the past few decades. Lisa Benton-Short and Melissa Keeley deftly discuss one of the most important issues of our time—urban sustainability, and how cities make decisions to decrease our impacts in urban spaces and come to a more maintainable future of urban planning and development. So many of these changes are further developed throughout this section—the impacts of automobiles, tourism, and especially the shift from industry to services.

As our first chapter posits, the response of urban and economic landscapes to the automobile's commanding presence in our lives has changed our cities perhaps more than any other factor. Ellen Hostetter takes readers to Conway, Arkansas, to see how that city has evolved due to the automobile. She also takes us outside of Albuquerque, New Mexico, to comment on how rural landscapes appear to us as we take to today's superhighways. Dydia DeLyser buttresses this chapter by providing greater context concerning the types of data we collect about our landscapes and how we interpret the human condition based on their interpretation.

As Hostetter details, automobiles move us away from home. While we built cities around that fact, we also take our cars on vacation as tourists. A significant portion of the American economy—over \$1 trillion worth—is in tourism. This activity increasingly impacts both rural and urban America. Velvet Nelson takes a broad brush to help us see how tourism landscapes have changed over the past few decades. While racial segregation or sexist policies may not directly influence our vacation sites anymore, serious questions still exist about how we travel, where we travel, and how these behaviors have created new landscapes in places that rely on the economic activity as a source of economic survival.

As these tourism habits attest, we cannot ignore the millions of citizens who live in rural America. Andrew Husa takes us to the rural Plains and overviews how the less populated spaces between our cities have changed over the past quarter century in response to population loss, cultural diversification, and continued reliance on agricultural productivity. Cheryl Morse adds to this investigation by looking specifically at how the youth of Rural America create their own meaningful places and landscapes amidst continued urbanization.

Our last three chapters in this section focus more on the economic side of these intertwined themes. Emily Fekete focuses on the development of retail landscapes from the foundational perspective of Walter Christaller to how today's consumption is guided by digitization and supposed "ease." Retail malls have shuttered, and Amazon lockers now dot our cities. Fekete details this transformation and its relationship to our consumptive behavior.

Mark Rhodes and Sarah Scarlett investigate deindustrialization's impacts on our collective memory and identity with our industrial past. How do we treat the history of labor, industry's environmental impacts, and the economic development it offered places large and small across the country? Rhodes and Scarlett ask how we interact with those sites today.

Andrew Herod concludes this section by detailing two crucial economic foci. His chapter on labor expresses how shifts in our economic activity have changed the way we work, unionize, and earn a living. The landscape of the workplace—the assembly line, presence of common areas, doors, and the new gig economy—have radically changed the way we socialize, collectivize, and endure a day's work. Herod also contributes a special essay on how geography's interpretation of economic landscape has developed over the past century, concluding with how Marxist theory influences much of today's work.

As a collection these chapters detail a continually shifting economy and associated urban and rural landscapes. Changes experienced years ago with deindustrialization continue even further with the digitization of our lives. Still, and despite these changes, our impact on the urban landscape continues to revolve around our use of the automobile as we chase the dream of individualized mobility. With different foci, these chapters synthesize our modern urban places as economic engines seeking to become more attractive and sustainable.

Reference

Egan, D. 2021. "A Battle Between a Great Lake and a Great City: The Climate Crisis Haunt's Chicago's Future." *New York Times*. July 7, 2021. www.nytimes.com/interactive/2021/07/07/climate/chicago-river-lake-michigan.html.

22

URBAN SUSTAINABILITY

Lisa Benton-Short and Melissa Keeley

For many, the term "American landscape" conjures images of bucolic countrysides and charming small towns. In reality, about 85 percent of the US population lives in urban areas. We cannot fully understand the American landscape without understanding the changing urban landscape.

Scholars have recognized that the contemporary urban landscape is not accidental. It is socially produced; the result of social, economic, political, and physical systems that create diverse material forms—from the built environment of skyscrapers and freeways to the natural environment of waterfronts, parks, and wildlife.

For a long time urban scholars largely ignored the physical nature of cities; instead the emphasis was on the social, political, and economic rather than the ecological. However, in recent years there has been renewed attention on the complex relationship between the city and its natural systems and how this impacts the urban landscape. Scholars in urban environmental history such as William Cronon (1991) and Matthew Klingle (2007) have looked at how cities such as Chicago or Seattle reflect a story of continued landscape modification. Other scholars examine the degree to which cities impact the environment. Ian Douglas (1981) has shown that cities rely on the physical environment for resources (water, food, energy) and have a tremendous impact on that environment (globally, cities account for 78 percent of greenhouse gas emissions, generate air and water pollution, and even reconfigure natural environments). Scholars also examine the social context of the city–nature dialectic in a body of work characterized as urban political ecology. Matthew Gandy (2002), for example, examined the relationships between nature, the city, and social power in the creation of New York City's water supply. Eric Keys and colleagues looked at changes over 20 years to the spatial structure of land use in Phoenix and found that the desert was increasingly fragmented, with implications on landscape change and biodiversity (Keys et al. 2007). Today a significant force shaping the contemporary urban landscape is the way in which urban sustainability is changing the city–nature dialetic and transforming the landscape.

This chapter critically surveys recent trends and developments in urban sustainability. Understanding how cities conceptualize, plan, and implement sustainability is critical in understanding recent changes in the urban landscape. The urban–environment dynamic *is* shifting as cities make efforts to implement sustainability. We highlight three key ways that the urban landscape is being transformed by efforts to address sustainability challenges. First, cities are adapting both built infrastructure and natural systems as they seek to adapt to the local

DOI: 10.4324/9781003121800-28

impacts of climate change. Second, cities are creatively redeveloping and re-purposing urban spaces to promote economic development through denser, more centralized, and mixed-use development, and facilitate more sustainable transportation choices. Third, as cities confront issues of inequality and environmental injustice, this emerging focus on equity seeks to reduce the existing segregation by race and class in US cities as well as alter the distribution of amenities and disamenities throughout the urban landscape.

To understand these changes, this chapter first introduces the trend towards urban sustainability. Next, we examine how cities plan for sustainability—underscoring the fact that this concept moves beyond simply "greening" the city and requires considerations of the economy and equity that shape both the natural and the built environment. Finally, the essay next explores how cities' sustainability plans and actions are changing the urban landscape in the three ways mentioned above.

Cities and Sustainability

The idea of sustainability coalesces around the three issues of environmental protection, economic development, and the advancement of social equity. These Three "E's" are often depicted as three interconnected circles (as shown in Figure 22.1). Importantly, sustainability is not concerned only with the environment: at its root it is concerned with the quality of human life over the long term. The concept emerges from a long and rich intellectual evolution of ideas that redefined nature–society relationships. Early framing came when the 1987 United Nations World Commission on Environment and Development issued a report called *Our Common Future* (also referred to as the *Brundtland Report*). The report defined sustainable development as "development that meets the needs of the present without compromising the ability of future generations to meet their own needs" (World Commission on Environment and Development United Nations 1987).

Sustainability moves beyond simply protecting the environment by embracing economic development that invests in people, places, and processes that generate jobs and maintains livelihoods. While equity has long been the least operationalized element of sustainability, its importance is increasingly understood. Genuine sustainability involves ensuring that all members of a community have their voices heard in decision-making processes and that environmental and economic benefits are distributed more equitably among all. Addressing underlying problems of economic and social inequity could be one of the biggest challenges to a more sustainable future and are also seen as essential to addressing environmental challenges.

Urban Sustainability Plans

In recent years and for a variety of reasons, cities have taken the lead in sustainability planning and action in the United States (Portney 2003; Zhenghong et al. 2010). As federal leadership in the creation of environmental policies has faltered, there has been growing support for local initiatives, often referred to as a "new localism" in environmental policies (Parker and Rowlands 2007). The city scale can be a beneficial starting point for local activism and community involvement around sustainability (Boone and Moddares 2006). Cities have tremendous control over numerous tools for change such as land use, public education, and economic development and are positioned to address unique local environmental, social, and economic issues in the process (Wheeler 2013).

The vision of cities as leaders in sustainability is not unique to the US. In fact, US cities have lagged behind European cities in sustainability planning. Further, the 2015 United Nations

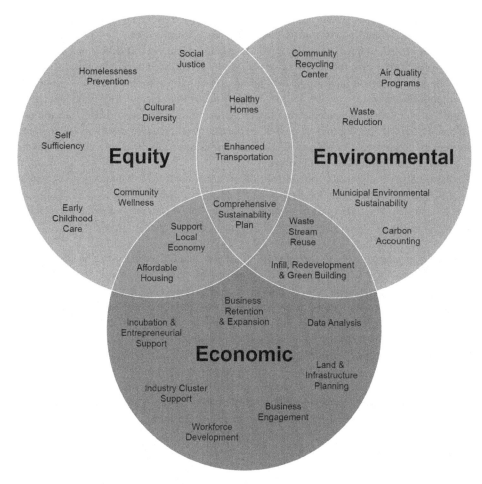

Figure 22.1   Three Es in the context of Urban Sustainability Plans. Figure by authors.

Sustainable Development Goals call for a substantial increase in the number of cities and human settlements adopting and implementing integrated policies and plans towards inclusion, resource efficiency, mitigation, and adaptation to climate change, and resilience to disasters.

As an approach to planning, sustainability has the power to transform the urban landscape in promoting projects and policies that simultaneously consider environmental, economic, and social equity concerns. The outcomes can be both highly visible (reducing air pollution), and less visible (reducing rates of childhood asthma caused by air pollution).

Municipal sustainability plans are holistic and multi-departmental documents that outline a city's goals, visions, and priorities for a sustainable future. These plans often inventory current problems and standings, identify solutions and priorities, and set indicators for measuring progress (Evenson et al. 2009). Sustainability plans are more holistic than most other planning documents and consider multiple goals relating to improving environmental, economic, and social equity conditions simultaneously. For this reason, sustainability plans often serve as an "umbrella" or connector between initiatives and they both prioritize and highlight ways in which projects and programs can provide multiple benefits to the city. However, there is great variety in the aspects of sustainability that cities prioritize or address first, and great differences in the roles that cities take in furthering the sustainability agenda that they outline, and the changes that implementing these plans may have on the landscape (Keeley and Benton-Short 2019).

Many cities have conducted lengthy public engagement exercises in developing their sustainability plans, often over the course of two or more years. Involving a diversity of voices in prioritizing and planning is increasingly recognized as essential. Some cities have also innovated to reach populations that may otherwise have not participated in planning by providing childcare, holding meetings in languages other than English, and partnering with local institutions like places of worship. Such efforts to include minority, low-income, or others who have been under-represented in previous planning efforts may mean sustainability plans will more effectively address the needs of these populations.

A sustainability plan intends to provide both short-term and long-term guidance for current and future decision makers, city employees, city leaders, city residents, and other community groups and entities. They are a starting point for change. And visions, even general ones, can be powerful. Consider the vision statement in Washington, DC's sustainability plan:

> In just one generation—20 years—the District of Columbia will be the healthiest, greenest, and most livable city in the United States. An international destination for people and investment, the District will be a model of innovative policies and practices that improve quality of life and economic opportunity. We will demonstrate how enhancing our natural and built environments, investing in a diverse clean economy, and reducing disparities among residents can create an educated, equitable, and prosperous society. (Sustainable DC 2012, 5)

A great variety in topical priorities are emphasized in the sustainability plans across US cities. Many of these plans are comprehensive in that they address a range and diversity of issues that include the "three E's: ecological protection, economic development, and social equity" (See again Figure 22.1). Among the most common issues addressed in sustainability plans are:

- Food
- Health
- Air quality
- Climate change
- Water quality
- Water supply
- Parks and recreation
- Social justice and equity
- Green economy/green jobs
- Transportation
- Energy use
- Housing
- Garbage and recycling
- Risk and resilience to hazards

While topical breadth is a feature of most sustainability plans, US cities emphasize different priorities, sometimes focusing on specific local or regional problems or connecting to broader stewardship around global issues like climate change, and highlighting local impacts and mitigation strategies.

Sustainability planning and action is complex and challenging, and many cities are challenged by limited funds and competing priorities. Other cities are much more focused on changing public behaviors, sometimes through the creation of new infrastructures, like bike lanes. While infrastructure projects are particularly time consuming, New York City describes the importance of engaging the public in sustainability efforts in this way:

> Changes by individuals and households… have the potential to be quicker and more cost-effective than policy initiatives. While it could take years to implement a new capital project or pass and implement a law, it only takes months to develop and execute a marketing campaign and seconds for a person to decide to switch off a light or choose to bike to work. (City of New York 2017)

Finally, planning is an iterative process, not an end result, in which plans are implemented and outcomes are assessed so that this information can be used to reassess strategies, goals, and actions for the future. In doing so it links back to how cities conceptualize sustainability. The pace of innovation in the realm of sustainability is rapid, and many cities frequently reassess and integrate emerging best practices. We argue that the sustainability planning process sets the stage for investment in a variety of diverse projects that change the physical, social, and economic urban landscape. Cities are using their sustainability plans to take actions and implement initiatives that are transforming the urban landscape. Such actions vary widely and impact the current and future urban landscape in countless ways. For example, in some cities, building code updates will promote a new generation of greener buildings or require green infrastructure (like rain gardens, street trees, and green roofs) to manage stormwater and provide other urban amenities. Similarly, in some cities stream restoration and wetlands reconstruction restores natural processes while connecting large-scale parkway systems provides corridors for wildlife.

Urban Sustainability Changes the Urban Lanscape

In this section we consider three ways that sustainability planning is transforming the contemporary urban landscape. First, cities are adapting their built infrastructure and natural systems in the face of the local impacts of climate change. Second, cities are creatively re-using and redeveloping spaces in order to promote economic development and facilitate more sustainable transportation choices. Third, cities' emerging focus on equity seeks to reduce the existing segregation by race and class in US cities as well as alter the distribution of amenities and disamenities throughout the urban fabric.

Addressing Local Climate Change Impacts

Climate change already impacts cities and is a major motivator behind many cities' plans for sustainability. Currently about 100 US cities have developed plans specifically for climate adaptation. This is in part because the impacts of climate change are increasingly visible in the landscape. One only needs to turn on the news to know that cities are repeatedly experiencing record-breaking floods, tornadoes, droughts, heat waves, and wildfires—sometimes in the same month. According to the National Oceanic and Atmospheric Administration, in the last three years (2017–2020) the US sustained some 44 climate disasters that cost a total of $460 billion (NOAA 2020). It is clear that our society, infrastructure, and ecological systems are unprepared for the effects of current climate realities.

Cities respond to climate change in two main ways: mitigation and adaptation (Bulkeley and Betsill 2005). While mitigation addresses the causes of climate change, adaptation instead prepares us for the changes ahead. Because of geography, US cities have different vulnerabilities to climate change that will no doubt impact their physical landscape. For example, western cities such as Las Vegas and Denver must prepare for an increased frequency of drought. Cities in the Midwest and Northeast may experience more frequent large storm events, wetter winters, and heavier snowfall. Some cities located on rivers such as Kansas City, Cincinnati, Memphis, and Sacramento, may see rivers flood in the spring as more snowpack melts sooner. Reduced snowpacks and the shifting melt times also impact cities like Denver, located in a semi-arid climate. On the other hand, coastal cities such as Miami, Boston, and New York City must prepare for sea level rise that causes storm surges, increased flooding, coastal erosion, and even salt water intrusion in groundwater supplies. Figure 22.2 summarizes some of the major expected impacts on the physical landscape, which vary by region.

Many adaptation measures, especially those focusing on built infrastructure and natural systems, are likely to visibly change the physical landscape. They may include relocating critical infrastructure to higher ground, creating and preserving networks of green infrastructure and wetlands to manage stormwater and prevent flooding, and developing distributed energy grids and improving transmission lines in areas vulnerable to wildfire. Other examples are described in Table 22.1.

Table 22.1 Examples of Adaptation for Climate Change

Cities	*Adaptation Strategies Implemented*
Phoenix, AZ, Boston, MA, Philadelphia, PA and New York City, NY	These cities have integrated climate change impacts into public health planning and implementation activities that include creating more community cooling centers, neighborhood watch programs, and reductions in the urban heat island effect.
San Diego, CA	Local governments around the metro area partnered with the port, the airport, and more than 30 organizations with direct interests in the Bay's future to develop the San Diego Bay Sea Level Rise Adaptation Strategy. The strategy identified key vulnerabilities for the Bay and adaptation actions that can be taken by individual agencies, as well as through regional collaboration.
Chicago, IL	Through a number of development projects, the city has added 55 acres of permeable surfaces since 2008 and has more than four million square feet of green roofs planned or completed.
Portland, OR	Updated the city code to require on-site stormwater management for new development and re-development. Provides a downspout disconnection program to help promote on-site stormwater management.
Boulder, CO, New York City, NY and Seattle, WA	Water utilities are using climate information to assess vulnerability and inform decision-making.
Philadelphia, PA	The Philadelphia Water Department began a program to develop a green stormwater infrastructure, intended to convert more than one-third of the city's impervious land cover to "Greened Acres": green facilities, green streets, green open spaces, green homes, etc., along with stream corridor restoration and preservation.

Source: National Climate Assessment Report. 2014 nca2014.globalchange.gov/report/response-strateg ies/adaptation.

Miami Beach launched a $400 million Sea-Level Rise Plan that consists of a series of stormwater pumps, improved drainage systems, elevated roads, and higher seawalls. Currently the city is raising 100 miles of roads by two feet to reduce the damage done by periodic flooding. While plans such as those in Miami Beach seem exorbitantly expensive, these investments may prove smart and cost-effective in the face of a rapidly changing climate, by averting future storm damage.

Former Chicago Mayor Richard M. Daley bemoaned that, "as a nation, we recycle aluminum, glass, and paper, but we don't recycle our most valuable commodity, our land" (US Conference of Mayors 2017). Land is one of the most valuable commodities in any city. Greenfields (undeveloped land) tend to be located on the urban periphery. In contrast, virtually all of the developable land in a city center has been developed before, and some is abandoned, neglected, and sometimes contaminated. These lands do not add to the tax base and, in many cases, are actually a burden on city coffers. By redeveloping land, cities create jobs, revitalize neighborhoods, draw tourists, increase the tax base, clean up a legacy of pollution, and create more dense urban development (which brings a host of environmental benefits, such as less reliance on cars).

A brownfield is property—often a former industrial or commercial site— in which redevelopment is complicated by the presence of actual hazardous substances or by potential contamination. Although brownfields can be found anywhere, many are concentrated in cities in the Northeast and Midwest that were home to manufacturing activities (and, in the 1970s and 1980s, deindustrialization). Some brownfields are large abandoned industrial sites, others are smaller, including abandoned gas stations or other vacant commercial properties. The US EPA estimates there are more than 450,000 brownfields in the US.

New York City's Highline, a linear park created on an elevated rail line running through Chelsea and the Meatpacking District in Manhattan, presents one of the more innovative brownfield redesigns of the urban landscape. As warehouses and factories closed, the thirteen-mile line was unused after 1980, and fell into disrepair until, in 1999, the Regional Plan Association, commissioned by the railway line owners, CSX Railroad, suggested turning the narrow railroad into a pedestrian promenade. The first section of the linear pedestrian park was opened in 2009 and instantly became popular amongst locals and visitors. The route retains an industrial aesthetic with railroad tracks and movable seats on railroad car wheels yet softens this experience with extensive plantings of native vegetation and provides unique views of the city (Figure 22.3).

The conversion of an abandoned railroad track into an urban public space and walking route is one of the more successful transformations in the shift from industrial to postindustrial. The Highline has been lauded for reimagining and reclaiming green spaces as both vertical and linear possibilities. It has helped to influence the design of other linear parks, such as Atlanta's Beltline Park.

Cities reuse land in other ways as well. At the height of the industrial era, waterfront locations hosted warehouses, factories, and docks. Water quality was abysmal, and rivers often stank from a mix of untreated sewage and dumped chemicals. Rail yards and highways frequently cut off these areas from the reast of the city and were classified "dangerous" and polluted. Deindustrialization and new shipping technologies left many cities with abandoned warehouses, buildings, and unused port facilities on their waterfronts. Progress was made in implementing the Clean Water Act, and water quality improved. Beginning in the 1980s, waterfront redevelopment became a widespread strategy in urban planning and is now among the most prominent urban efforts to simultaneously revitalize both the economy and the environment (Gordon 1997).

Northwest
- Reduced snowpack and lower summer streamflows impact water supplies
- Sea level rise may increase erosion of coastlines, damaging infrastructure and ecosystems.
- Higher temperatures, may increases pests and diseases impacting forests, agriculture, and fish populations

Midwest
- Warmer and wetter winters, springs with heavier precipitation, hotter and drier summers
- Warmer temperatures may decrease agricultural yields
- Reduced air quality and increased allergens

Northeast
- More frequent heat waves will impact human health
- Sea level rise and more frequent heavy rains may increase flooding and storm surge

Southwest
- Increases in temperatures and more frequent and severe droughts
- Greater water demand in growing cities could stress drinking water supplies
- High temperatures will increase poor air quality, particularly ground-level ozone
- High temperatures will increase residents' vulnerability to heat-related illnesses
- Drought, wildfire, invasive species, pests, and changes in species' geographic ranges will increase

Great Plains
- Increases in temperature and drought frequency will further stress the High Plains Aquifer
- An increase in frost-free days will lengthen the pollen season for common allergens impacting public health

Southeast
- Sea level rise may increase salinity of estuaries, coastal wetlands, tidal rivers, and swamps
- Higher temperatures will strain water resources
- Increase in extreme weather including hurricane activity

Figure 22.2 The impacts of climate change vary by region but will no doubt change the physical geography of many cities. As a result, cities are planning for adaptation in diverse ways. Map by authors, adapted from data in "Climate Change Impacts by Region," the US EPA, accessed April 1, 2018, https://archive.epa.gov/epa/climate-impacts/climate-change-impacts-region.html >
Promoting Economic Development and a More Sustainable Transportation Network.

Waterfronts, once the city's "back door," have become focal points and are regaining "their rightful role as prized public treasures" (Project for Public Spaces 2008). Cities in the United States, and around the world, have transformed their waterfronts into vibrant, public spaces that offer a wide range of activities for people to enjoy, attracting both locals and tourists. Examples range from San Antonio and Boston to Pittsburgh and Charleston. Waterfront redevelopment takes a number of forms, some focused on increasing parkland and recreational trials, others on mixed use retail and housing, and still others on creating tourist destinations or revitalized commercial space. Figure 22.4 shows waterfront redevelopment in Tampa, Florida, which features a large convention center, linear green space with water access, and high density residences. Yet the city's industrial past remains visible as shown by the oil tanks and a tanker ship in the background.

Cities are reconsidering land use and repurposing land. The automobile has been a dominant force shaping urban design and infrastructure created for single-occupancy vehicles—roads, highways, and parking garages—consuming vast amounts of land (Figure 22.5). A new, more

Figure 22.3   The New York City Highline. Photo by Lisa Benton-Short.

Figure 22.4   Tampa's mix of new developments and historic industrial sites. Photo by Geoffrey L. Buckley.

Figure 22.5 Interstate 405 in Los Angeles. Infrastructure for automobiles is considerable; this stretch of the 405 is 16 lanes across. Photo by Lisa Benton-Short.

sustainable transportation hierarchy prioritizes self-locomotion like walking and cycling above all else, and favors public transportation and even shared vehicles over private vehicles. Efforts to overcome automobile dependence are beginning to reshape our urban landscapes. In contrast to Los Angeles is Boston, which transformed a section of Interstate 93 that ran through the heart of the city into a 1.5-mile tunnel underground. In 2008, after many years of construction, the city celebrated a new linear green space that features gardens, promenades, plazas, fountains, and art.

Another of these changes can be seen in New York City's adoption of "Complete Streets" priorities in street design projects. This approach recognizes that street design in the US has prioritized the experience of motorists above all other users, and seeks to change this by "enabling safe and convenient travel for all users including pedestrians, bicyclists, bus riders, motorists, children, older adults, and the disabled" (Smart Growth America 2018). Best practices include separated bike lanes, bus lanes, transit prioritization, wider sidewalks, and pedestrian bump outs for safe street crossing. Other changes include: designing streets to be closed sometimes to allow for public gatherings or enhanced recreational opportunities; designing streetscapes to help define neighborhood character and reflect context, history, and nearby landmarks; and to contribute to climate resiliency by incorporating street trees and other green infrastructure (New York City Department of Transportation 2020). During the 2020 COVID-19 pandemic, many cities experimented with "streateries"—allowing restaurants to apply for permits to use expanded sidewalk space, alleys, parking lanes, and travel lanes for table seating. It remains to be seen if these changes will be permanent, but for many months, the look, feel, and sound of many downtown neighborhoods was significantly changed.

A final way that cities repurpose urban land is through Transit Oriented Development (TOD), a practice that promotes dense, connected, walkable neighborhoods around public transportation nodes. This planning tool recognizes that dense, mixed-use communities around public transportation nodes are essential to attracting ridership and vital to the economic sustainability of these systems. Cities around the US are implementing TOD projects—that is, permitting and incentivizing high density development around public transportation nodes—including greater Washington DC. There, for example, a new stop was constructed on an existing subway line in 2004 at a cost of $110 million. Since that time, development in that neighborhood (called NoMa, short for North of Massachusetts Avenue) had sparked $3 billion in new development including over 7,000,000 square feet of new office, residential, hotel, and retail space (MacCleery and Stone 2021). Studies indicate that not all projects are having the desired consequences of encouraging walkability, promoting public transportation ridership, and reducing vehicle miles traveled, though we are learning more about what elements of these projects promote these ends (Jacobsen and Forsyth 2008).

A More Just and Equitable Future

Income inequality has been sharply increasing in the US since the 1970s, and the country currently has one of the most inequitable distributions of wealth in the Global North. Inequities extend beyond just income, and the patterns that we see in the US today—in affluence, education, occupation, and health. These inequities have molded our cities and spatially concentrated poverty within them. Socioeconomic and racial segregation remains an intractable problem, particularly in large cities in the Northeast and Midwest. This spatial distribution reinforces existing patterns of inequity in that amenities like high-quality schools, full service grocery stores, banks, doctors and hospitals, and well-maintained parks are more available to privileged communities. In the US, racial and economic inequity have shaped the urban landscape in innumerous ways: concentrated poverty, insufficient affordable housing, health disparities, educational gaps, and the burden of environmental inequities or "disamenities" such as proximity to hazardous waste on low income or minority communities (Logan 2013). Alarmingly, given continued segregation and concentration of poverty, many people remain unaware of conditions existing within their own communities.

A first step in increasing equity within cities is by making existing *inequities* more visible to planners, public officials, and citizens alike. GIS and mapping are powerful geospatial tools cities use to identify and understand inequity in health, income, and the distribution and access to amenities and disamenities. Cities can use maps as tools for priority setting; they serve as starting points in understanding community assets or problems. Mapping amenities or disamenities can identify an inequity, and start a community conversation and discussion, with the hope that more people are drawn into the conversation to continually strengthen data and identify place-specific solutions.

Environmental inequities occur not only through the presence of environmental disamenities, but also when there is an absence of environmental amenities. Washington, DC has effectively mapped food deserts in the city—areas where residents lack access to full-service supermarkets, and fresh fruits, vegetables, and affordable healthy food options. Food deserts have significant public health ramifications, and are directly connected to the location of supermarkets, public transportation networks, and personal car access. In another example, geographers mapped trees in Milwaukee and found fewer trees in poorer and blacker communities. Because trees absorb and filter pollution, an uneven distribution of trees reinforces uneven social and health outcomes (Heynen et al. 2006).

If done right, sustainability planning challenges cities to incorporate equity into every goal within their plans. For example, instead of setting a goal to increase the tree canopy city-wide to 35 percent, a city could instead increase the tree canopy to 35 percent in each neighborhood,

Table 22.2 Examples of Investing in Equity—What Cities Could and Should Do

Procedural Equity	• Name, identify, and locate communities that might require particular outreach and attention • Remove barriers to participation: hold meetings in languages other than English, at houses of worship and other community facilities, and provide childcare. • Develop a municipal government staff that reflects the demographics of the city's residents
Distribution and access	• Use GIS and other spatial technologies to identify problems of distribution and access of services, amenities and disamenities • Prioritize investment and programs in underserved neighborhoods
Education	• Cultivate a community of lifelong learners • Invest in STEM curriculum that prepares students for high-tech jobs and clean energy • Develop culturally relevant curriculum and classrooms • Develop environmental education and outdoor education curriculum
Health	• Develop health in all policies • Recognize the co-benefits of health in other areas of sustainability • Provide subsidies for full-service grocery stores to locate in food deserts • Work with corner grocery stores to provide more fresh food options • Provide financial incentives for electronic bank transfer (EBT) users to buy produce at farmers markets
Housing	• Create mixed income neighborhoods • Locate affordable housing near public transportation amenities • Remove regulatory barriers to affordable housing construction • Promote inclusionary zoning • Confront the challenge of homelessness

Source: by authors

more evenly providing urban forest benefits across the city. Or, they could target tree planting specifically in neighborhoods experiencing the most heat stress or air quality challenges. Other strategies that integrate equity into sustainability efforts are outlined in Table 22.2.

Unequal urban landscapes have been created by forms of discrimination woven into public policies and everyday attitudes. These have been accepted and reinforced by our society for generations. Rectifying this situation will likely take the same level of effort that created it, and will not happen overnight, but might well affect the most dramatic changes within the urban landscape of all sustainability actions. Denver's commitment to equity is already shaping development and transportation in the city. The city is actively expanding its public transit system, investing $7.8 billion in 122 miles of new light rail, 18 miles of bus rapid transit, and enhanced regional bus service to the region (West 2016). While this might seem like a straightforward sustainability win, studies have shown that unless carefully executed, expansion of public transit systems may prioritize improving service for more affluent communities or unintentionally exacerbate gentrification and displacement of low-income residents. In Denver, a coalition of local nonprofits together with public and private actors are collaborating in the Mile High Connects initiative, which works to make the city's investments in public transit improve the lives of low-income residents and residents of color (MileHighConnects 2021). Among other initiatives, the group has developed the Denver Regional Transit Oriented Development Fund, spending $32.8 million to create or preserve "1,354 affordable homes, a new public library, and well over 100,000 square feet of supportive commercial and non-profit space, all near public transit" (Enterprise Community Investment 2020). This example shows that some cities are beginning to make changes that address equity, and reshape the physical and cultural landscape, although much work is needed to achieve this goal.

Summary

The American urban landscape is being transformed as cities plan and implement sustainability. This is because a call to sustainability fundamentally changes the status quo and the urban fabric along with it. In preparation for localized changes in precipitation patterns resulting in floods or droughts, rising sea levels, and more extreme heat wrought by climate change, cities must take steps such as moving critical infrastructure and enhancing green spaces to absorb water. Cities are also reusing and repurposing urban land. Some are reclaiming contaminated areas along waterfronts or industrial areas; still others are reconfiguring streets, public spaces, and the land around transportation nodes as they seek to reduce automobile dependence. Finally, cities' emerging focus on equity holds the potential to reduce segregation by race and class as well as alter the distribution of amenities and disamenities throughout the urban fabric. If cities are able to meaningfully improve equity, this could represent a transformational change in the urban landscape. Sustainability is both literally and figuratively changing the urban landscape in important ways.

References

Boone, C. and A. Moddares. 2006. *City and Environment*. Philadelphia: Temple University Press.

Bulkeley, H. and M. Betsill. 2005. *Cities and Climate Change: Urban Sustainability and Global Governance*. New York: Routledge.

City of New York. 2017. "Small Steps, Big Strides, Insights from GreenNYC: The City of New York's Behavior Change Program," n.d., 18, www1.nyc.gov/assets/sustainability/downloads/pdf/publicati ons/greenyc_lessons_2017_online_final.pdf.

Cronon, W. 1991. *Nature's Metropolis: Chicago and The Great West*. New York: Norton.

Douglas, I. 1981. "The City as an Ecosystem," *Progress in Physical Geography* 5: 315–367.

Enterprise Community Investment. 2020. "Denver Regional Transit-Oriented Development (TOD) Fund." www.enterprisecommunity.org/financing-and-development/community-loan-fund/denver-regional-tod-fund.

Evenson, K., Aytur, S., D. Rodriguez, and D. Salveson. 2009. "Involvement of Parks and Recreation Professionals in Pedestrian Plans," *Journal of Park and Recreation Administration* 27(3): 132–142.

Gandy, M. 2002. *Concrete and Clay: Reworking Nature in New York City*. Cambridge MA: MIT Press.

Gordon, D. 1997. "Managing the Changing Political Environment in Urban Waterfront Development," *Urban Studies* 34(1) 61–83.

Heynen, N., H. Perkins, and P. Roy. 2006. "The Political Ecology of Uneven Urban Green Space: The Impact of Political Economy on Race and Ethnicity in Producing Environmental Inequality in Milwaukee," *Urban Affairs Review* 42 (1) 3–25. doi:10.1177/1078087406290729.

Keeley, M. and L. Benton-Short, L. 2019. *Urban Sustainability: Cities Take Action*. Cham, Switzerland: Palgrave Macmillan

Keys, E., E.A. Wentz and C.L. Redman 2007. "The spatial structure of land use from 1970–2000 in the Phoenix, Arizona Metropolitan Area," *The Professional Geographer* 59: 131–147.

Klingle, M. 2007. *Emerald City: An Environmental History of Seattle*. New Haven: Yale University Press.

Jacobson, J. and A. Forsyth. 2008. "Seven American TODs: Good Practices for Urban Design in Transit-Oriented Development Projects," *Journal of Transport and Land Use* 1 (2): 51–88. www.jstor.org/stable/26201614. Accessed December 30, 2020.

Logan, J. 2013. "The Persistence of Segregation in the 21st Century Metropolis," *City Community*, 12, no. 2. www.ncbi.nlm.nih.gov/pmc/articles/PMC3859616/.

MacCleery, R. and A. Stone 2021, "NoMa: The Neighborhood That Transit Built," *Urban Land Magazine, Urban Land Institute*, urbanland.uli.org/development-business/noma-the-neighborhood-that-transit-built/.

MileHighConnects. 2021. "About Us," milehighconnects.org/

National Oceanic and Atmospheric Administration and National Centers for Environmental Information (NCEI). 2020. "US Billion-Dollar Weather and Climate Disasters." www.ncdc.noaa.gov/billions/, doi:10.25921/stkw-7w73

New York City Department of Transportation. 2020. "Transportation Design Manuel," Third Edition. www.nycstreetdesign.info/sites/default/files/2020-04/.pdf.

Parker, P. and I. Rowlands 2007. "City Partners Maintain Climate Change Action Despite National Cuts: Residential Energy Efficiency Programme Valued at Local Level," *Local Environment*, 12 (5): 505–517.

Portney, K. 2003. *Taking Sustainable Cities Seriously: Economic Development, the Environment and Quality of Life in American Cities*. Cambridge, MA: The MIT Press.

Project for Public Spaces. 2008. "The Global Waterfront Renaissance," August 1, www.pps.org/reference/theglobalwaterfrontrenaissance/.

Saha, D. 2009. "Empirical Research on Local Government Sustainability Efforts in the USA: Gaps in the Current Literature," *Local Environment*, 14 (1).

Smart Growth America. 2018. "National Complete Streets Coalition," accessed March 20, 2018, smartgrowthamerica.org/program/national-complete-streets-coalition/.

Sustainable DC. 2012. "Sustainability DC," 5, sustainable.dc.gov/sites/default/files/dc/sites/sustainable/page_content/attachments/DCS-008%20Report%20508.3j.pdf.

The United States Conference of Mayors. 2017. "Written Testimony of Elizabeth Mayor J. Christian Bollwage For The U.S. Conference of Mayors Before the House Transportation and Infrastructure Subcommittee on Water Resources and Environment on 'Building a 21st Century Infrastructure for America: Revitalizing American Communities through the Brownfields Program.'" transportation.house.gov/uploadedfiles/2017-03-28_-_bollwage_testimony.pdf.

West, D. 2016. "Achieving Healthy Communities Through Transit Equity," Stanford Social InnovationsReview. ssir.org/articles/entry/achieving_healthy_communities_through_transit_equity

Wheeler, S. 2013. *Planning for Sustainability: Creating Liveable, Equitable, and Ecological Communities,* 2nd edn. New York: Routledge Press.

World Commission on Environment and Development United Nations. 1987. "Our Common Future."

Zhenghong T., S. D. Brody, C. Quinn, L. Chang & T. Wei. 2010 "Moving From Agenda to Action: Evaluating Local Climate Change Action Plans, *Journal of Environmental Planning and Management*," 53 (1) 41–62. doi: 10.1080/09640560903399772.

23

THE ORDINARY-EXTRAORDINARY AUTOMOBILE LANDSCAPE

Ellen Hostetter

This chapter starts at the intersection of Oak Street and Harkrider Street in Conway, Arkansas, a city of approximately 66,000 located roughly 30 miles north of Little Rock. Figures 1 and 2 capture the view from this intersection. Face south and turn to the right and you'll see the start of "historic downtown" Conway (Figure 23.1). Face south and turn to the left and you'll see a major commercial throughway, East Oak Street (Figure 23.2). The chapter starts here, in part, because it is convenient. I live in Conway. But I also invite you to stand with me at this intersection so we can first, dissect the substantial ways the automobile has shaped American landscapes and American lives and second, survey the automobile landscape literature of the past thirty years.

For the second part of this chapter, we will hit the road. As Jean Baudrillard reminds us, to really understand America, one must *drive* (2010, 16). We'll pause on an overpass in New Mexico, overlooking Interstate 40, with two goals in mind. First, to explore the standardized similarity between Interstate 40 and Oak Street. Second, to review three key perspectives coming out of the "mobilities turn"—access, meaning, and practice—that are driving (pun intended) contemporary work on the automobile landscape.

On the Street

West Oak and East Oak exhibit ordinariness and mundanity. West Oak is one manifestation of a landscape type found throughout the United States: a downtown streetscape of tightly packed, two-story commercial buildings, built in the early 1900s, directly fronting a sidewalk and on-street parking. East Oak is the ubiquitous feeder highway that provides access to stand-alone commercial boxes, fronted by swaths of parking. Even the kinds of businesses found on each are predictable: downtown Conway is for local coffee, antiques, boutique clothing, and special-feel dining while East Oak is largely for everyday needs—the drugstore, grocery, bank, brake-and-lube, gas station. For most of us living in Conway, West Oak is the quaint heart of town, but relatively unimportant in our daily routine, while East Oak is the "seen one, seen them all" backdrop to busy lives.

To cultivate an appreciation of this street corner, we'd do well to adopt the perspective of J.B. Jackson (1997) and Robert Venturi, Denise Scott Brown, and Steven Izenour ([1972] 2001), some of the first scholars to step back from knee-jerk reactions to the American automobile landscape—quaint, shrug—and approach it "on its own terms" (Davis 2003, 62) with a "non-chip-on-the-shoulder view" (Venturi, Brown, and Izenour ([1972] 2001, 3; see also Raitz 1998). Pair their

DOI: 10.4324/9781003121800-29

Figure 23.1 West Oak Street, gateway to "historic downtown" Conway, Arkansas. Photograph by author.

Figure 23.2 East Oak Street, one of many commercial throughways in Conway, Arkansas. Photograph by author.

perspective with the scholarship of historians who provide broad overviews of auto-induced change in the organization and layout of American towns and cities (Rae 1971; Liebs 1985; Longstreth 1997) and we can understand and even revel in the logical beauty of the scene before us.

West Oak looks the way it does because it was built when locals got around on foot or on a horse. The density of buildings and intimate relationship between store-front and sidewalk made it efficient for pedestrians to go about their daily business. Fire-insurance maps of Conway from 1897 show that drugstores, groceries, banks, and pre-automobile blacksmiths and harness shops used to be concentrated downtown. It was the rapid increase in automobile ownership that created East Oak Street, a new way to shop born of the new reality: shoppers were no longer pedestrians, but drivers. The everyday life of Conway leaked out onto East Oak where "the façade of the traditional retail street has been fragmented by parking lots, drive-in facilities, conspicuous signage... either for better visibility or for more convenient access" (Jackson 1997, 246). Historian of suburbia Kenneth T. Jackson (1985) would add another storyline: basic retail and services left downtown as more and more people moved to auto-centric suburbs in search of the "good life" over the course of the twentieth century (for more on auto-centric homes see Jennings 1990; Goat 1989). Jackson would take you on a tour of the post-World War II subdivisions that flank East Oak, past the interstate interchange to Conway's brand-new brick-box developments. When we turn our heads from West Oak to East Oak, we bear witness to all the ways the automobile has fundamentally altered where and how we live, work, and shop.

We should not be seduced, however, into accepting the scene before us as inevitable or normal. We should be shocked by what we see. Geographer Nigel Thrift gives a grand summary: "[a]round a relatively simple mechanical entity... a whole new civilization has been built" (2004, 46). Let us take this statement and run it through a conceptual frame I call "landscape backstory" (Hostetter 2015 and see Mitchell 2008). *Every single component* of the automobile landscape has a backstory, a detailed history of how it came to be not readily apparent from mere observation. *Everything* in the scene before us *had to be invented*. West Oak and East Oak should take our breath away.

It is one thing to say that the parking lots fronting East Oak are a logical response to the needs of shoppers-turned-drivers. It is quite another to go back to the turn of the twentieth century and realize that the backstory of parking lots starts with the backstory of curbside parking, which itself not only had to be figured out, but was fought over in city council chambers, courtrooms, and on the pages of trade journals and newspaper editorials. Haphazard parking of automobiles along city streets raised critical questions about what (and who) streets were for and how far municipal governments could go in telling individual motorists where, when, how, and for how long they could park their machines (Jakle and Sculle 2004, 19–46). We live in places shaped by the automobile to such an extent that we *experience* two-hour parking and parking lots as natural. And yet these banal components of West Oak and East Oak are not simply *there*, they were "*actively made*" (Mitchell 2008, 34) and are still being debated today (Henderson 2009).

Thanks to the scholarship of the past 30 years, we have a rich body of literature that details the backstories of individual components of our automobile landscape. Many tell the backstory of roadside businesses devoted to selling and servicing the automobile and meeting the needs of travelling motorists—auto showrooms and dealerships (Liebs 1985, 75–94), gas stations (Jakle and Sculle 1994; Liebs 1985, 95–116), motels (Jakle, Sculle, and Rogers 2002; Belasco 1997; Beecher 1990; Liebs 1985, 169–192), and roadside restaurants (Jakle and Sculle 1999; Langdon 1986). These authors dig into the minutiae of trial and error, conflict and negotiations, profit and loss.

Recent work by John Jakle and Keith Sculle has turned to the more mundane aspects of the automobile landscape—parking (2004), referenced above, and garages (2013)—but it is scholars working in crosscurrents of history, technology and science studies, and legal geography that give us the conflict-ridden backstories of the most ubiquitous and standardized components of our automobile landscape—traffic lights, traffic signs, crosswalks (see McShane 1999; Norton 2011; Vinsel 2019), sidewalks (Blomley 2011), and even your dreaded Department of Motor Vehicle office (Hostetter 2019). They do not share one theoretical frame or viewpoint, but they offer a different perspective from that of J.B. Jackson (1997), Venturri, Brown, and Izenour ([1972] 2001), and even historians Rae (1971), Liebs (1985), and Longstreth (1997). The automobile and the people who drove them "arrived in American city streets as intruders" (Norton 2011, 7; although see Flink 1972, 453). The safety and regulatory infrastructure we take for granted today is the product of heated protest against the disruption, damage, and death wrought by the automobile.

Scholars working in "energy geography" would not want us to forget that our automobile landscape is also a petroleum landscape. Products derived from crude oil are ingrained in everyday life—in our case, gasoline, plastic components, asphalt—and, yet "six generations back, oil was a bit player in an emerging lubricant market" (Bridge and Le Billion 2017, v). How petroleum became central to the scene before us is a sprawling epic of geopolitical strategy and violence, global financial power plays, and domestic "struggles over the sociospatial stuff of everyday life" (Huber 2013, xii). The story told by West Oak and East Oak, therefore, is not a neat narrative of logical, orderly adaptations, but a cacophony of resistance and contestation.

We have covered a lot of ground, but a question might nag: why do these landscape backstories matter to *us*, two people standing at a busy intersection? The concept of automobility provides one answer, defined by sociologists Mimi Sheller and John Urry in a seminal article "The City and the Car" (2000). Automobility is the "complex amalgam of interlocking machines, social practices and ways of dwelling" (Sheller and Urry 2000, 739) that "dominates how both car-users *and* non-car users organize their lives through time-space" (Sheller and Urry 2000, 745). We've been thinking about how the automobile has shaped our built environment. Sheller and Urry ask us to think about how the automobile landscape shapes—even dictates—the simplest of everyday experiences and interactions, whether we're in a car or not.

Turning our head from West Oak to East Oak we can trace how experiences and interactions have changed over time. Simply put, West Oak concentrated the everyday life of Conway in downtown while East Oak fragments interactions. More and more time is spent behind the wheel. Sheller and Urry join a wide range of social commentators who have mounted a sustained, scathing critique against this fragmentation. Everyone from Jane Jacobs (1961) to James Kunstler (1993), Richard Sennet (1977) to Henri Lefebvre (1991) recognize a profound loss: "the 'coming together of private citizens in public space' is lost to a privatization of the mechanized self moving through emptied non-places" (Sheller and Urry 2000, 746). Don Mitchell, in *Mean Streets: Homelessness, Public Space, and the Limits of Capital*, takes this one step further (2020). As the title indicates, his book is not about automobility and yet he connects how we get around, symbolized by the Sport Utility Vehicle, to court decisions in the US that are redefining citizenship and public space: "what US courts are pushing us toward, is a model of citizenship and urban life that matches the cars we drive" (2020, 123). The individual—on the street or sidewalk, in a park or plaza—has the right to be left alone and move unimpeded, with "the sense of inviolability that a couple of tons of steel and fiberglass can instill" (Mitchell 2020, 123). Navigating our automobile landscape has not just inured us to the generic act of driving, it has seeped into our understanding of rights and relationships.

What of the future? Subtle indications of change are in the scene before us. The CVS on the south corner of East Oak Street sits flush to the sidewalk and parking is in back, a configuration that mimics the pedestrian-oriented layout of West Oak. The city rejected CVS's proposal to turn this corner into a parking lot and required the national chain to conform to a new set of design standards inspired by New Urbanism, a nation-wide architecture and urban planning movement (see Duany, Plater-Zyberk, and Speck 2000; Congress for New Urbanism, n.d.). New Urbanists translate the scathing critiques of our automobile landscape, outlined above, into new ordinances

Figure 23.3   Overlooking Interstate 40 from Route 6, approximately 40 miles west of Albuquerque, New Mexico. Photograph by author.

and codes. The CVS represents a sliver of hope for a future Conway designed "around the true needs of individuals" (Duany, Plater-Zyberk, and Speck 2000, xiii) instead of the demands of cars.

On the Road

Time to hit the road. We arrive at an overpass overlooking Interstate 40 in New Mexico, just west of Albuquerque (Figure 23.3). We stop here to first appreciate the similarity of Interstate 40 and Oak Street in Conway. Second, we consider three theoretical perspectives underpinning what's been labeled the "mobilities turn" in the social sciences (Sheller and Urry 2006; Adey et al. 2014, 21–24): access, meaning, and practice. The goal here is to highlight scholarship on the road coming out of not just geography and sociology, but American studies, literature, history, media studies, and transportation policy that is easily lost in the voluminous citation lists following any discussion of the mobilities turn.

Connecting Interstate 40 and Oak Street

Interstate 40 and Oak Street are both engineered ribbons of asphalt whose markings communicate the same information to motorists. To appreciate this banal observation, consider the roughly 880 miles between these roads and the fact that we travelled this distance on a vast network of similarly standardized roadbeds. What we are looking at and what we experienced is nothing less than the materialization of a strategic political vision that predates the automobile: to connect the nation physically and, therefore, economically, socially, and ideologically with a system of "good roads." Karl Raitz gives us the long, winding backstory of this materialization through an edited treatise focused on the National Road, the first federally funded highway in the United States (1996). The National Road was conceived and begun in 1808 during the wagon-canal era, replaced by US 40 in the 1920s, which was then bypassed by Interstates 68 and 70 in the late 1950s in an ever-evolving reformulation of what a "good road" should look like and how it should function. Each stage represents a tremendous, contentious investment of resources. We experienced our trip as one continuous piece of asphalt, but the story of how we got here is layered and halting (see also McShane 1979; Jakle and Sculle 2008).

Interstate 40, Three Theoretical Views

Access

Looking down onto Interstate 40, at the vehicles speeding into and out of view, it is tempting to conclude that we're looking at unfettered mobility, *America* on the move. No. From the perspective of the mobilities literature, "[m]obility is a resource that is differentially accessed" (Cresswell 2010, 21). While these ribbons of asphalt are legally defined as *public* infrastructure and funded for the *public* good, they are highly regulated, controlled, and exclusionary. Codes and manuals outline the rules and standards intended to get *those in motorized vehicles* where *they* need to go (see Prytherch 2018), while driver's licenses specify whose minds and bodies are "fit" to be behind the wheel (see Adar 2019, 585). The most mundane part of our automobile landscape—paved roadbeds—has become disturbingly important for determining one's ability to function and flourish in the United States. The opportunities for those without cars, in places with underfunded, inefficient, or nonexistent public transportation is diminished and curtailed (Bullard 2004), while voter identification laws have made the consequences of living without a driver's license equally disturbing (Adar 2019, 575). Equitable access to the road is framed as a "basic right, a right worth fighting for" (Bullard 2004, 2) and some of the most exciting work being done on the road today is moving "beyond critique of traffic laws and roadway design to changing their orientation for the better" (Prytherch 2018, 14–15; see also Lucas and Jones 2012; Golub and Martens 2014; Cook and Butz 2019).

Meaning

Let's now look up and out, contemplate the asphalt ribbons of Interstate 40 as they converge on the horizon. I'm tempted to skip the conclusions, get in the car, and *go*—drive to the horizon and escape into the nowhere freedom and possibility of the open road. If we gave in to that temptation, we could explore meaning, a key mobilities perspective. Getting from point A to point B is not just a physical act. All movement is loaded with meaning (Cresswell 2006, 2) and the meaning I ascribed to driving down Interstate 40 is decidedly American. The promise of America is wrapped up in the promise of mobility—"the freedom to go anywhere and become anyone" (Brigham 2015, 3)—a promise made before the auto-age (see Lewis 1996; Cresswell 2001) that is now fundamentally tied to the road trip, that "quintessential expression of Americanness" (Brigham 2015, 3).

Practice

And yet the work of American Studies scholar Cotten Seiler shows us that the "quintessential American" is a particular kind of American—white and male—and driving on the open road expresses a particular brand of Americanness: that of the consumer-citizen whose individualism is tethered to the mass production of automobiles and federally funded roads (2008, 13). Seiler's dissection of the open-road-as-symbol gets us close to practice, our third conceptual frame from the mobilities turn. Movement is *always* particular, *never* universal even when—as symbol—it is framed as such. Movement is a lived, embodied experience (Cresswell 2006, 3). The American idea of the road is, therefore, always enacted through the bodies and identities of particular people, both fictional and real.

We might assume how these enactments unfold on the "big screen," on the pages of novels, or in someone's account of their experience: those with power and privilege have a much easier time experiencing "Americanness" on the road than the marginalized and disenfranchised (see Hague 2010). There is unsettling truth to this assumption. The scholarship on Black drivers, for instance, documents a history of movement denied or restricted, made fraught or terrifying. There are "countless stories of trouble on the road that have informed a black 'highway consciousness' distinct from that of white drivers" (Seiler 2006, 1094), everything from Jim Crow era anxieties, humiliations, and dangers of securing services and conforming to local racial customs (Foster 1999), to the racialized traffic stops, fines, arrests, and license suspensions that began in the early 1900s and continues today (Albert 2001; Meehan and Ponder 2002). And yet there are countless stories of Black "mastery, elegance, self-possession, and decorum" (Seiler 2006, 1095) on the road (and see Franz 2004; Foster 1999) as well as stories of resistance, such as Black activist Jo Ann Robinson, living in Montgomery, Alabama in the mid-1950s, using her Chrysler to organize rides for Black bus riders during the city's now iconic bus boycott (Scharff 2003, 139–156).

What should we conclude from these stories? Our vast network of engineered asphalt carries more than just the load of automobiles and trucks, it carries the load of a pressing question—what does it mean to be an American?—that we are always in the process of answering in simultaneously liberating and disciplining, inclusionary and exclusionary ways.

Box 23.1 Landscape Interpretation and Qualitative Research
Dydia DeLyser

When we study landscapes like the automobile landscapes that form the focus of Ellen Hostetter's chapter—landscapes often referred to as "ordinary"—we nearly always do so qualitatively. This is not because we can't count, or resist applying quantitative methods, but rather because of what qualitative methodology (the philosophical underpinnings of the qualitative methods we employ in our research) opens to research in landscapes like these.

Whether our interests are in interpretations driven by Marxist political economy or those open to ideas that cannot be apprehended representationally, whether our interests are historical or contemporary (or both), qualitative research seeks to open out rather than limit analytical and interpretative possibilities of studying ordinary landscapes.

We can review core tensions and differences between qualitative and quantitative methodologies in order to understand how qualitative methods open such analytical purchase for the study of ordinary landscapes (Table 23.1).

From the Space Age of the 1960s to the era of Big Data we now find ourselves in there have been compelling reasons to turn to quantitative research: answering important questions globally, for example about the speed and pace of climate change, demands such approaches. Yet, at the same time, we can see how across different periods where quantitative techniques offered compelling

Table 23.1

Qualitative Research	Quantitative Research
Inductive	**Deductive**
• Theory emerges from the data	• Data is applied to existing theory
• Research concludes with theory and ideas	• Research begins with a hypothesis to test
"Naturalistic"	**"Experimental"**
• Research is undertaken in place and/or directly with those involved	• Research often takes place offsite, often through data analysis and data transformations
• Fieldwork often a significant component	
Intensive	**Extensive**
• Research involves in-depth research with a limited number of participants	• Research is broad based and includes large numbers of participants/samples
• "Small N" even N=1, or undertaking research with just one participant, is valid and important	• For research to be valid it must include large populations or sample sizes
Contextual	**Generalizable**
• Research is context dependent and cannot be applied directly any place else	• Research results can be applied directly to other sites and populations
Subjective	**Objective**
• Grounded in situated knowledges that validate individual perspectives	• Grounded in measurement and the idea that humans can be neutral
• "cold"	• 65 degrees Fahrenheit
• "tall"	• 5' 10"
• "sweet"	• 200 grams of sugar
Focused on meaning	**Focused on measurement**

solutions, qualitative approaches have predominated in the study of landscapes: understanding local responses and local realities calls for them.

Thus, underlying the study of a unique-but-seemingly insignificant landscape (one intersection in one town in Arkansas) lies the validation of uniqueness, not only for its own sake, but as a window onto something greater—here revealed as the dividing line between older and newer forms of development, between an artisanal and small-business past, and a mass-manufactured and corporate present.

For Hostetter, every landscape we see has multiple "backstories"—elements that cannot be known until they are studied. Here we can see the draw of qualitative methodology: in inductive research that builds ideas from the data itself, in "naturalistic" research that takes place in the field, in intensive research that can illuminate America through one intersection, in a contextual approach that validates uniqueness even as it seeks to draw connections, and in a subjective perspective grounded in researcher positionality (Hostetter, after all, lives in this town) and emplaced knowledge. Combining these aspects in a qualitative approach to research, the geographers (and others) who study landscapes, seek out such backstories for each one adds a dimension that, once seen, can no longer be unseen. Each element reveals new stories of power, struggle, resistance, and empowerment. Learning them helps us understand the world around us in richer, more complex ways.

Coda

To reinforce this final point, I offer two snapshots from the road. An emerging lifestyle trend, van-life: people opting out of a traditional fixed home, as well as the largesse of recreational vehicles, and converting a standard commercial van into a place to live. Van-lifers talk about multiple freedoms: individual self-expression, adventure, and rebellion (Rubey 2019). For those, however, on the edge of homelessness who sleep in their vehicle out of necessity, not choice, vehicle-as-home does not tap any easy notions of freedom; a symbol, perhaps, of survival. The US cities that have enacted laws making it illegal to sleep in your vehicle have made this tenuous symbol an even more tenuous reality (National Law Center on Homelessness and Poverty 2018, 25).

That you can find on Instagram #vanlifeisawesome and in a city's municipal code a prohibition on the use of streets for habitation is *not* a contradiction of American life. It is the product of choices being made in and about the Automobile landscape.

References

Adar, C. 2019. "Licensing Citizenship: Anti-Blackness, Identification Documents, and Transgender Studies." *American Quarterly* 71 (2): 569–594.

Adey, P., D. Bissell, K. Hannam, P. Merriman, and M. Sheller, eds. 2014. *The Routledge Handbook of Mobilities*. London: Routledge.

Albert, D. M. 2001. "Primitive Drivers: Racial Science and Citizenship in the Motor Age." *Science as Culture* 10: 327–351.

Baudrillard, J. 2010. *America*. Brooklyn, NY: Verso.

Beecher, M. A. 1990. "The Motel in Builder's Literature and Architectural Publications: An Analysis of Design." In *Roadside America: The Automobile in Design and Culture*, edited by J. Jennings, 115–124. Ames, IA: Iowa State University Press.

Belasco, W. J. 1997. *Americans on the Road: From Autocamp to Motel, 1910–1945*. Baltimore, MD: The Johns Hopkins University Press.

Blomley, N. 2011. *Rights of Passage: Sidewalks and the Regulation of Public Flow*. New York: Routledge.

Bridge, G. and P. Le Billon. 2017. *Oil*. Malden, MA: Polity Press.

Brigham, A. 2015. *American Road Narratives: Reimagining Mobility in Literature and Film*. Charlottesville, VA: University of Virginia Press.

Bullard, R. D. 2004. "Introduction." In *Highway Robbery: Transportation Racism & New Routes to Equity*, edited by R. D. Bullard, G. S. Johnson, and A. O. Torres, 1–14. Cambridge, MA: South End Press.

Congress for New Urbanism. n.d. "Who We Are." Accessed February 8, 2021. www.cnu.org/who-we-are.

Cook, N. and D. Butz, eds. 2019. *Mobilities, Mobility Justice and Social Justice*. New York, NY: Routledge.

Cresswell, T. 2001. *The Tramp in America*. London: Reaktion Books.

———. 2006. *On the Move: Mobility in the Modern Western World*. New York, NY: Routledge.

———. 2010. "Towards a Politics of Mobility." *Environment and Planning D: Society and Space* 28: 17–31.

Davis, T. 2003. "Looking Down the Road: J.B. Jackson and the American Highway Landscape." In *Everyday America: Cultural Landscape Studies after J.B. Jackson*, edited by C. Wilson and P. Groth, 62–80. Berkeley, CA: University of California Press.

Duany, A., E. Plater-Zyberk, and J. Speck. 2000. *Suburban Nation: The Rise of Sprawl and the Decline of the American Dream*. New York, NY: North Point Press.

Flink, J. J. 1972. "Three Stages of American Automobile Consciousness." *American Quarterly* 24 (4): 451–473.

Foster, M. S. 1999. "In the Face of 'Jim Crow': Prosperous Blacks and Vacations, Travel and Outdoor Leisure, 1890–1943." *The Journal of Negro History* 84 (2): 130–149.

Franz, K. 2004. " 'The Open Road': Automobility and Racial Uplift in the Interwar Years." In *Technology and the African-American Experience*, edited by D. Miller, 131–154. Cambridge, MA: MIT Press.

Goat, L. G. 1989. "Housing the Horseless Carriage: America's Early Private Garages." *Perspectives in Vernacular Architecture* 3: 62–72.

Golub, A. and K. Martens. 2014. "Using Principles of Justice to Assess the Modal Equity of Regional Transportation Plans." *Journal of Transport Geography* 41: 10–20.

Hague, E. 2010. "'The Right to Enter Every Other State'–The Supreme Court and African American Mobility in the United States." *Mobilities* 5 (3): 331–347.

Henderson, J. 2009. "The Spaces of Parking: Mapping the Politics of Mobility in San Francisco." *Antipode* 41 (1): 70–91.

Hostetter, E. 2015. "Landscape Backstories." American Association of Geographers Annual Meeting, Session 2210. Chicago, IL.

———. 2019. "The Landscapes of Early Automobile Registration and Licensing Laws: Creating New Jersey's Department of Motor Vehicles, 1903 to 1957." *Journal of Urban History* 45 (3): 452–482.

Huber, M. T. 2013. *Lifeblood: Oil, Freedom, and the Forces of Capital*. Minneapolis, MN: University of Minnesota Press.

Jacobs, J. 1961. *The Death and Life of Great American Cities*. New York, NY: Random House.

Jackson, J. B. 1997. *Landscape in Sight: Looking at America*. Edited by H. L. Horowitz. New Haven, CT: Yale University Press.

Jackson, K. T. 1985. *Crabgrass Frontier: The Suburbanization of the United States*. New York: Oxford University Press.

Jakle, J. A. and K. A. Sculle. 1994. *The Gas Station in America*. Baltimore, MD: The Johns Hopkins University Press.

———. 1999. *Fast Food: Roadside Restaurants in the Automobile Age*. Baltimore, MD: The Johns Hopkins University Press.

———. 2004. *Lots of Parking: Land Use in a Car Culture*. Charlottesville, VA: University of Virginia Press.

———. 2008. *Motoring: The Highway Experience in America*. Athens, GA: University of Georgia Press.

———. 2013. *The Garage: Automobility and Building Innovation in America's Early Auto Age*. Baltimore, MD: The Johns Hopkins University Press.

Jakle, J. A., K. A. Sculle, and J. S. Rogers. 2002. *The Motel in America*. Baltimore, MD: The Johns Hopkins University Press.

Jennings, J. 1990. "Housing the Automobile." In *Roadside America: The Automobile in Design and Culture*, edited by J. Jennings, 95–106. Ames, IA: Iowa State University Press.

Kunstler, J. H. 1993. *The Geography of Nowhere: The Rise and Decline of America's Man-Made Landscape*. New York, NY: Touchstone.

Langdon, P. 1986. *Orange Roofs, Golden Arches: The Architecture of American Chain Restaurants*. New York, NY: Knopf.

Lefebvre, H. 1991. *The Production of Space*. Cambridge, MA: Blackwell.

Lewis, P. 1996. "The Landscapes of Mobility." In *The National Road*, edited by K. Raitz, 3–44. Baltimore, MD: Johns Hopkins University Press.

Liebs, C. H. 1985. *Main Street to Miracle Mile: American Roadside Architecture*. Baltimore, MD: Johns Hopkins University Press.

Longstreth, R. W. 1997. *City Center to Regional Mall: Architecture, the Automobile, and Retailing in Los Angeles, 1920–1950*. Cambridge, MA: MIT Press.

Lucas, K. and P. Jones. 2012. "Social Impacts and Equity Issues in Transport: An Introduction." *Journal of Transport Geography* 21: 1–3.

McShane, C. 1979. "Transforming the Use of Urban Space: A Look at the Revolution in Street Pavements, 1880–1924." *Journal of Urban History* 5 (3): 279–307.

———. 1999. "The Origins and Globalization of Traffic Control Signals." *Journal of Urban History* 25 (3): 379–404.

Meehan, A. J. and M. C. Ponder. 2002. "Race and Place: The Ecology of Racial Profiling African American Motorists." *Justice Quarterly* 19 (3): 399–430.

Mitchell, D. 2008. "New Axioms for Reading the Landscape: Paying Attention to Political Economy and Social Justice" In *Political Economies of Landscape Change: Places of Integrative Power*, edited by J. L. Wescoat, Jr. and D. M. Johnston, 29–50. Dordrecht, The Netherlands: Springer.

———. 2020. *Mean Streets: Homelessness, Public Space, and the Limits of Capital*. Athens, GA: University of Georgia Press.

National Law Center on Homelessness and Poverty. 2018. "Housing Not Handcuffs: Ending the Criminalization of Homelessness in US Cities." nlchp.org//wp-content/uploads/2018/10/Housing-Not-Handcuffs.pdf.

Norton, P. D. 2011. *Fighting Traffic: The Dawn of the Motor Age in the American City*. Cambridge, MA: The MIT Press.

Prytherch, D. L. 2018. *Law, Engineering, and the American Right-of-Way: Imagining a More Just Street*. Cham, Switzerland: Palgrave Macmillan.

Rae, J. B. 1971. *The Road and the Car in American Life*. Cambridge, MA: The MIT Press.

Raitz, K., ed. 1996. *The National Road*. Baltimore, MD: John Hopkins University Press.

———. 1998. "American Roadside, Roadside America." *Geographical Review* 88 (3): 363–387.

Rubey, P. 2019. "Van-Life: The Realities of Living in a Rolling Home." Undergraduate Thesis. Norbert O. Schedler Honors College, University of Central Arkansas.

Scharff, V. 2003. *Twenty Thousand Roads: Women, Movement, and the West*. Berkeley, CA: University of California Press.

Seiler, C. 2006. "'So That We as a Race Might Have Something Authentic to Travel By': African American Automobility and Cold-War Liberalism." *American Quarterly* 58 (4): 1091–1117.

———. 2008. *Republic of Drivers: A Cultural History of Automobility in America*. Chicago, IL: University of Chicago Press.

Sennet, R. 1977. *The Fall of Public Man*. New York, NY: Alfred A. Knopf.

Sheller, M. and J. Urry. 2000. "The City and the Car." *International Journal of Urban and Regional Research* 24 (4): 737–757.

———. 2006. "The New Mobilities Paradigm." *Environment and Planning A* 38: 207–226.

Thrift, N. 2004. "*Driving* in the City." *Theory, Culture & Society* 21 (4/5): 41–59.

Venturi, R., D. S. Brown, and S. Izenour. (1972) 2001. *Learning from Las Vegas: The Forgotten Symbolism of Architectural Form*. Cambridge, MA: MIT Press.

Vinsel, L. 2019. *Moving Violations: Automobile, Experts, and Regulations in the United States*. Baltimore, MD: The Johns Hopkins University Press.

24

TOURISM IN THE AMERICAN LANDSCAPE

Velvet Nelson

Tourism is undeniably a part of the American landscape. In 2019, the United States (US) registered 2.3 billion domestic tourists and 79 million international tourists. These tourists collectively spent 1.1 trillion US dollars. As a result, direct travel and tourism employment in the country accounted for nine million jobs, while nearly seven million additional jobs were indirectly supported (US Travel Association 2020c). Yet, the significance of tourism goes beyond its economic effects. The ever-changing patterns of tourism also have clear social and environmental impacts that warrant closer attention.

Academic interest in tourism increased in the 1960s. Historians and sociologists took a leading role in the critique of tourism and tourists (Boorstin 1961; Cohen 1979; McCannell 1973), whereas geographers played an integral role in investigating the relationship between tourism and landscape. Richard Butler (1980) first proposed the idea of a tourist area life cycle to understand the development of tourism and the changes it brings to a place over time. Although the study of tourism was too often dismissed as frivolous (Gibson 2008), research on the subject continued to increase. In tourism geography specifically, scholarly publications have experienced the most marked growth since 2008 (Müller 2019).

This chapter provides a brief overview of the evolution of American tourism and follows key tourism trends. It examines the transition from the beginnings of modern tourism as an exclusive activity for the upper classes through the developments that led to a democratization of travel. It also considers issues such as gender bias and racial discrimination that have shaped if and how social groups have been able to participate in tourism. The chapter continues by looking at contemporary problems created by an abundance of tourism (i.e., overtourism) as well as the lack of tourism during the COVID-19 pandemic and considers movement toward a more sustainable American tourism landscape. Finally, the chapter brings these issues into focus by looking at examples from California, the most visited state in the US.

Overview of Tourism Development

Although travel has taken place throughout human history, tourism is a more recent concept. In England in the late eighteenth century, the word "tourist" emerged to describe individuals who traveled for observation or pleasure (Simmons 1984). The practice of "tourism" became a popular activity among Britain's elite upper classes that had sufficient disposable income and

DOI: 10.4324/9781003121800-30

leisure time. Tourism was slower to develop in the US. Practically, the young nation lacked an aristocratic class with sufficient free time for long pleasure journeys (Nelson 2013) as well as a national transportation network that would facilitate these journeys (Shaffer 2001). Conceptually, the US lacked the type of history and culture that had long provided the basis for European attractions (Gassan 2008). At this time, tourism was largely spatially concentrated in the northeastern US and centered around resorts based on mineral springs (Shaffer 2001). Such places had long been popular attractions in England (Kevan 1993). Ballston Spa in Saratoga County, New York emerged as an elite retreat in the 1790s and held the distinction of having the country's first major hotel outside a city in 1804. Within two decades, the area was firmly established with the country's highest non-urban concentration of hotels (Figure 24.1) (Gassan 2008).

In the nineteenth century, US tourism development continued as Americans experienced an increase in leisure time and desire for leisure activities, as well as an ever-expanding transportation infrastructure (Nelson 2013). Americans also began to revise their attitudes towards the natural landscape. Where wild nature was formerly seen as dangerous and uncivilized, intellectual interpretations of the beautiful, sublime, and picturesque allowed America's mountains, valleys, and forests to be viewed as scenic attractions (Gassan 2008; Shaffer 2001; Young 2017). For some, witnessing such attractions called for a certain solemnity that would allow them to experience the appropriate emotions. These tourists perceived the steady growth of tourism as a threat to this experience. After a visit to Niagara Falls in 1831, French political theorist Alexis de Tocqueville reportedly warned that visitors should hasten to visit the site before it was too late (Löfgren 1999).

Figure 24.1 Upstate New York had some of the US's earliest destinations. This is a scene from Ballston Springs, New York. "Ballston Springs" by W.H. Bartlett. 1839. Collection of Brookside Museum, Saratoga County Historical Society.

Travel was disrupted by the Civil War, but afterwards opportunities for travel in the US again increased through the end of the century. Further expansion of the national transportation infrastructure opened new places for visitation, particularly in the western US. However, the railroad remained an important mediator of scenic tourism experiences in the region. Until the tourism infrastructure could be more fully developed tourists were limited to viewing the landscape from passenger cars while the dramatic scenery lying outside of the rail corridor remained inaccessible (Shaffer 2001). For example, Yellowstone became the country's first national park and "a quintessentially American tourist attraction" in 1872. Yet, spur tracks to the southern entrance were not completed until 1883, which was followed by the hotel and road construction that would allow tourists to easily access the park (Shaffer 2001, 44).

American tourism historian Marguerite Shaffer (2001, 4) writes:

> As the tourist infrastructure expanded, public and private tourist advocates worked together to develop a canon of American tourist attractions that manifested a distinct national identity. They encouraged white, native-born middle- and upper-class Americans to reaffirm their American-ness by following the footsteps of American history and seeing the nation firsthand.

This ritual gained new importance with the outbreak of World War I in Europe in 1914. Then, the explosion of automobile ownership and highway construction that took place during the interwar years allowed even more people to participate (Nelson 2013).

During this time, resort tourism saw a resurgence, but a host of new attractions also began to emerge. While earlier attractions were primarily based on natural features, these attractions were expressly built for tourism. Disneyland opened in southern California in 1955. Critics complained about issues ranging from chronic traffic around the park to the corruption of American culture (Marling 1991). Yet, for the public, Disneyland "was a place where plastic crocodiles were better than live ones, since half the fun came from noticing that the beasts were *almost* real" (Marlin 1991, 174, original emphasis). This supported Daniel Boorstin's (1961) critique that tourists were those who were satisfied with such artificial experiences. Dean McCannell (1973) took this opposing view that tourists sought authenticity from their experiences, while Erik Cohen (1979) suggested that tourists should not be viewed as a single type.

The US has the potential to appeal to many types of tourists with its extensive and diverse attractions. These attractions range from spectacular scenes of nature (e.g., Yosemite National Park, California), to major urban centers (e.g., Boston, Massachusetts), places with distinctive cultural patterns (e.g., the French Quarter in New Orléans, Louisiana), popular family beaches (e.g., Daytona Beach, Florida), and even adult entertainment districts (e.g., the Strip in Las Vegas, Nevada). Despite the potential for significant and widespread interest in the American tourism landscape, it must be noted that, over time, there have been barriers that have prevented some tourists from effectively engaging in tourism.

Barriers to Tourism

While women have long traveled, it was rare that a woman would travel alone until the nineteenth century. Even then, a woman who traveled by herself had to walk a fine line between personal autonomy and propriety (Anderson 2006). Yet, as tourism opened up to upper and subsequently middle-class women, many found it liberating, not only because it provided a break from everyday life but also an opportunity to experience new things in new ways. Male tourists, however, complained that female tourists, and the infrastructure that allowed them

Figure 24.2   Niagara Falls, New York is an early example of "wild" nature that was turned into an attraction for all types of tourists. "Niagara Falls from the American Side" by Jacques Gérard Milbert. Date Unavailable. Digital Public Library of America, https://dp.la/item/982a038914ca6ef2f9fec529c26 13a17. (Accessed September 15, 2021.)

to visit places, threatened the "wild" nature of places such as Niagara Falls (Figure 24.2). The desire to protect masculinized spaces prompted a reaction against female tourists as well as a push to find new spaces free of these women (Löfgren 1999).

Tourism opened up to women throughout the twentieth century, but female tourists— particularly solo female tourists—continued to experience places differently than their male counterparts. The prospect of negotiating an unfamiliar place can create anxiety for any traveler, but these anxieties may be magnified by media representations of violence against women (see, for example, Specia and Mzezewa 2019). In light of real or perceived security issues, female tourists are more likely to be constantly aware of their surroundings to avoid potentially unsafe situations. As such, they may not be able to immerse themselves in the experience, to visit some places, or participate in certain activities (Nelson 2013). For female tourists traveling within the US as well as those traveling from other countries, they may not always be treated with respect or dignity. For example, women of color in particular report higher rates of verbal harassment and/or being mistaken for sex workers in public spaces, such as hotel lobbies. Women in unfamiliar places may not feel that they are able to react in the ways they would if they were comfortable in their surroundings (Nelson 2021).

Despite these, and other barriers, more women are traveling today than ever. The US tourism industry is starting to experience a shift with more women in positions of power to better recognize and meet the needs of this market. In addition, there is an increasing number of women who are disrupting the industry in ways that will empower female tourists (CNT Editors 2019).

Figure 24.3 During the Jim Crow era, African Americans travelers and tourists relied on resources like *The Negro Motorist Green Book* to identify places where they could eat, stay, and recreate like the Lewis Mountain Negro Area in Shenandoah National Park. "Lewis Mountain Negro Area" by Unknown. 1930s. National Park Service, https://www.nps.gov/shen/images/20070117113507.jpg. (Accessed September 15, 2021.)

Upper-class African Americans began to participate in tourism in the late nineteenth century. These professionals visited the same resorts as white tourists (Armstead 2005). However, this was relatively short-lived, as the Jim Crow system of segregation changed patterns of travel and tourism particularly in the southeastern US but also throughout the country. Even public spaces and facilities, supported by everyone's tax dollars, were segregated (Figure 24.3) (Young 2017). Existing resorts closed to African American tourists, but in time, new resorts opened. From the early to mid-twentieth century, rural Idlewild, Michigan offered a retreat for African Americans who came from midwestern cities such as Detroit, Cleveland, Chicago, Indianapolis, and even St. Louis (Stephens 2001). While tourists were able to relax at resorts such as this, the act of travel was fraught with challenges, such as whether they would be able to obtain products and services on the road. In 1936, Victor Green produced the first *Negro Motorist Green Book* to provide a listing of businesses that welcomed African American customers (Townsend 2016; Figure 24.3).

Green continued to publish this guide until the mid-1960s. The Civil Rights Act of 1964 outlawed discrimination, which opened up greater opportunities for African American travel and tourism. Although places are no longer segregated, they are still racialized. According to Briona Lamback (2019), an African American travel blogger, "Though that dark chapter of our history [Jim Crow] has passed, the woes of being a black traveler haven't completely disappeared. There is still much anxiety felt around being a black person and visiting a new state for the first time." While many countries warn their citizens considering travel to the US of dangers due to gun violence, countries with predominantly black populations further warn about police brutality. Following the fatal police shootings of Philando Castile and Alton Sterling in 2016, the Bahamas issued a travel advisory as well as recommendations, particularly for young male travelers, to exercise extreme caution in interactions with police offers in the US (Davis 2016). While this violence continues to affect the everyday lives of African Americans, it can also add to the anxiety of black tourists in unfamiliar circumstances.

African American tourists and tourist spending have been on the rise over the past decade. In 2018, African American tourists spent an estimated 63 billion US dollars. This represents an increase of $20 billion USD from 2010 (Diskin 2018). The tourism industry is starting to recognize these tourists as a significant market. While African American representation in traditional industry roles or media remains small, social media has allowed for the rise of influencers such as Jessica Nabongo (@thecatchmeifyoucan—the first documented black woman to visit every country in the world) and Phil Calvert (@philwaukee—host of the YouTube show Phil Good Travel). Virtual communities, like Nomadness Tribe with over 20,000 members, are also increasingly exerting influence on the industry (Nelson 2021).

From Overtourism to (almost) No Tourism

Concerns about growing numbers of tourists, and their effects on the landscape, are not new. However, these concerns—expressed through the idea of overtourism—came to dominate conversations about tourism in the late 2010s. The United Nations World Tourism Organization (UNWTO 2018) describes overtourism as the impacts of tourism on a destination that negatively affects the perceived quality of life for residents and/or the perceived quality of experiences for visitors. Overtourism is often associated with mass tourism, in which standardized experiences are provided to large numbers of tourists. Yet, mass tourism does not necessarily result in the problems associated with overtourism. These problems are primarily attributed to poor tourism planning and/or poor destination management. Thus, overtourism occurs in destinations ranging from major urban areas like New York City, New York to rural locations such as the T.A. Moulton Barn in Wyoming, where photographers line up to take pictures because of its reputation as the most photographed barn in the US.

Although expressions of overtourism are place-specific, some key factors contribute to the issue. More people, both within the US and from countries around the world, have been enabled to travel and to travel more often. This is due to the combination of increasing socioecomic development and decreasing travel costs, including low-cost airlines (e.g., Spirit), budget cruises (e.g., Carnival), and vacation rental platforms (e.g., Airbnb). Tourism stakeholders in positions of power have prioritized tourism growth with a short-term focus on economic gain. The priorities of other stakeholders, such as community residents, are ignored until the situation becomes unsustainable.

Overtourism has brought new attention to the role of tourism in changing places. Scholars have proposed the term *touristification* to describe the physical and social changes associated with tourism development in urban destinations. Tourist accommodations (i.e., hotels and investor-purchased apartments for short-term rentals on platforms like Airbnb), entertainment venues, and non-essential retail shops expand into residential areas. Local retail shops close, which makes it harder for residents to get necessities and changes long-standing cultural patterns. Affordable residential houses become scarce, and particularly low-income residents are vulnerable to displacement (Nelson 2021). This is not necessarily a new problem. For example, Charleston, South Carolina experienced *touristification* in the 1980s and 1990s. Since that time, approximately half of the city's African American residents have been displaced from the downtown area (Lyles 2013). However, additional issues have arisen in the past decade with the growth of short-term rentals. Despite a city-wide ban, these rentals grew to over 2,000 and threatened the continued existence of Charleston's residential neighborhoods. The full ban was lifted in 2018 and replaced with strict regulations requiring that the homeowner inhabit the house while hosting short-term rental guests (Darlington 2018).

Tourists' actions also play an integral role in overtourism. In recent years, social media has driven people to travel for the sake of posting photographs of themselves in distinctive—and often trending—places. This leads to overcrowding in specific places, which has negative social and environmental effects. In just one example, the author of a popular media article writes, "Social media has been blamed for ruining Kanarraville Falls [Utah], a once hidden gem but now featured in countless Instagram posts. Bottlenecks can back up for an hour or more at the ladders, rescue teams are dispatched regularly to retrieve injured hikers, and stream banks are eroding and littered with trash" (Simmonds et al. 2018).

In 2019, the co-founder of *Lonely Planet*, Tony Wheeler, wrote the introduction to a text on overtourism. While he discussed these issues, he also wrote, "remember that time-worn truism that what goes up can equally easily come down" (Wheeler 2019). Within a year, the COVID-19 pandemic drastically changed the tourism landscape in the US and around the world. The US began the process of suspending incoming international flights. Passengers were quarantined on a cruise ship off the coast of California before the Centers for Disease Control and Prevention (CDC) issued a No Sail Order. The CDC further recommended that individuals avoid all non-essential travel while the US Department of State issued a Global Level 4 Health Advisory: Do Not Travel.

National weekly travel spending fell from over 20 billion US dollars in February to a low of 2.6 billion in March (US Travel Association 2020d). At this time, major national travel and tourism stakeholders began to raise concerns about the impact of COVID-19 on the nine million workers whose livelihoods directly depend on tourism (US Travel Association 2020a). By October 2020, the US Travel Association was predicting that 50 percent of all travel-supported jobs would be lost by the end of the year (US Travel Association 2020b). In the same month, the organization estimated that the losses in the US travel economy from the pandemic accounted for 415 billion US dollars. Losses were highest for states such as Hawaii, New York, and Washington, DC while western states such as Colorado, Utah, and Wyoming fared better with summer tourists traveling to participate in outdoor activities (US Travel Association 2020d).

People and places have had—and will continue to have—varied responses to tourism under these circumstances. In 2019, Hawaii's tourism industry accounted for 17.8 billion US dollars and 216,000 jobs (Hawai'i Tourism Authority 2019). Yet, in April 2020 the state's Visitors and Convention Bureau pleaded with media outlets to stop promoting tourism to the state, citing concerns that the pandemic would overwhelm their health care system (Monahan 2020). The state was able to keep cases to a minimum, but the economic cost was devastating. In the same year, Florida's tourism industry accounted for 91.3 billion US dollars and 1.5 million jobs (Visit Florida 2020a). By March 2020, weekly hotel revenue in the state was down 416.1 million US dollars over the same time in 2019 (Visit Florida 2020b). To regain some of this income and employment, the state was one of the first to "re-open" as early as May. By June, new COVID-19 infections were on the rise.

Although scholars predict strong political and industry pressure to "restart" the economy, which will mean a push to return to "business-as-usual" in tourism (Hall, Scott, and Gössling 2020), the crisis generated by the COVID-19 pandemic should be viewed as an opportunity to critically evaluate the tourism system. We should reevaluate the longstanding emphasis on tourism growth and the idea that more tourists will lead to more economic benefits (Gössling, Scott, and Hall 2021). We should consider ways in which tourism can be more socially inclusive and equitable (Benjamin, Dillette, and Alderman 2020). Finally, we should find ways to better address tourism's growing environmental impacts, from localized degradation to contributions to climate change (Prideaux, Thompson, and Pabel 2020). With these aims in mind, both tourism stakeholders and tourists need to work together to move towards a more sustainable American tourism landscape.

Towards a Sustainable Tourism Landscape

Tourism scholars and practitioners have used the framework of sustainable development to weigh the economic, social, and environmental costs and benefits of tourism and to work towards sustainable tourism. Sustainable tourism should ensure long-term economic opportunities that are fairly distributed among stakeholders. It should respect local cultural heritage and contribute to inter-cultural understanding. Finally, sustainable tourism should promote efficient resource consumption and support the conservation of natural heritage and biodiversity (UNWTO nd). The UNWTO (nd) reminds stakeholders that "Sustainable tourism development guidelines and management practices are applicable to all forms of tourism in all types of destinations, including mass tourism and the various niche tourism segments."

The US does not have a federal tourism policy to promote or certify sustainable tourism development. However, various initiatives are in place through federal agencies, such as the National Park Service (Bricker and Schultz 2011). Individual states and specific destinations have undertaken a variety of sustainability measures as well. Major hospitality brands have also made sustainability commitments. US-based Hilton Worldwide's corporate social responsibility program supports environmental objectives (e.g., reducing energy and water use intensity, responsibly sourcing products) and social objectives (e.g., human rights, community investment) (Hilton nd).

A market research study found that one-third of the US population would consider themselves sustainable travelers. However, the study also found that there is little consensus about what sustainable tourism means for these travelers (Peltier 2016). Nonetheless, tourists also have a part to play in supporting sustainable tourism businesses and destinations as well as voluntarily making changes in their behavior. Actions that tourists can take include supporting local businesses, buying locally made products, respecting the lives and culture of local people, traveling with refillable water bottles or reusable bags, minimizing food waste, and walking or using public transportation at a destination (Nelson 2021).

California Tourism

California's tourism industry embodies many of the points made here, evidenced in part by the above discussion of Disneyland. US tourism to California dates to the late nineteenth century. At this time, tourism was predicated partly on infrastructure development (e.g., inter-state and transcontinental railroads) and partly on place promotion supported by writers, painters, and photographers (Davis 1999). For example, Carleton E. Watkins's photographs of Yosemite were displayed in Washington, DC in 1864 and played a role in the campaign to preserve the valley as a national park (Runte 1990). While California continued to capitalize on its diverse natural attractions, the post-World War II era saw the development of built attractions, such

Figure 24.4   California is well-known for its built attractions, most notably, Disneyland. "Sleeping Beauty Castle Disneyland Anaheim" by Tuxyso. 2013. Licensed under CC BY-SA 3.0, https://commons.wikimedia. org/wiki/File:Sleeping_Beauty_Castle_Disneyland_Anaheim_2013.jpg. (Accessed September 15, 2021.)

as Disneyland (1955) (Figure 24.4), SeaWorld (1964), Six Flags Magic Mountain (1971), San Diego Zoo Safari Park (1972), and more.

Although the state has a strong destination image among domestic and international audiences, the destination marketing organization Visit California actively promotes tourism. The organization uses its website, social media platforms, and even commercials with celebrity appearances to create a demand for California's experiences. In 2019, the state saw an estimated 268 million US visitors and nearly 18 million international visitors (Tourism Economics 2019). These tourists contributed 144.9 billion US dollars to the state's economy and directly supported 1.2 million jobs (Visit California 2021b).

Although tourism is integral to the state's economy, it is not immune to the problems of overtourism. When southern California experienced a "super bloom" of wildflowers in 2019, thousands of people flocked to the area daily. In the quest for an attention-grabbing photograph, tourists strayed from designated walking paths, trampled the flowers, laid down in the flowers, and picked the flowers. With such destruction of the landscape and disruption of life, the nearby city of Lake Elsinore was forced to declare a public safety emergency (Morton 2019).

Various initiatives work to combat the negative environmental and social effects of tourism. For example, in 2019, California's governor signed a bill prohibiting hotels from providing personal care products in small plastic bottles. These products must be phased out by 2023 for hotels with more than 50 rooms and by 2024 for those with fewer rooms (California Legislative Information 2019). In 2020, Visit California launched a Responsible Travel Code that promotes travel with RESPECT (i.e., **R**oam responsibly, **E**ducate oneself, **S**afety first, **P**reserve California, **E**mbrace community, **C**elebrate culture, and **T**each others) (Visit California 2021a).

Conclusion

The American tourism landscape is diverse, complex, and ever-changing. Over the past two hundred years, many factors have created opportunities for Americans to travel—from expanding transportation infrastructure and lowered the cost to increasing disposable incomes and leisure time. While this is sometimes described as the democratization of travel, there have been (and continue to be) many barriers to travel as well—from a lack of accessible tourism infrastructure to discriminatory practices that have prevented people from participating in tourism. This chapter briefly considered barriers to travel for women and African Americans, but this discussion could also be expanded to include lower socioeconomic classes, Muslim tourists, the LGBTQIA+ community, persons with disabilities, and others.

With the rise of travel, places across the United States have looked for opportunities to engage in tourism. While some places were able to capitalize on natural resources, others specifically created attractions that would draw in people who could contribute to the local economy. However, the recent, and opposing, crises created by overtourism and the COVID-19 pandemic have prompted various stakeholders to rethink the current tourism system. As tourism recovers, tourism stakeholders and tourists themselves need to work together to promote a more equitable and a more sustainable American tourism landscape.

References

Anderson, M. 2006. *Women and the Politics of Travel, 1870–1914.* Madison: Fairleigh Dickinson University Press.

Armstead, M. B. Y. 2005. "Revisiting Hotels and Other Lodgings: American Tourist Spaces through the Lens of Black Pleasure-Travelers, 1880–1950." *The Journal of Decorative and Propaganda Arts* 25: 136–159.

Benjamin, S., A. Dillette, and D. H. Alderman. 2020. " 'We Can't Return to Normal': Committing to Tourism Equity in the Post-Pandemic Age." *Tourism Geographies* 22 (3): 476–483. doi:10.1080/14616688.2020.1759130.

Boorstin, D. J. 1961. *The Image: A Guide to Psuedo-Events in America.* New York: Harper and Row.

Bricker, K. S. and J. Schultz. 2011. "Sustainable Tourism in the USA: A Comparative Look at the Global Sustainable Tourism Criteria." *Tourism Recreation Research* 36 (3): 215–229.

California Legislative Information. 2009. "AB-1162 Lodging Establishments: Personal Care Products: Small Plastic Bottles." October 10, 2019. leginfo.legislature.ca.gov/faces/billTextClient.xhtml?bill_id=2019 20200AB1162.

CNT Editors. 2019. "The Women Who Travel Power List." *Condé Nast Traveler*, March 8, 2019. www.cnt raveler.com/gallery/the-women-who-travel-power-list.

Cohen, E. 1979. "A Phenomenology of Tourist Experiences." *Sociology* 13: 179–202.

Darlington, A. 2018. "Charleston's New Short-Term Rental Rules Enforceable So Far, But Debate Rages On." *The Post and Courier*, October 19, 2018. www.postandcourier.com/business/real_estate/charles ton-s-new-short-term-rental-rules-enforceable-so-far/article_1cb2672c-cca5-11e8-bec8-6f83f60ef 0fb.html.

Davis, A. C. 2016. "The Bahamas' New U.S. Travel Advisory: Use 'Extreme Caution' Around Police." *The Washington Post*, July 9, 2016. www.washingtonpost.com/news/morning-mix/wp/2016/07/09/ the-bahamas-travel-advisory-for-the-u-s-use-extreme-caution-around-the-police/.

Davis, S. G. 1999. "Landscapes of Imagination: Tourism in Southern California." *Pacific Historical Review* 68 (2): 173–191.

Diskin, E. 2018. "African-Americans Spent $63 billion on Travel in 2018." *Matador Network*, December 26, 2018. matadornetwork.com/read/african-americans-spent-billion-travel-2018/.

Gassan, R. H. 2008. *The Birth of American Tourism: New York, the Hudson Valley, and American Culture, 1790–1830.* Amherst: University of Massachusetts Press.

Gibson, C. 2008. "Locating Geographies of Tourism." *Progress in Human Geography* 32 (3): 407–422.

Gössling, S., D. Scott, and C. M. Hall. 2021. "Pandemics, Tourism and Global Change: A Rapid Assessment of COVID-19." *Journal of Sustainable Tourism* 29 (1): 1–20. doi.org/10.1080/09669582.2020.1758708.

Hall, C. M., D. Scott, and S. Gössling. 2020. "Pandemics, Transformations and Tourism: Be Careful What You Wish For." *Tourism Geographies* 22 (3): 577–598. doi.org/10.1080/14616688.2020.1759131

Hawai'i Tourism Authority. 2019. "Fact Sheet: Benefits of Hawai'i's Tourism Economy." www.hawaiitourismauthority.org/media/4167/hta-tourism-econ-impact-fact-sheet-december-2019.pdf.

Hilton. ND. "Travel with Purpose." cr.hilton.com.

Kevan, S. M. 1993. "Quests for Cures: A History of Tourism for Climate and Health." *International Journal of Biometeorology* 37: 113–124.

Lamback, B. 2019. "The 6 Best US Travel Destinations for Black Travelers in 2019." *Matador Network*, May 20, 2019. matadornetwork.com/read/best-travel-destinations-black-travelers/.

Löfgren, O. 1999. *On Holiday: A History of Vacationing*. Berkeley: University of California Press.

Lyles, J. 2013. "Gentrification Breaks a Neighborhood Down from the Inside Out." *Charleston City Paper*, April 24, 2013. www.charlestoncitypaper.com/story/gentrification-breaks-a-neighborhood-down-from-the-inside-out?oid=4615103.

Marling, K. A. 1991. "Disneyland, 1955: Just Take the Santa Ana Freeway to the American Dream." *American Art* 5 (1/2): 168–207.

McCannell, D. 1973. "Staged Authenticity: Arrangements of Social Space in Tourist Settings." *Journal of Sociology* 79 (3): 589–603.

Monahan, J. 2020. "Request to Suspend Hawaii Travel-Focused Editorial Coverage." April 6, 2020. www.hawaiitourismauthority.org/media/4424/request-to-suspend-hawaii-travel-focused-editorial-coverage.pdf.

Morton, C. 2019. "Tourists are Already Destroying California's Super Bloom." *Condé Nast Traveler*, March 19, 2019. www.cntraveler.com/story/how-to-see-californias-super-bloom.

Müller, D. K. 2019. "Tourism Geographies: A Bibliometric Review," In *A Research Agenda for Tourism Geographies*, ed. D. K. Müller, 7–22. Cheltenham: Edward Elgar Publishing.

Nelson, V. 2013. *An Introduction to the Geography of Tourism*. Lanham: Rowman & Littlefield Publishers, Inc.

———. 2021 *An Introduction to the Geography of Tourism*, 3rd Edition. Lanham: Rowman & Littlefield Publishers, Inc.

Peltier, D. 2016. "U.S. Travelers Like Sustainable Tourism but Love Transparency From Brands." May 10, 2016. skift.com/2016/05/10/u-s-travelers-like-sustainable-tourism-but-love-transparency-from-brands/.

Prideaux, B., M. Thompson, and A. Pabel. 2020. "Lessons from COVID-19 Can Prepare Global Tourism for the Economic Transformation Needed to Combat Climate Change." *Tourism Geographies* 22 (3): 667–678. doi:10.1080/14616688.2020.1762117.

Runte, A. 1990. "Introduction: The California National Parks Centennial." In *Yosemite and Sequoia: A Century of California National Parks*, eds. R. J. Orsi, A. Runte, and M. Smith-Baranzini, 1–5. Berkeley: University of California Press.

Shaffer, M. S. 2001. *See America First: Tourism and National Identity, 1880–1940*. Washington, D.C.: Smithsonian Institution Press.

Simmonds, C., A. McGivney, P. Reilly, B. Maffly, T. Wilkinson, G. Canon, M. Wright, and M. Whaley. 2018. "Crisis in Our National Parks: How Tourists are Loving Nature to Death." *The Guardian*, November 20, 2018. www.theguardian.com/environment/2018/nov/20/national-parks-america-overcrowding-crisis-tourism-visitation-solutions.

Simmons, J. 1984. "Railways, Hotels, and Tourism in Great Britain 1839–1914." *Journal of Contemporary History* 19 (2): 201–222.

Specia, M. and T. Mzezewa. 2019. "Adventurous. Alone. Attacked." *The New York Times*, March 25, 2019. www.nytimes.com/2019/03/25/travel/solo-female-travel.html

Stephens, R. J. 2001. *Idlewild: The Black Eden of Michigan*. Charleston, S.C.: Arcadia.

Tourism Economics. 2019. "California Travel & Tourism: Overview of Key Drivers and Outlook." October 22, 2019. industry.visitcalifornia.com/-/media/industry-site/pdfs/research/ca-travel-forecaststate-oct-2019.pdf.

Townsend, J. 2016. "How the Green Book Helped African-American Tourists Navigate a Segregated Nation." *Smithsonian Magazine*, April 2016. www.smithsonianmag.com/smithsonian-institution/history-green-book-african-american-travelers-180958506/?no-ist.

United Nations World Tourism Organization. ND. "Sustainable Development." Accessed November 13, 2020. www.unwto.org/sustainable-development.

———. 2018. "'Overtourism'? Understanding and Managing Urban Tourism Growth Beyond Perceptions." www.e-unwto.org/doi/pdf/10.18111/9789284420070.

US Travel Association (a). 2020. "About COVID-19." April 28, 2020. www.ustravel.org/sites/default/files/media_root/document/Coronavirus_Specific.pdf.

———— (b). 2020. "US Travel Applauds Introduction of Hospitality Jobs Bill." October 15, 2020. www.ustravel.org/press/us-travel-applauds-introduction-hospitality-jobs-bill.

———— (c). 2020. "US Travel and Tourism Overview (2019)." March 2020. www.ustravel.org/system/files/media_root/document/Research_Fact-Sheet_US-Travel-and-Tourism-Overview.pdf.

———— (d). 2020. "Weekly Coronavirus Impact on Travel Expenditures in the US." October 15, 2020. www.ustravel.org/sites/default/files/media_root/document/Coronavirus_WeeklyImpacts_10.15.20.pdf.

Visit California. 2021a. "Safe Travels in California." www.visitcalifornia.com/things-to-do/travel-california-respect-california/.

————. 2021b. "Why Travel Matters." industry.visitcalifornia.com/partner-opportunities/programs/why-travel-matters.

Visit Florida. 2020a. "About Us." www.visitflorida.com/en-us/about-us.html.

————. 2020b. "COVID-19 FL Tourism Impacts." October 22, 2020. www.visitflorida.org/resources/crisis-preparation/covid-19-resources-and-information-for-businesses/covid-19-fl-tourism-impacts/.

Wheeler, T. 2019. "Forward." In *Overtourism: Excesses, Discontents and Measures in Travel and Tourism*, edited by C. Milano, J. M. Cheer, and M. Novelli, xv–xviii. Oxfordshire: CABI.

Young, T. 2017. *Heading Out: A History of American Camping*. Ithaca: Cornell University Press.

25

THE CHANGING RURAL LANDSCAPE

Andrew Husa

What is America's rural landscape? The United States Census Bureau does not specifically define "rural"; instead, the absence of that which is urban defines a rural place. Therefore, "rural" encompasses all population, housing, and territory not included in an urban area, defined in the US as having at least 2,500 people (United States Census Bureau 2020). However, census categories do not always adequately describe the lived experience of a place. Across the country, demographers, geographers, and government leaders struggle to develop definitions of "rural" that are both uniform and conform to the personal experience of living in a place. How "rural" is experienced as an identity, a place, and a representation varies between individuals and groups (Halfacree 2006).

In order to define "rural," Keith Halfacree (2006) suggests using a model that intertwines three facets: rural localities, formal representations of the rural, and everyday lives of those in rural areas. Individuals and groups inscribe rural localities through spatial practices, which may be related to production or consumption. Formal representations of the rural refer to the way that the rural is framed within the production process. Everyday lives of the rural incorporate individual and social elements in interpretation and negotiation. Together, these three facets comprise rural space and help us define rural places outside of census categories and individual interpretations (Halfacree 2006).

Those who call rural America home have found that the landscape around them has changed dramatically over the second half of the twentieth century and the first two decades of the twenty-first century. This chapter examines the evolution of America's rural landscape, including changes in rural economics, rural communities, rural population trends, and rural diversity. The chapter closes with a consideration of new areas of research in rural geography.

Rural Economics

Across the country, the rural economic landscape is changing. One of the most striking changes is manifested in agriculture: the size of farms continues to increase while the number of farmers and farm families decreases. We can trace this shift towards large, consolidated farms back to the Great Depression and the Dust Bowl years of the 1930s, when banks forced struggling small-scale farmers to sell their land. An increase in both technological advancements and market

DOI: 10.4324/9781003121800-31

pressures has seen the number of small farms decline at an accelerated rate since the 1980s (Gardner 2002).

At the same time, the number of large farms has increased. The United States Department of Agriculture (USDA) has found that the number of large farms, farms with at least 2,000 acres of cropland, increased in each Census of Agriculture after 1987. In 1987, 66,786 farms with at least 2,000 acres of cropland operated; in 2017, there were 85,127 such farms (MacDonald, Hoppe, and Newton 2018).

The total number of farms has been declining in recent years. From 2007 to 2017, the total number of farms dropped by more than 150,000. In the same period, the average farm size increased by 23 acres (MacDonald, Hoppe, and Newton 2018). This increase echoes the words of John Fraser Hart, who wrote, "[t]he scale of farming has changed so dramatically that farmers have had to add a zero or two to the way they once thought, be it dollars or acres, crops or animals, bushels or head" (Hart 2003, 1–2).

Changes in scale and shifts in production are happening in other rural-based sectors as well. Consolidation among food processing and manufacturing companies, for example, has led to large plants dominating the industry. By the end of the twentieth century, plants with more than 400 employees accounted for most of the meat production in the US (MacDonald et al. 1999). As they have gotten larger, food processing and manufacturing companies have increasingly relocated their plants to rural areas in order to reduce livestock transportation and feed costs, and to ensure more consistent quantities of animals. A combination of industry consolidation, increasing plant size, and new plant locations has raised the demand for low-skilled workers in rural areas.

Manufacturing is another important source of employment in rural America. Rural manufacturing is no longer the driver of job growth that it once was, however. Many of these jobs have been lost to cost-saving measures, as companies outsource to countries with cheaper labor. Others have been lost to labor-saving measures, including the increased use of automation in manufacturing plants. Still, manufacturing provides employment that many rural residents and communities rely on. The survival of a manufacturing plant is crucial for the survival of the community where it is located (Low and Brown 2016; Low 2017). As a rule of thumb, first these communities lose the factories, and then second, they lose the people that supported them (Collins 2019).

The dairy landscape has also changed markedly in recent decades: the number of dairy farms is declining, while the average herd size is growing. In 2002, there were 74,100 licensed dairy farms in the US. This number dropped to 34,187 in 2019 (MacDonald, Low, and Mosheim 2020). Although the number of licensed dairy farms has declined, milk production has continued to grow as production has shifted toward larger farms. By 2017, nearly 2,000 dairy farms had herds of at least 1,000 cows, and those farms milked over half of the country's cows. In 1992, there were only 500 such farms, with those farms milking less than 10% of US cows (MacDonald, Low, and Mosheim 2020). These numbers include America's Dairyland (Figure 25.1), as Wisconsin's milk production increasingly comes from larger herds, while the number of small, family-operated dairy farms continues to dwindle (Cross 2001, 2012).

Rural Communities

While rural America's farms, factories, and feedlots are getting bigger, most of the small towns that dot the rural landscape are getting smaller. In many areas of the country, the rural population has plummeted since the early twentieth century. While these patterns of rural depopulation exist elsewhere, they are especially pronounced in the Great Plains. Perhaps no state

Figure 25.1 A creamery in Chaseburg, Wisconsin. Photo courtesy of Carissa Dowden.

provides a better example of ongoing rural depopulation than Nebraska, which lies in the heart of the Great Plains: 72 of Nebraska's 93 counties had their year of highest census population in 1930 or earlier (Husa 2020).

Although the majority of rural towns in the Great Plains have declined in population, they should not be considered failures. These towns were successful, and many still are, despite their declining populations, as they are still performing the role they were intended to perform as grain-collection points. In a speech given at Kansas State University in 1996, John C. Hudson argued that depopulation should not be equated with failure and that the Great Plains does not require a sedentary, stable, dispersed rural population to make it function (Hudson 1996). Although many rural towns themselves have lost hundreds of people and dozens of businesses, their grain elevators continue to thrive today.

Unfortunately, the same cannot be said for their main streets. In recent decades, people have largely abandoned the small-town main streets which were once the hearts and souls of rural America. In the Midwest and the Great Plains, the historical "T-town," whose main street and businesses were platted perpendicular to the railroad, has been replaced by a new version, where gas stations and fast-food chains line up along the interstate exits, while the local businesses remain miles away from the interstate itself. While the interstate has led to the decimation of many main streets, it has helped keep some of these towns on the map by allowing consumers to come to them, often in the form of motel and restaurant customers, or truck repairs and fuel sales (Husa 2020).

Along with their populations and businesses, many small towns have also lost their high schools. Rural depopulation, limited funding, poorer facilities, relatively low salaries, and few special programs have forced schools to close their doors and consolidate their districts (Blauwkamp, Longo, and Anderson 2011). These closures and consolidations force students to abandon their old relationships and place attachments while getting acclimated to their new schools. Communities are at a loss as they forfeit the benefits of having a high school. School consolidations, "can leave a small community empty and its citizens angry with a loss of connection to the newly formed district and to their own community" (Surface 2016, 62).

Many small-town high schools have found ways to fight back against declining enrollments. In the Midwest and the Great Plains, reduced-player football teams have become a mainstay as high schools have trouble finding enough students to field a traditional eleven-man team. Across these regions, young men play six- and eight-man football on eighty-yard football fields, which are twenty yards shorter than the traditional field, creating a distinct landscape in numerous small towns. The geography of reduced-player football will continue to reflect rural depopulation patterns; six-man football, which has a high number of teams in Texas, is growing in Nebraska, and is being either introduced or reintroduced in a handful of states, including Kansas, North Dakota, and South Dakota (Husa 2019).

The enduring problem of rural depopulation has led to state and community leaders exploring solutions as exemplified by population recruitment and retention strategies. The towns that are able to sustain their populations and recruit new residents have several things in common, including strong leadership and dedicated residents, diverse employment opportunities, strong schools and access to healthcare, good and affordable housing, and broadband internet access (Husa 2020). These towns have kept their residents happy and their main streets healthy.

In an effort to reverse decades of population loss, some rural communities are offering free land for people to settle. In fact, in the early 2000s, a handful of senators from states in the Great Plains tried to pass what were essentially new versions of the Homestead Act of 1862. Although the proposed bills never passed, many were enthusiastic about the free-land idea. In 2005, *USA Today* profiled various Kansas towns in which new settlers could receive free building lots (Shortridge 2004). Although the success of these programs varies from town to town, many of the participating communities have received population boosts, which has raised the tax base and provided the community with more money, which has been used to improve schools and infrastructure (Bauer 2016).

Rural Population Trends

Many rural towns have been able to sustain their populations, and some have even grown significantly. In 1998, Kenneth Johnson and Calvin Beale wrote about the "rural rebound" that occurred over the last decade of the twentieth century when, between 1990 and 1996, the population of America's rural counties grew by nearly three million. They largely attributed this rebound to an increase in Hispanic immigration into rural America to fill the high demand for low-skilled workers sought by manufacturing and food processing companies (Johnson and Beale 1998).

Rural population gains in the early 2000s were considerably smaller than they had been during the 1990s: the rural population grew by just 2.2 million between 2000 and 2010. The smaller gains were a result of the declining manufacturing economy and the subsequent reduced in-migration of workers. The rural counties with manufacturing jobs had been the leader in in-migration in the 1990s, but were surpassed by those with natural amenities, recreational opportunities, and/or those located near urban areas or major transportation corridors (Johnson 2012).

Net out-migration remains a serious problem: between 2010 and 2016, 462,000 more people left rural areas than moved in (Johnson and Lichter 2019). The map in Figure 25.2 shows that 1,351 of 1,976 (68.37%) rural (non-metro) counties lost population during these years. As part of this out-migration, many rural areas have seen younger generations move away, which has increased the percentage of the population of those aged 65 or over, resulting in rural populations becoming disproportionately old (Glasnow and Brown 2012). The aging

Population loss now widespread in the Eastern United States

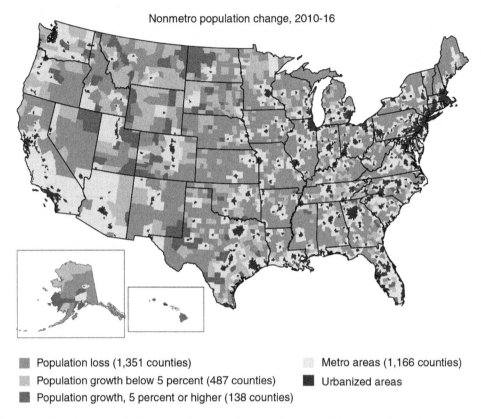

Nonmetro population change, 2010-16

- ■ Population loss (1,351 counties)
- ▨ Population growth below 5 percent (487 counties)
- ■ Population growth, 5 percent or higher (138 counties)
- ▨ Metro areas (1,166 counties)
- ■ Urbanized areas

Source: USDA, Economic Research Service using data from the U.S. Census Bureau.

Figure 25.2   Map showing Nonmetro Population Change, 2010–2016. Map from Cromartie 2017.

population may lead to rural community leaders pushing to offer new services, especially those that improve healthcare access (Thiede et al. 2016).

Rural Diversity

Although rural America is generally less diverse than urban areas, it is more racially and ethnically diverse than popular imagination often presumes (Brown and Schafft 2019). For example, since 1990, the Hispanic population in rural America has more than doubled, as migrant workers have sought out places with new job opportunities and lower costs of living (Kandel and Cromartie 2004). Although the majority of the research on the impact of Hispanic immigration seems to focus on the country as a whole, there are several local-scale case studies, including some that focus on rural communities in Georgia (Engstrom 2001), Kansas (Broadway and Stull 2006), and Kentucky (Shultz 2000). Holly Darcus has also written on the impact of new Hispanic immigrants in rural Kentucky (2006), as well as Appalachia (2007) and the Great Plains (2013). New industries that rely on low-skill and inexpensive labor have increasingly drawn Hispanic immigrants to these rural areas. In moving to these areas, Hispanic immigrants have stimulated local economies that are increasingly lacking consumers. This new stream of in-migration, "holds hope for changing the trajectory of depopulation" (Barcus and Simmons 2013, 150).

While this population influx has helped revive many small towns, it has often led to a culture clash between white residents and their new neighbors (Carr, Lichter, and Kafalas 2012). Along with facing language barriers and some forms of social discrimination, Hispanic immigrants may not be accepted by the white population. Some try to dissuade the Hispanic workers, especially those who have come to the US illegally, from living in and doing business in their towns.

A perceptive case study comes from the middle of the country in Fremont, Nebraska. Lincoln Premium Poultry (Figure 25.3), in collaboration with Costco, opened a chicken processing plant in Fremont in 2018. When the plant opened, it sought over 1,000 new workers. With the addition of the new plant next to the Hormel hog plant, the nation's largest producer of Spam, Fremont has become a popular destination for Hispanic immigrants seeking gainful employment. In 2014, however, the town of 26,397 people became the only one in the country where it is illegal to rent a house to an unauthorized immigrant (Grabar 2017).

With the influx of new Hispanic immigrants to the area, both documented and undocumented, nearby small towns such as Scribner, population 857, voted to take similar measures as Fremont in regard to unauthorized immigrants. In discussing the housing ordinance, Scribner's mayor said, "[w]e certainly are not going to discriminate against anyone… If you're here lawfully, you have nothing to worry about" (Hammel 2018).

These housing ordinances have divided neighbors. Those who oppose the ordinances argue that they make their community appear hostile and intolerant, which hurts development efforts. Those who support the ordinances don't believe that they hurt development efforts and that they have nothing to do with intolerance or racism (Bergin 2014). Several rural towns share the changes and tensions felt by Fremont and Scribner, where just about the only growth,

Figure 25.3 The Lincoln Premium Poultry in Fremont, Nebraska. Photo by author.

both in terms of employment opportunities and population, has come from food processing and manufacturing companies and the immigrant workers they attract. Moving forward, it will be interesting to see how rural communities respond to increased diversity.

New Areas of Research

Research on rural populations has often focused on why people leave a place, rather than why they stay. In recent years, however, there has been a rise in research on "stayers," those who are born and raised in an area and have never left (Stockdale and Haartsen 2018). These rural stayers are not stuck, but rather they have made the decision to remain in a place. Aileen Stockdale and Tialda Haartsen (2018) have called for the adoption of this new perspective on immobility in migration research as they consider stayers to be active participants in residential decisions and staying as an active process. Likewise, Cheryl Morse and Jill Mudgett (2018) have highlighted the importance of examining immobility as a choice.

To explore the various reasons why people choose to stay in a place, Cheryl Morse conducted the Vermont Roots Migration Survey in 2014, asking Vermonters to cite the reasons why they chose to remain in their home state (Figure 25.4). A statistical analysis of the survey's responses showed the importance of rural place-based factors and family ties as key factors in stayers' residential decision making. The research also showed that, by and large, rural Vermonters were content in their decision to stay (Morse 2015, 2017; Morse and Mudgett 2017, 2018).

Rural residential mobility research in Montana (Erickson, Sanders, and Cope 2018) and Nebraska (Husa and Morse 2020) provides similar results with the principal influences in the

Figure 25.4   The rural landscape of Glover, Vermont. Photo courtesy of Cheryl Morse.

decision to stay in rural areas being family ties, community attachment, community satisfaction, and many rural-based factors, including rural hospitality and a simpler lifestyle. The results from Nebraska, for example, suggest that people who grew up in a rural place and stayed as adults, "are more likely to enjoy living in, value raising their children in, and appreciate the culture, community, and landscape of their home state more than those who do not have rural living experience" (Husa and Morse 2020, 16). Other research on place attachment in the rural Great Plains (Smith and McAlister 2015; Wetherholt 2016) supports the notion that many of the region's rural residents are content in their decision to stay in their hometowns.

Box 25.1 Youth in Rural Landscapes
Cheryl Morse

Researchers know little about the everyday lives of young people who live in the vastly diverse cultural and physical landscapes of the rural North. The existing research on rural youth tends to center on pressing issues such as outmigration and mobility (Corbett 2007), substance use (Keyes et al. 2014), and educational attainment (Byun, Meece, and America Irvin 2012). In these accounts rural youth are often positioned as recipients of unfortunate circumstances. My own research fell into this "problem trap." A couple of decades ago I asked how rural youth craft their social lives in an impoverished region of Vermont. I found that gendered rules, a lack of built spaces, and a remote location in a cold climate created risky conditions for boys and a desire to leave among many of the girls (Dunkley 2004). The problem with my problem-oriented focus is what it overlooked: young people's placemaking activities, emotional attachments, and agency.

Now, looking through the lens of rural youth as active producers of place, I see countless examples of how young people are responding to challenges both large and small in my home state of Vermont. There is the group of young alumni from a regional high school who designed and delivered an anti-racism curriculum for students and teachers. They saw a gap in their education and decided to fill it through a collaborative and volunteer effort. A recent university graduate returned to her hometown of 700 residents and ran for a state legislative office. At twenty-three, she became one of the youngest state representatives in the country. Vermont's legislature is part-time; she works in a sawmill during the off months. She focuses on supporting rural healthcare programs, and bringing publicly owned, universal, high-speed internet access to rural Vermont. And there are the sixteen-year-old students in an outdoor program who built a picnic table along a snowmobile trail, a teeter totter for a mountain bike trail, and a campsite along a canoe trail. In each of these instances, youth demonstrate attentiveness to the needs of their communities. Such acts are modest, grounded, and practical; they do not tend to attract attention. They are the stuff of everyday life. In what ways do youth in the deserts of Arizona, the Everglades of Florida, and the coast of Maine experience and contribute to the making of their places? Society needs research on challenges rural youth face; we also will learn much from asking rural youth to share what they know about their home places.

References

Byun, S., J. L. Meece, and M. J. Irvin. 2012. "Rural-Nonrural Disparities in Postsecondary Educational Attainment Revisited." *American Educational Research Journal* 49 (3): 412–437. doi: 10.3102/0002831211416344.

Corbett, M. J. 2007. *Learning to Leave: The Irony of Schooling in a Coastal Community*: Fernwood Pub.

Dunkley, C. M. 2004. "Risky Geographies: Teens, Gender, and Rural Landscape in North America." *Gender, Place and Culture* 11 (4): 559–579. doi: 10.1080/0966369042000307004.

Keyes, K. M., M. Cerdá, J. E. Brady, J. R. Havens, and G. Sandro. 2014. "Understanding the Rural–Urban Differences in Nonmedical Prescription Opioid Use and Abuse in the United States." *American Journal of Public Health* 104 (2): e52-e59. doi: 10.2105/ajph.2013.301709.

Another development in research on rural populations comes from Holly Barcus and Stanley Brunn (2009, 2010), who coined the term "place elasticity" to describe how individuals can stay tied to a place through technology and communication while living elsewhere. Place attachment and mobility used to be seen as mutually exclusive, but their research shows that they can be complementary. For some, place attachment is expressed by feeling a connection to the place rather than physically residing there (Barcus and Brunn 2009, 2010). Research suggests that, along with missing family and friends, people who leave rural America still feel connected to the physical landscapes of the places where they grew up (Morse and Mudgett 2017).

An individual's experience of rural life is influenced by a range of identity factors, including ability, age, gender, and sexual orientation. Rural places are often stereotypically coded as being "for" certain groups of people and not for others (see Morse). As Cary W. DeWit (2001, 2004) has shown, women have long experienced the social and physical conditions of rural places much differently than their male counterparts, which has led to differing values and place attachments. Related research shows that women and men often consider much different ties in their residential decision making (Mulder and Malmberg 2014; Shepard 2014). Additionally, how gender identities and gender roles are constructed and contested, and sometimes even reinvented, is an emerging focus in rural geography (Little and Panelli 2003; Little and Morris 2005; Little 2017).

America's rural communities provide excellent opportunities for additional studies across a wide array of research topics. Although the country is becoming increasingly urbanized, a significant proportion of Americans remain in rural areas. While the rural landscape often occupies an empty space in the American geographical imagination, or perhaps one that is frozen in a past moment of time, it is anything but static. America's rural landscape is ever changing, with many farms getting bigger and populations getting smaller. However, new employment opportunities, changing demographic trends, and population recruitment and retention strategies will help many of the country's small towns survive and ensure that rural America is a place that many are happy to call home for years to come.

References

Barcus, H. R. 2006. "New Destinations for Hispanic Migrants: An Analysis of Rural Kentucky." In *The New South: Latinos and the Transformation of Place*, edited by O. J. Furuseth and H. A. Smith, 89–109. London: Ashgate Press.

———. 2007. "The Emergence of New Hispanic Settlement Patterns in Appalachia." *The Professional Geographer* 59 (3): 298–315.

Barcus, H. R. and S. D. Brunn. 2009. "Towards a Typology of Mobility and Place Attachment in Rural America." *Journal of Appalachian Studies* 15 (1 & 2): 26–40.

———. 2010. "Place Elasticity: Exploring a New Conceptualization of Mobility and Place Attachment in Rural America." *Geografiska Annaler: Series B, Human Geography* 92 (4): 281–295. doi:10.1111/j.1468-0467.2010.00353.x.

Barcus, H. R. and L. Simmons. 2013. "Ethnic Restructuring in Rural America: Migration and the Changing Faces of Rural Communities in the Great Plains." *The Professional Geographer* 65 (1): 130–152. doi:10.1080/00330124.2012.658713.

Bauer, K. F. 2016. "Free Land Programs Revisited: A Case Study of Four Kansas Communities." B.S., University of North Dakota.

Bergin, N. 2014. "Fremont Voters Overwhelmingly Affirm Anti-Illegal Immigration Ordinance." *Lincoln Journal Star*, February 11, 2014. journalstar.com/news/state-and-regional/nebraska/fremont-voters-overwhelmingly-affirm-anti-illegal-immigration-ordinance/article_91d84a16-e66b-5b30-8acc-bf741705f2b4.html.

Blauwkamp, J. M., P. J. Longo, and J. Anderson. 2011. "School Consolidation in Nebraska: Economic Efficiency vs. Rural Community Life." *Online Journal of Rural Research & Policy* 6 (1). doi:10.4148/ojrrp.v6i1.1309.

Broadway, M. J. and D. D. Stull. 2006. "Meat Processing and Garden City, KS: Boom and Bust." *Journal of Rural Studies* 22 (1): 55–66. doi:10.1016/j.jrurstud.2005.06.001.

Brown, D. L. and K. A. Schafft. 2019. *Rural People and Communities in the 21st Century: Resilience and Transformation.* Cambridge: Polity Press.

Carr, P. J., D. T. Lichter, and M. J. Kefalas. 2012. "Can Immigration Save Small-Town America? Hispanic Boomtowns and the Uneasy Path to Renewal." *The Annals of the American Academy of Political and Social Science* 641 (1): 38–57. doi:10.1177/0002716211433445.

Collins, M. 2019. "The Abandonment of Small Cities in the Rust Belt." *Industry Week*, October 10, 2019. www.industryweek.com/talent/article/22028380/the-abandonment-of-small-cities-in-the-rust-belt.

Cromartie, J. 2017. "Rural Areas Show Overall Population Decline and Shifting Regional Patterns of Population Change." *Amber Waves*, September 5, 2017. www.ers.usda.gov/amber-waves/2017/september/rural-areas-show-overall-population-decline-and-shifting-regional-patterns-of-population-change/.

Cross, J. A. 2001. Change in Americas Dairyland. *Geographical Review* 91 (4): 702–714. doi:10.2307/3594727.

———. 2012. "Changing Patterns of Cheese Manufacturing in Americas Dairyland." *Geographical Review* 102 (4): 525–538. doi:10.1111/j.1931-0846.2012.00173.x.

DeWit, C. W. 2001. "Women's Sense of Place on the American High Plains." *Great Plains Quarterly* 21 (1): 29–44.

———. 2004. "Gender and Sense of Place." In: *Encyclopedia of the Great Plains*, edited by D. J. Wishart, 329–330. University of Nebraska Press.

Engstrom, J. D. 2001. "Industry and Immigration in Dalton, Georgia." In: *Latino Workers in the Contemporary South*, edited by A. D. Murphy, C. Blanchard, and J. A. Hill. Athens: University of Georgia Press.

Erickson, L. D., S. R. Sanders, and M. R. Cope. 2018. "Lifetime Stayers in Urban, Rural, and Highly Rural Communities in Montana." *Population, Space and Place* 24 (4). doi:10.1002/psp.2133.

Gardner, B. L. 2002. *American Agriculture in the Twentieth Century: How it Flourished and What it Cost.* Cambridge, MA: Harvard University Press.

Glasgow, N., and D. L. Brown. 2012. "Rural Ageing in the United States: Trends and Contexts." *Journal of Rural Studies* 28 (4): 422–431. doi:10.1016/j.jrurstud.2012.01.002.

Grabar, H. 2017. "Who Gets to Live in Fremont, Nebraska?" *Slate*, December 6, 2017. slate.com/business/2017/12/latino-immigrants-and-meatpacking-in-midwestern-towns-like-fremont-nebraska.html.

Halfacree, K. 2006. "Rural Space: Constructing a Three-Fold Architecture." In: *Handbook of Rural Studies*, edited by P. J. Cloke, T. Marsden, and P. H. Mooney, 44–62. London: Sage. doi:10.4135/9781848608016.n4.

Hammel, P. 2018. "Scribner Voters Approve Ordinance barring Illegal Immigrants from Housing, Jobs." *Omaha World-Herald*, November 7, 2018. omaha.com/state-and-regional/scribner-voters-approve-ordinance-barring-illegal-immigrants-from-housing-jobs/article_59f8a313-86d2-515d-b0d8-22687fd91f13.html.

Hart, J. F. 2003. *The Changing Scale of American Agriculture.* Charlottesville: University of Virginia Press.

Hudson, J. C. 1996. "The Geographer's Great Plains." *Occasional Publications in Geography*. Kansas State University, Manhattan: Department of Geography.

Husa, A. 2019. "Six and Eight-Man Football and Community Identity in Nebraska." *Nebraska History Magazine* 100 (3): 210–219.

———. 2020. "Population Sustainability in Rural Nebraska Towns." Ph.D., University of Nebraska-Lincoln.

Husa, A. and C. E. Morse. 2020. "Rurality as a Key Factor for Place Attachment in the Great Plains." *Geographical Review*. doi:10.1080/00167428.2020.1786384.

Johnson, K. M. 2012. "Rural Demographic Change in the New Century: Slower Growth, Increased Diversity." *The Carsey School of Public Policy at the Scholars' Repository* 159. scholars.unh.edu/carsey/159.

Johnson, K. M. and C. L. Beale. 1998. "The Rural Rebound." *The Wilson Quarterly* 22 (2): 16–27.

Johnson, K. M. and D. T. Lichter. 2019. "Rural Depopulation: Growth and Decline Processes over the Past Century." *Rural Sociology* 84 (1): 3–27. doi:10.1111/ruso.12266.

Kandel, W. and J. Cromartie. 2004. "New Patterns of Hispanic Settlement in Rural America." *Rural Development Research Report* 99. www.ers.usda.gov/webdocs/publications/47077/rdrr-99.pdf?v=0.

Little, J. 2017. *Gender and Rural Geography*. Oxford: Taylor & Francis.

Little, J. and C. Morris. 2005. *Critical Studies in Rural Gender Issues*. Aldershot: Ashgate.

Little, J. and R. Panelli. 2003. "Gender Research in Rural Geography." *Gender, Place and Culture* 10 (3): 281–289. doi:10.1080/0966369032000114046.

Low, S. A. 2017. "Rural Manufacturing Resilience: Factors Associated with Plant Survival, 1996–2011." *Economic Research Report* 230. www.ers.usda.gov/publications/pub-details/?pubid=83540.

Low, S. A. and J. P. Brown. 2016. "Manufacturing Plant Survival in a Period of Decline." *Growth and Change* 48 (3): 297–312. doi:10.1111/grow.12171.

MacDonald, J. M., et al. 1999. "Consolidation in U.S. Meatpacking." *Agricultural Economic Report* 785. www.ers.usda.gov/webdocs/publications/41108/18011_aer785_1_.pdf?v=0.

MacDonald, J. M., R. A. Hoppe, and D. Newton. 2018. "Three Decades of Consolidation in US Agriculture." *Economic Information Bulletin* 189. www.ers.usda.gov/publications/pub-details/?pubid=88056.

MacDonald, J. M., J. Law, and R. Mosheim. 2020. "Consolidation in U.S. Dairy Farming." *Economic Research Report* 274. www.ers.usda.gov/publications/pub-details/?pubid=98900.

Morse, C. E. 2015. "The Risks and Rewards of Using Social Media in Rural Migration Research: Findings from the Vermont Roots Project." *The Northeastern Geographer* 7: 72–88.

———. 2017. "The Emotional Geographies of Global Return Migration to Vermont." *Emotion, Space and Society* 25: 14–21. doi:10.1016/j.emospa.2017.09.007.

Morse, C. E. and J. Mudgett. 2017. "Longing for Landscape: Homesickness and Place Attachment Among Rural Out-Migrants in the 19th and 21st Centuries." *Journal of Rural Studies* 50: 95–103. doi:10.1016/j.jrurstud.2017.01.002.

Morse, C. E. and J. Mudgett. 2018. "Happy to Be Home: Place Attachment, Family Ties, and Mobility of Contented Rural Stayers." *The Professional Geographer* 70 (2): 261–269. doi:10.1080/00330124.2017.1365309.

Mulder, C. H. and G. Malmberg. 2014. "Local Ties and Family Migration." *Environment and Planning A: Economy and Space* 46 (9): 2195–2211. doi:10.1068/a130160p.

Shepard, R. 2014. "The Role of Gender in Rural Population Decline in Kansas and Nebraska, 1990–2010." *Great Plains Research* 24 (1): 1–12. doi:10.1353/gpr.2014.0010.

Shortridge, J. R. 2004. "A Cry for Help: Kansasfreeland.com." *Geographical Review* 94 (4): 530–540. doi:10.1111/j.1931-0846.2004.tb00187.x.

Shultz, B. J. 2008. "Inside the Gilded Cage: The Lives of Latino Immigrant Males in Rural Central Kentucky." *Southeastern Geographer* 2: 201–218. doi:10.1353/sgo.0.0024.

Smith, J. S. and J. M. McAlister. 2015. "Understanding Place Attachment to the County in the American Great Plains." *Geographical Review* 105 (2): 178–198. doi:10.1111/j.1931-0846.2014.12066.x.

Stockdale, A. and T. Haartsen. 2018. "Editorial Introduction: Putting Rural Stayers in the Spotlight." *Population, Space and Place* 24 (4). doi:10.1002/psp.2124.

Surface, J. L. 2016. "Losing a Way of Life: The Closing of a Country School in Rural Nebraska." *Publications of the Rural Futures Institute* 2.

Thiede, B. C., et al. 2016. "A Demographic Deficit? Local Population Aging and Access to Services in Rural America, 1990–2010." *Rural Sociology* 82 (1): 44–74. doi:10.1111/ruso.12117.

United States Census Bureau. 2020. "Urban and Rural." www.census.gov/programs-surveys/geography/guidance/geo-areas/urban-rural.html.

Wetherholt, W. A. 2016. "Exploring Rootedness in the Very Rural Great Plains Counties of Kansas and Nebraska. Ph.D., Kansas State University.

26

ROOTED IN PLACE—
RETAIL ON THE AMERICAN
LANDSCAPE

Emily Fekete

The Urban Hierarchy and the Retail Landscape

The origins of the exploration of retail geographies can be traced to the historic development of urban hierarchy theory in the work of the renowned German geographer, Walter Christaller (1933 reprint 1966), ultimately giving rise to a generation of work concentrated on Central Place Theory. Central Place Theory focuses on retail and shopping, essentially arguing that places fall into an order that reflects the hierarchy of goods and services that each place offers. From this work, Christaller claims retail trade as the central dynamic driving urban growth.

Drawing from Zipf's law, Berry (1970) expanded on Christaller's Central Place Theory and introduced the theory of city-size relationships (Lake 2009), famously conceptualizing "cities as systems within systems of cities" (Lake 2009, 306). Using Berry's framework, urban hierarchy theory offers a means to evaluate and compare metropolitan environments. Both Alan Pred (1977) and later Peter Hall (1998), also used Christaller's theory to make the claim that urban spatial development is uneven and that those cities occupying lower levels of the urban hierarchy are disappearing as global cities grow in economic significance.

Building on these examples of previous work, geographers today understand the concept of the urban hierarchy to be a ranking of cities in which those with larger populations have a greater range of financial activities and superior access to information, which in turn leads them to exert disproportionate economic influence over other urban areas and the whole economy. Viewing the urban hierarchy from the standpoint of retail, larger populations can support niche markets and provide access to a wide variety of goods and services. Retail businesses can afford to offer specialized items in locations found further up the urban hierarchy, while cities with smaller populations must rely on larger ones for access to particular products.

Under recent rounds of globalization, emphasis on the study of urban hierarchies has shifted to world cities and networks of the international economy. Due to growing linkages between world cities such as New York City, London, or Tokyo, smaller cities within countries are gradually becoming disconnected, leading to an increase in inequality among cities worldwide (Alderson et al. 2010). Though goods and services are becoming more globally available, the experience of retail in places in the US on lower levels of the urban hierarchy are strikingly similar to one another as most rely on big box stores—like Walmart (Brunn 2006)—and chain

DOI: 10.4324/9781003121800-32

restaurants, illuminating the lack of specialty goods and services available in lower order cities (Fekete 2014). This absence reinforces the notion that people in small communities have a gradually more analogous lived experience because of limited variety and uniform commercial choices in these locations, exemplifying the idea of placelessness. As Cresswell (2004) explains, consumer culture and mass communication have led to a homogenization of the world. Our lives increasingly:

> take place in spaces that could be anywhere—that look, feel, sound, and smell the same wherever in the globe we may be. Fast food outlets, shopping malls, airports, high street shops, and hotels are more or less the same wherever we go. These are spaces that seem detached from the local environment and tell us nothing about the particular locality in which they are located.
>
> *(Cresswell 2004, 43)*

Cities at the bottom of the hierarchy not only lack access to the global economy and the wide range of retail options it offers, but also share the same few available options (Fekete 2014).

While the urban hierarchy still holds up as an effective model for understanding the interactions and competition between cities, the theory is not without critique. Many geographers have criticized early studies of retail and economic geographies for ignoring individual culture, dynamics of labor markets, and other socioeconomic processes. Recently, economic geography has started to reinvent itself amid ongoing research on culture, inequality, and globalization (Barnes 2001).

Retail as an Expression of Culture and Identity

Today, geographers who study landscapes of retail recognize that there are a variety of cultural factors that affect economic decisions in addition to geographic location and the urban hierarchy. New research areas geographers began to study in the 1990s include marketing (Hartwick 1998), spaces of retail consumption (Goss 1999), and the consuming body (Valentine 1999). The types of things that we consume, whether food, clothing, electronics, or other household items, "help us create social identities and relationships with others" (Kneale and Dwyer 2003, 300).

What people consume and where ties into a variety of aspects of individual identity creation such as ethnicity (Williams et al. 2001), age (Thomas 2005; Morrison, Nelson, and Ostry 2011), gender (Leslie 2002; Ren and Kwan 2009; Morrison, Nelson, and Ostry 2011), income, and disability (Shaw 2006). For example, Slocum (2008) identifies the ethnic differences in participating in the Minneapolis Farmers' Market, concluding that the demand for local, alternative, and cleaned (no roots and in packaging) foods is largely a white phenomenon while other foods are purchased by immigrant populations (Slocum 2008). Grocery shopping can also be viewed as an ethnic and cultural practice. A study by Wang and Lo (2007) found that among Chinese immigrants in Toronto, many were willing to drive longer distances to be able to shop at ethnic grocery stores because of the unique social experience these retail outlets provide. Even the most mundane of consumer activities is deeply rooted in social and cultural experience.

Aside from farmers' markets and grocery stores, food is one of the more commonly studied topics in geography because eating is a very embodied practice—everyone must eat somewhere—and what people eat is linked to identity. Geographers Bell and Valentine emphasize the links between consumption, identity, and food "in a world in which self-identity and place identity are woven through webs of consumption, what we eat (and where, and why) signals ... who we are" (2013, 3). The foods that people choose to purchase and eat are often proclamations of cultural identity, metaphors for other aspects of culture or society (Shortridge and Shortridge 1998). As Bell and Valentine (2013) further remark, food consumption also marks different passages of time (through seasonal offerings), distinct moments in a person's life, and specific traditions with family and friends. Consuming food is not simply about the physical act of eating, but rather about performing identity, culture, and social relations.

New Spaces of Retail—Growth of the Internet

The landscape of retail and consumption has further changed throughout the past several decades because of the development of the internet. Online retailing, also sometimes referred to as e-tailing or ecommerce, has altered the spaces in which retail activity takes place, bringing shopping to the internet and bypassing traditional offline storefronts. Even though customers can purchase goods online, most items bought online must be consumed offline. Therefore, online retail is not geographically even. Studying the online sites of retail requires knowledge of internet access and education (often associated with class, age, ethnicity, and other aspect of identity), as well as the offline geographies of the physical world to understand how goods and services are shipped to those who purchase them.

Development of Online Retail

Throughout the nineteenth and twentieth centuries there are many examples where retail took place outside of physical storefronts such as catalogues and television shopping channels (Williams 2009). The rapid growth of the internet at the start of the twenty-first century represents another change in how people purchase and consume goods and services based on new technology. Though there are some retail corporations existing completely online (known as a pure-play establishment), most retail corporations follow a click-and-mortar, otherwise known as bricks-and-clicks, model, meaning that they are hybrids of traditional retailing in physical storefronts (brick-and-mortar stores) and online institutions.

The development of online retail has taken longer to catch on than initially predicted, partially due to the burst of the dot com bubble in the late 1990s. Seattle-based Amazon.com is currently the largest online retailer in the US. Launched in 1995, Amazon has grown into a company that reported annual revenue of $280.52 billion in 2019 (Statistica 2016). Though Amazon became the fifth company in the world to be valued at over $1 trillion it had a long journey to profitability, not turning a profit for the first five years of its existence and reporting a negative net income of $241 million as recently as 2014. Initial challenges were largely because of the desire to decrease shipping and fulfillment costs and time (Mangalindan 2015). Amazon's innovative Amazon Prime membership, a subscription service where users pay an annual fee to receive free two-day shipping on products as well as access to free streaming movies and television shows, loses Amazon money but the company maintains it to increase brand loyalty among their user base (Mangalindan 2015).

As demonstrated by Amazon, online retail is directly bound to the geographies of distribution, shipping, and transit as physical goods must be delivered to the customer. Many early online retail enterprises failed or were slow to turn a profit because of the lack of infrastructure required to be able to ship items directly to consumers at a relatively quick pace (Wrigley et al. 2002). Brick-and-mortar establishments already have supply chains in place, whereas online retailers need to ship directly to consumers and must rely on established shipping companies like United Parcel Service (UPS) or the US Postal Service (USPS). Success in the world of online retail is driven by the ability to develop fulfillment centers and distribution networks.

Looking again to Amazon as an example, the company has worked to develop additional distribution methods aside from the traditional business-to-customer's home model to increase the speed and efficiency of moving products over distance. Innovations such as Amazon lockers where customers go to a central location (for example, a local convenience store or college campus) to pick up and drop off packages is one attempt to consolidate shipping and distribution. Recently, due to rising shipping costs incurred by Amazon as it continues to offer free or discounted shipping to its members, the company developed their own shipping service in an attempt to undercut traditional shipping services like FedEx, UPS, and the US Postal Service (Figures 26.1a and 26.1b). Amazon has even been investing in the possible use of drones to deliver packages as a quicker and more efficient way to reach customers' homes. The future of online retail is largely tied up in the offline geographic networks of shipping and distribution where successful companies will be those able to best move products across space.

In order to compete with growing online retailers like Amazon, brick-and-mortar establishments now offer services such as in store pick up, where a customer can purchase an item online and pick it up directly from a store instead of having it shipped to their home. This model allows businesses to take advantage of both online sales and consolidated, pre-established shipping networks. Big box store giant Walmart has been increasing its efforts to directly compete with online retailers such as Amazon by investing in the development of their online and mobile platforms. Posting a 40 percent growth in online sales during the second quarter of 2018, Walmart now holds second place for most online sales in the US. In order to compete with Amazon, Walmart has offered free two-day shipping without having to purchase a membership, free in store pick up, and has been testing the use of driverless cars for delivery services. Walmart's online sales still lag far behind, however, making up only 5.3% of all online retail while Amazon makes up 38.7% of all online retail (emarketer.com).

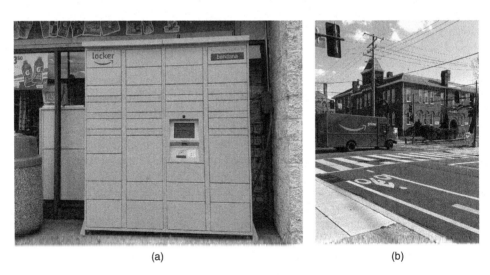

(a)   (b)

Figures 26.1a and b Innovations from Amazon have been introduced to increase the efficiency of moving products from warehouses to customers. Amazon lockers (Figure 26.1a) are often located in public places like in front of convenience stores as shown here or on college campuses. Amazon has also introduced its own shipping services (Figure 26.1b) to move goods faster to customers. Photos taken by the author in Baltimore, MD (Figure 26.1a) and Washington, DC (Figure 26.1b).

The food and grocery business has been one area of retail that online venues have found difficult to crack. Olive oil appears to be one of the few foodstuffs easily sold online to date. Due to the nature of fulfillment and shipping, it is difficult as of now to establish completely online grocery businesses. Complications with food spoilage and unknown delivery times make delivering groceries problematic outside of urban areas. Distribution networks and the need for extensive warehouses also have prevented many pure-play grocery establishments from succeeding thus far, whereas several grocery stores have successfully adopted a bricks-and-clicks model available in large cities (Murphy 2007). Groceries is one area where Amazon competitor Walmart has dominated, ousting Amazon as the top seller of groceries over the internet. To maintain its position, Walmart has introduced express delivery which guarantees two-hour delivery service to residential areas in almost 70% of the US (Ward 2021). The ubiquitous presence of Walmart in many cities lower on the urban hierarchy makes it easier to reach a large proportion of the US population. Amazon's recent purchase of Whole Foods Market is an attempt for the company to develop a successful grocery component, though with a less wide-spread network of stores, Whole Foods may not be able to reach as many customers.

Despite the difficulties of selling groceries over the internet, food retail in general has been altered by the online presence of stores as well as online restaurant delivery services. People are increasingly looking to services such as Yelp and Google for restaurant reviews before they venture out to the restaurant, making it more pertinent than ever for restauranteurs to maintain an online presence (Fekete 2017). Some online retailers allow customers to skip the grocery store all together, providing the option to buy pre-packaged and pre-portioned meal prep kits that are delivered to a customer's home and cooked and prepared on the customer's own time. While Walmart has started to break into the grocery delivery market, middleman companies such as Instacart allow for users to create and purchase a grocery list of items from their local supermarket which are then shopped for and delivered by the employees of the shopping service, not of the grocery store.

Similarly, restaurant food delivery services have grown. Services like UberEats and Grubhub hire contract workers who pick up food from a restaurant and deliver it to customers who have purchased the food online through their mobile phone. Food delivery services have affected the offline restaurant market through analysis of search history data. In one example, UberEats noticed that their users were looking for hamburgers in a location where there were no restaurants selling hamburgers. The company approached some local establishments, encouraging them to start offering hamburgers online only, to which the restaurants complied. The only way to order a hamburger from these physical restaurants is by using an online sale and delivery service (Garsd 2018). While the internet may open up new retail spaces, it also serves to buttress consumption patterns already in place.

Though online retail has been growing at higher rates than offline retail over the last two decades, most consumers still purchase goods and services offline (Figure 26.2). In 2018, the most recent data available at the time of writing, estimated online retail sales (e-commerce) was $5.19 billion USD. However, total retail sales in the US accounted for $5,269.47 billion USD. Overall, these numbers represent the fact that online retail represents only a fraction of total sales in the US despite its sustained growth.

Certain goods and services are more likely to be purchased online over others. Statistics from the US Census Bureau show that clothing and clothing accessories have the highest percentage of online sales as compared to total retail sales. This category is followed by motor vehicles and parts dealers, and books, music, and sporting goods. Categories of goods with the lowest percentage of online purchases compared to total retail purchases include health and personal care stores as well as food and beverage; building materials and gardening equipment and supplies dealers; and furniture and home furnishings.

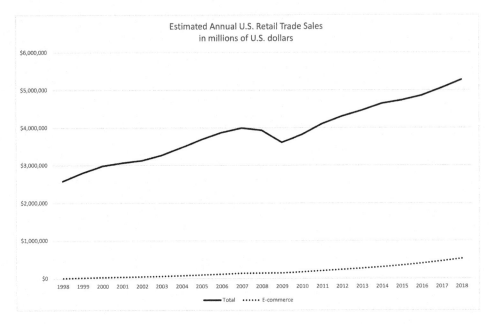

Figure 26.2 Estimated Annual US Retail Trade Sales in millions of US Dollars 1998–2018. Source: US Census Bureau. Graph by the author.

Identity and Online Retail

Despite a lower use of internet shopping in the mid-to-late 2000s, the impact online shopping has had on society and geography has been comprehensively explored (Hjorthol 2009; Ren and Kwan 2009). Online shopping has largely been an urban phenomenon, reflecting internet use patterns and availability of distribution networks (Farag et al. 2006). Internet connectivity is higher in urban areas, making it easier for people living in cities to purchase items online. Aspects of identity like gender and social class also play a role in online shopping with more women than men likely to adopt e-shopping (Ren and Kwan 2009). E-shopping is often most utilized to save time, money, and travel and is also largely driven by familiarity with the internet as well as access to cars and proximity to physical retail outlets (Ren and Kwan 2009). According to Ren and Kwan (2009), the internet does not alter consumption patterns significantly in that people are not *substituting* their shopping behavior with e-commerce, but rather *supplementing* it.

The internet also influences where people consume goods and services because of who developed internet applications or the way that internet programs were coded. For example, a search on Google Maps for services is dependent on the keywords used in the search as well as on the overall web presence of a retail establishment. Those establishments with less of an internet presence will show up less frequently or towards the bottom of a search on Google Maps regardless of whether the venue is physically located nearer to the person conducting the search than other retail outlets. Similarly, national chain businesses are more likely to appear near the top of search results as opposed to local retailers because of the ability of these large corporations to distort their online presence through marketing techniques and monetary means (Zook and Graham 2007). While Google Maps and Open Street Map are not sites of online retail activity per se, they do drive people to specific brick-and-mortar retail establishments and affect where people go to shop.

Gender bias also exists in the creation of online maps of and navigation to retail establishments. In Open Street Map, for instance, there are several categories for places that commodify women (e.g., strip clubs, escort services), but suggested categories for traditionally feminized spaces of care (e.g., childcare, hospices) have been routinely rejected as necessary additions to the mapping platform by other users, creating a gender-biased space (Stephens 2013). The omission of these types of businesses poses a concern as many applications including Apple products, are abandoning paying for mapping services in favor of the free Open Street Map application (Stephens 2013).

Though online shopping has not come close to surpassing traditional consumption patterns, the internet has an ever-increasing effect on the types of businesses people frequent and where these establishments are located. There are a multitude of retail services that require an offline space because they deal in provisions that are embodied and necessitate a physical presence— restaurants or hair salons, for example. Certain types of goods and services cannot be wholly consumed online, but still need to maintain an online presence in order to compete with other retailers.

Alternative Online Spaces of Retail Activity

New spaces of sale or the exchange of goods and services associated with individuals rather than major corporations have also been made possible with the internet. Websites such as Craigslist, Facebook, and Etsy provide online marketplaces for individuals to sell products and goods while third party platforms like Airbnb provide a space for online point of sale exchanges. Craigslist, for example, is a mostly free website, established in 1995, that allows locals to list an item for sale or trade as well as property rentals or jobs (craigslist.org). Though Craigslist is an online site, it relies heavily on physical geography because its listings are location dependent (Zook et al. 2004). Over 50 million pages are viewed each month in the approximately 700 local sites for cities, towns, or counties found in over 70 countries worldwide (craigslist.org). Craigslist online activity also has offline consequences: Craigslist acts as a facilitator of trade between two unknown local parties, coordinating offline economic activity in an online space, and taking over traditional forms of classified advertising such as newspapers (Seamans and Zhu 2013). One study found that the effect of a new Craigslist local site on local newspaper classified advertisements led to newspapers dropping the cost of their advertisement space by 20.7% while increasing their subscription prices by 3.3% (Seamans and Zhu 2013). The presence of Craigslist has also been shown to lead to a reduction in real estate vacancy rates (Kroft and Pope 2014). Like Craigslist, Facebook market-place offers ways to help individuals buy, sell, trade, or rent their goods and services. Etsy, an online site that started in 2005 for "crafters, artists and makers" to "sell their handmade and vin-tage goods and craft supplies" (etsy.com), is largely dominated by female crafters who are looking to sell their homemade items as a supplement to their incomes as students or stay at home mothers (Luckman 2013). Similarly, commission-based services like Airbnb offer a new way for property owners to list their short-term rentals. While transactions from individuals not associated with corporations existed before the internet, primarily through newspaper classifieds or local craft fairs, the Internet has further popularized these alternative spaces of economic activities in ways that may not have been possible prior to the growth of online sales.

Unknown Impacts of COVID-19 on the Retail Landscape

The COVID-19 pandemic radically uprooted lives throughout the US and the world including the spaces where people purchased goods and services. As many people were pushed to spend more time at home or limit their travel to reduce the spread of the novel coronavirus, retail outlets adapted to meet this new socially distant demand. Though exact numbers are not yet known, experts estimate that online retail saw several years of growth in a six-month window. During the second quarter of 2020, when most pandemic-driven lockdowns occurred in the US, online retail sales were 44.5% higher than during the second quarter of 2019 while overall retail in the US dropped by 3.6% (US Census 2020). Online grocery delivery services, traditionally an area with low sales numbers, saw massive growth throughout 2020. Instacart even reported that customer demand for their middleman service was up 500% year over year (Petrova 2020). Restaurant businesses, which are physically rooted in place, pivoted to increase their take-out offerings and bring the restaurant experience into diners' homes by offering prix fixe meals, links to curated music playlists, and at-home cocktail making classes. As local delivery increased to meet the social distancing demand, inequalities have also been exposed. For many who work in grocery or delivery services, the lockdown was not a prominent feature of their 2020 experience. Instead, postal workers, grocery store employees, and food delivery service providers became front-line workers as they navigated a virus-infected environment, altering the standard way many in the US were used to obtaining goods and services. Despite the reported rapid growth of online spaces of sale, the long-term effects of the COVID-19 pandemic on the retail landscape of the US remain to be seen.

Retail Geographies—Rooted in Place

Geographers have studied the landscapes of retail for over 100 years, developing theories and models as to why certain goods and services are available in particular locations. While early research on retail landscapes helped geographers understand that goods and service availability is largely a result of population size and urban areas, today geographers also know that factors of individual identity play a role in where people go to purchase and consume products. The internet has started to shift the location of the point of sale—online through a computer or mobile phone. However, these online sales are still rooted in space as goods and services must move from the manufacturer or warehouse to the consumer. Shipping and delivery services are becoming a battleground area where corporations are striving to be the leader in time and efficiency. During 2020, the COVID-19 pandemic had an immediate impact on online purchasing and mobility of goods through home delivery. Though the effects of the pandemic may be short-lived, landscapes of retail will continue to change.

References

Alderson, A. S., J. Beckfield, and J. Sprague-Jones. 2010. "Intercity Relations and Globalisation: the Evolution of the Global Urban Hierarchy, 1981–2007. *Urban Studies* 47 (9): 1899–1923.

Barnes, T. 2001. "Retheorizing Economic Geography: From the Quantitative Revolution to the 'Cultural Turn." *Annals of the Association of American Geographers* 91 (3): 546–565.

Bell, D., and G. Valentine. 2013. *Consuming Geographies: We Are Where We Eat.* New York: Routledge.

Berry, B. and F. Horton. 1970. *Geographic Perspectives on Urban Systems.* Eaglewood, NJ: Prentice-Hall Press.

Brunn, S. 2006. *Wal-Mart World: The World's Biggest Corporation in the Global Economy.* New York, NY: Routledge.

Christaller, W. 1966. *Central Places in Southern Germany.* Translated by C. Baskin. Eaglewood, NJ: Prentice-Hall Press.

Cresswell, T. 2004. *Place: An Introduction.* West Sussex: John Wiley.

Farag, S., J. Weltevreden, T. van Rietbergen, and M. Dijst. 2006. "E-shopping in the Netherlands: Does Geography Matter?" *Environment and Planning B: Planning and Design* 33: 59–74.

Fekete, E. 2014. "Consumption and the Urban Hierarchy in the Southeastern United States." *Southeastern Geographer* 54 (3): 249–269.

———— 2017. "Foursquare in the City of Fountains: Using Kansas City as a Case Study for Combining Demographic and Social Media Data." In J. Thatcher, J. Eckert, and A. Shears, eds. *Thinking Big Data in Geography: New Regimes, New Research*, Lincoln, NE: University of Nebraska Press: 165–188.

Garsd, J. 2018. Uber's Online-Only Restaurants: the Future, or the End of Dining Out? *NPR*. www.npr. org/sections/thesalt/2018/10/23/658436657/ubers-online-only-restaurants-the-future-or-the-end-of-dining-out Accessed November 24, 2018.

Goss, J. 1999. Once-upon-a-time in the Commodity World: an Unofficial Guide to the Mall of America. *Annals of the Association of American Geographers* 89 (1): 45–75.

Hall, P. 1998. *Cities in Civilization: Culture, Technology, and Urban Order*. London: Weidenfeld and Nicolson.

Hartwick, E. 1998. Geographies of Consumption: a Commodity-chain Approach. *Environment and Planning D: Society and Space* 16: 423–437.

Hjorthol, R. 2009. Information Searching and Buying on the Internet: Travel-related Activities? *Environment and Planning B: Planning and Design* 36 (2): 229–244.

Kneale, J. and C. Dwyer. 2003. Consumption. In *A Companion to Cultural Geography*, edited by J.S. Duncan, N.C. Johnson, and R.H. Schein, 298–315. Oxford: Blackwell.

Kroft, K. and Pope, D. G. 2014. Does Online Search Crowd Out Traditional Search and Improve Matching Efficiency? Evidence from Craigslist. *Journal of Labor Economics* 32 (2): 259–303.

Lake, R. 2009. Berry, B. In R. Kitchin and N. Thrift eds. *International Encyclopedia of Human Geography*. Oxford: Elsevier: 305–307.

Leslie, D. 2002. Gender, Retail Employment and the Clothing Commodity Chain. *Gender, Place and Culture: A Journal of Feminist Geography* 9 (1): 61–76.

Luckman, Susan. 2013. The Aura of the Analogue in a Digital Age: Women's Crafts, Creative Markets and Home-based Labour After Etsy. *Cultural Studies Review* 19 (1): 249–270.

Mangalindan, J.P. 2015. Inside Amazon Prime. *Fortune*. fortune.com/2015/02/03/inside-amazon-prime/

Morrison, K. T., T. A. Nelson, and A. S. Ostry. 2011. Mapping Spatial Variation in Food Consumption. *Applied Geography* 31 (4): 1262–1267.

Murphy, A. 2007. Grounding the Virtual: The Material Effects of Electronic Grocery Shopping. *Geoforum* 38: 941–953.

Petrova, M. 2020. Coronavirus is Making Grocery Delivery Services Like Instacart Really Popular and They Might Be Here to Stay. CNBC News Online. www.cnbc.com/2020/05/13/coronavirus-making-grocery-delivery-services-like-instacart-popular.html. Accessed February 14, 2021.

Pred, A. 1977. *City Systems in Advanced Economies*. London: Hutchinson.

Ren, F. and M-P Kwan. 2009. The Impact of Geographic Context on e-Shopping Behavior. *Environment and Planning B: Planning and Design* 36 (2): 262–278.

Seamans, R. and F. Zhu. 2013. Responses to Entry in Multi-Sided Markets: The Impact of Craigslist on Local Newspapers. *Management Science* 60 (2): 476–493.

Shaw, H. J. 2006. Food Deserts: Towards the Development of a Classification. *Geografiska Annaler: Series B, Human Geography* 88 (2): 231–247.

Shortridge, B.G. and J. R. Shortridge 1998. *Taste of American Place*. Rowman & Littlefield.

Slocum, R. 2008. Thinking Race Through Corporeal Feminist Theory: Divisions and Intimacies at the Minneapolis Farmers' Market. *Social and Cultural Geography* 9: 849–869.

Statistica A. 2016. Statistics and Facts about Amazon. www.statista.com/topics/846/amazon/

Stephens, M. 2013. Gender and the Geoweb: Divisions in the Production of User-generated Cartographic Information. *GeoJournal* 78 (6): 981–996.

Thomas, M. E. 2005. Girls, Consumption Space and the Contradictions of Hanging Out in the City. *Social & Cultural Geography* 6 (4): 587–605.

U.S. Census. 2020. Quarterly Retail E-commerce Sales 3rd Quarter 2020. www.census.gov/retail/mrts/www/data/pdf/ec_current.pdf.

Valentine, G. 1999. A Corporeal Geography of Consumption. *Environment and Planning D: Society and Space* 17: 329–351.

Wang, L. and L. Lo. 2007. Immigrant Grocery-Shopping Behavior: Ethnic Identity Versus Accessibility. *Environment and Planning A* 39 (3): 684–699.

Ward, T. 2021. Walmart Drops $35 Minimum for Express Delivery. Walmart, Inc. online press release. corporate.walmart.com/newsroom/2021/03/01/walmart-drops-35-minimum-for-express-delivery. Accessed May 1, 2021.

Williams, D. E. 2009. The Evolution of etailing. *The InternationalReview of Retail, Distribution and Consumer Research* 19 (3): 219–249`.

Williams et al. 2001. Consumption, Exclusion, and Emotion: The Social Geographies of Shopping. *Social & Cultural Geography* 2 (2): 203–220.

Wrigley, N., M. Lowe, and A. Currah. 2002. Progress Report: Retailing and e-Tailing. *Urban Geography* 23 (2): 180–197.

Zook, M., M. Dodge, Y. Aoyama, and A. Townsend. 2004. "New Digital Geographies: Information, Communication, and Place." In *Geography and Technology*, edited by S. Brunn, S. Cutter, and J. Harrington, 155–176. Springer: Netherlands.

Zook, M. and M. Graham. 2007. Mapping DigiPlace: Geocoded Internet Data and the Representation of Place. *Environment and Planning B: Planning and Design* 34: 466–482.

27

POST-INDUSTRIAL LANDSCAPES

Mark Alan Rhodes II and Sarah Fayen Scarlett

The term "post-industrial" appeared as a fearful spectre shortly after the 1973 energy crisis and has largely come into acceptance for describing the shift in the US and elsewhere from manufacturing to service economies (High, MacKinnon, and Perchard 2017; Storm 2016, 8–10). In its most obvious sense a post-industrial landscape is a location that once supported but no longer supports industrial activity. But the term forces emphasis on the past—on the activity of which we are now "post." It labels today based on yesterday and leaves little room for the future. It emphasizes what's missing. But just as any "post-" demarcated landscape (post-colonial, poststructural, etc.) a post-industrial landscape is not untouched or even stricken of its industrial influences. As Jo Sharp (2009, 4–5) explains of postcolonialism, the very use of the hyphen itself holds significance. Not only does the hyphen represent a clear break within a linear temporality but it also grants further power to the colonial, or in this case, the industrial. Fong (2019, 5) even accentuates this difference through the deliberate use of the hyphenated "post-industrial" to indicate a clear temporal break from the industrial allowing both "post" and "industrial" to "stand on their own."

In this chapter we expand upon "post-" industrial landscape engagement in order to best identify a broader status of post-industrial landscapes rooted in an understanding of landscape as both a way of seeing and the accretion of tangible and intangible layers of cultural meaning. Collective memory and industrial heritage, in particular, become embedded into both the communities who inhabit these (post-) industrial landscapes as well as the tourists and residents whose interpretation of these spaces and places lend power to processes of industrial heritage—both in place and in the academy from a primarily US perspective. This brief review reveals, within practice, scholarship, *and* the landscapes themselves, significant themes of absent presence—"the physical, emotional, or spiritual presence or representation of absence" (Rhodes 2020, 49). Processes of development, waste, and ruination all actively remove, forget, obfuscate, and even violently displace communities and memories within the post-industrial landscape, yet reinvention, reanimation, and resistance linger on and even draw power at times from these structural forces. We draw upon these themes for the remainder of the chapter, first through a discussion of absent presence and racialized othering and then turning to the dialectical relationship of built absences and reanimated presence to offer what we hope is a nuanced and comprehensive, albeit brief, reflection upon the status of post-industrial landscape in an American context.

DOI: 10.4324/9781003121800-33

The potential for a landscape approach to post-industrial memory emerges from its adaptability and breadth. Rather than focus only on the temporal, sociological, or spatial elements, we treat post-industrial landscapes as the embedded layers of memory and cultural meaning that they are, allowing us to step back and ask necessary questions of both space and place. The places of community, identity, heritage, and emotion which develop alongside the processes of industrialization and deindustrialization are simultaneously unique and connected to the spatial networks of multiscalar, economic, cultural, and political influences.

A Landscape Approach to the Post-Industrial

Del Pozo and González (2012, 447) identify an absence of both geography and landscape approaches within industrial heritage scholarship and the prevalence of a "naive and object-oriented approach to industrial sites" which disregards broader spatial narratives and complex cultural landscapes. Of course, there are clear exceptions. Caignet (2019) and Gobin (2011) both draw upon the visual arts, whereby the land shapes art and artist, but simultaneously, post-industrial aesthetics shape multiple layers of deeply buried meaning. Others, such as Mitchell (2013) or Proulx and Crane (2020) draw from Schein's (1997) conceptualization of "landscape as discourse" to advance theoretical frames. Further, we identify an additional pattern of methodological framing rooted within a landscape epistemology. While diverse in the specific employment of methods, coming from geography (DeLyser 1999; Rhodes 2021), archaeology (McAtackney and Ryzewski 2017; Sweitz 2010), geography *and* archaeology (Del Pozo and González 2012), and architecture (Scarlett 2021, 2013; Heatherington, Jorgensen, and Walker 2019), each case mobilizes a mixed methods approach to offer a unique perspective upon reading post-industrial landscapes.

Post-industrial landscapes are clearly—from the very inclusion of "post-"—memorial devices. As others note, despite the historical focus on their material presence, recent scholarship acknowledges the power of the spectral and absent presence of the past (Rhodes 2020). Every post-industrial landscape, and memorial landscape more broadly, presents a dialectic relationship of remembering and forgetting through cultural, economic, and political decision-making. In contrast to the narratives present in landscape, those which have been erased often never-the-less emerge in the form of a noticeably present absence creating haunted or obfuscated landscapes (Coddington 2011; DeSilvey 2017; Hill 2013). More broadly, this absence often follows in the wake of colonialism and its resistance. Spectral presence of such radical memory permeates the post-industrial landscape.

The many racialized landscapes of industrialized, deindustrializing, and post-industrial communities must be centered (Figure 27.1). We have found, however, not only an absence of Black geographies and decolonial discourses in the landscape itself, but also in the very scholarship meant to establish and designate these post-industrial spaces and places. Attention towards the racialized absence and presences of post-industrial landscapes contextualizes the ongoing and historical legacies of exclusion and inclusion, racism and power built into both urban and rural landscapes, and the continued embeddedness of such systemic and structural violence today.

The racialization of these spaces and places in the American context is inescapable as we examine the urban spatial structures of othering and dispossession (Ranganathan 2016; Inwood 2018). Likewise, we find clear cooperation between the militarized state and private industry in the genocidal actions taken in Hawaii (Tengan 2008; MacLennan 2014), the Cherokee Nation (Knapp 2018), or on southern plantations (McKittrick 2011; Rapson 2020). The processes of racialized industrialization and deindustrialization manifest in absences within post-industrial heritage, as well (Gobin 2011; Stanton 2007; Hallock 2019). Combined, the post-industrial landscape faces a doubled racialization: first, within the processes of dispossession and

Figure 27.1 The "Living Timeline: Paul Robeson" mural by Cory Stowers and Andrew Katz sponsored by and painted on the wall of the Hung Tao Choy Mei Leadership Institute in the Cardozo neighborhood of Washington, D.C. Photo by Mark Rhodes 2019.

exploitation and second, within the heritage industry. However, this also offers a wide lens from which to view resistance, resilience, and agency. As many will know, social justice has likewise been built into, used, and countered industrial landscapes. In the following sections we will further deconstruct first these acts of dispossession and built violence and absence in the post-industrial landscape before returning to means of resistance and reinvention.

Absent Presence in the Post-industrial Landscape

Development as Othering

Development has often been centralized within post-industrial landscapes: tourism development, redevelopment, heritage development, urban development, and more recently sustainable and community development. Within these contexts, development itself maintains its own complicated history in the context of Christianity, colonialism, capitalism, and imperialism whereby the morality of the development masks the underlying narratives which paint the giver as politically and culturally superior. In this way, development, even with the best of intentions, "serves to differentiate and map social difference" or in other words, to Other (Nally 2014, 41). In the 1970s, American cities responded to their vacated manufacturing districts by engaging planning professionals in "urban rebranding" campaigns, which reinvented post-industrial landscapes into newly trendy loft apartments, arts centers, shopping districts, and public spaces. These approaches throughout North America repeatedly leveraged public–private funding to attract young professionals to urban downtowns. Seen as a way to counteract white flight, this gentrification transformed formerly working-class manufacturing spaces into middle-class service spaces (Neumann 2016). Gentrification manufactures post-industrial landscapes as spaces and places of displacement and disenfranchisement for the working-class and racial minorities from these former industrial neighborhoods (Herstad 2017; High 2003; Mathews and Picton 2014).

Urban scholars regularly draw from Neil Smith's (1996) work which reframes Marx's frontiers of capitalist extraction in the context of gentrifying cities and the "surplus labor" dispossessed in the process. Thomas (2004, 278) further identifies that the histories of these places are of little interest to redevelopment agencies beyond the burden of addressing them, and that this ignorance does not just lead to absences upon the landscape of certain narratives and peoples but "more or less guarantees the reproduction of injustice and the categories—in this case racial—which have been used to justify them." This reproduction of dispossession can be seen through many lenses. Addie and Fraser (2019) point out the neo-settler colonialism occurring in the rapidly gentrifying Over-the-Rhine neighborhood of Cincinnati where German heritage narratives are displacing Black spaces (Figure 27.2). In Ferguson, now infamous for the murder of Michael Brown, Derickson (2017) also identifies similar interconnected economic, political, and community-led structures of racism built into the urban post-industrial landscape.

Gentrification ultimately creates landscapes of access within post-industrial spaces leading to two ultimate questions: who has access and what do they have access to? Gentrified landscapes which follow a neoliberal ethic in redevelopment more often than not lead to exclusionary housing costs, built exclusions upon the landscape, and a general decrease in housing stock and accompanying decrease in affordable food and services. Their benefits—jobs, tourism, repopulation, environmental rehabilitation—rarely provide the existing community the jobs and housing promised and instead tend to price out the few existing businesses or heritage

Figure 27.2   The Cincinnati skyline with the National Underground Railroad Freedom Center in the foreground. Photo by Mark Rhodes 2016.

Figure 27.3 Parking Garage at Heritage Square, Phoenix, Arizona. Photo by Sarah Fayen Scarlett 2016.

spaces extant in the community. Weedon and Jordan (2011, 844) also frame the absences of gentrification through means of resistance as sites of contest for those excluded. Collective memory can find "meaningful and often empowering forms of identity" in the face of lost histories and/or experiences.

Tourism and heritage more broadly both too often fall into similar neoliberal traps that are both exclusionary and dismissive of local, Indigenous, and/or political diversity. In these cases, development equates to state- or sub-state-wide economic prosperity with little regard for fine scale economic inequality, linguistic or cultural context, or political negotiations where local authorities have a substantive voice in heritage and tourism development (Figure 27.3).

Furthermore, when landscapes of industrial heritage and tourism do fall to local authorities, that does not guarantee either representation or sustainability. However, just as urban gentrification can spur a means of resistance, the development of a rural idyll and the picturesque can also be an arena for radical ecological and economic resistance from industrial communities (Scarlett and Rouleau 2022, MacLennan 2014). Just as any community is defined through processes of othering, power dynamics of a community can lead to further absences in the landscape. Ecologically speaking, industrialization has had *the* significant role in the climate crises since the 1700s, however this narrative is also too often omitted, obfuscated, or ignored by heritage and tourism developers and/or visitors and tourists (Price and Ronck 2018; Price and Rhodes 2020).

Landscapes of Waste and Ruin

Capitalistic systems make "Other" that which is not profitable—including human and non-human actors. As a necessity, industrial capitalism generates waste from resource, including labor. When considered from a landscape point of view which points towards a nexus of racialization, exploitation, and representation with layers of past, present, and future processes, we can observe wasteland as a process. Some historians demonstrate that the process often includes conflicting valuations that change over time with implications for historic sites. For example, Fredric L. Quivik argues that tailings from hard rock mining, while known today for their chemical toxicity, carried positive associations about the value of work and labor activism for generations of mining families in Idaho's Coeur D'Alene region (Quivik 2013).

In mining regions where the waste piles are less toxic, such as our own Copper Country in northern Michigan, the National Park Service and its allies struggle to overcome the overwhelmingly negative public opinion about mine tailings in order to retain innocuous waste piles in the landscape for interpreting the technological, geological, and social processes that drive capitalistic systems (Gohman 2013). While we see commodities break their chains—such as old tires being used to "develop" wildlife sanctuaries in Florida (Andrzejewski 2020)—in other cases we see the same capitalistic system generating waste with no intention of reuse or remediation, instead leaving industrial waste for public or non-profit organizations to address (Thompson 2021). While industrial heritage and redevelopment are both common responses, too often they fail to escape the capitalist confines which drove these landscapes into "waste" in the first place. Dillon (2014), coining the term "waste formation" to describe these socio-ecological structures in San Francisco, argues that by examining the social relations rooted within these processes we might best challenge neoliberal progress-oriented development. Others have focused particularly on the social process of making "human-as-waste" in the context of capitalist, industrial, deindustrial, and post-industrial processes (Cowell and Thomas 2002; Yates 2011; Bauman 2004; Kern 2015). Yates (2011, 1688–1689), in particular, outlines the processes by which humans labor within the industrial capitalist system to generate capital which is reinvested to improve efficiency, which then renders laborers as redundant under the guise of further capital accumulation.

Used in innumerable ways, waste and wastelands have been forgotten, romanticized, commodified, and weaponized (DeSilvey 2017). Millington (2013, 282), in particular, describes the ways in which media representations of Detroit's industrial ruins through a binary understanding of humans and the environment obscure the innovative ecological ways the environment (including humans) responded to industrialization. Abstracting these processes down to a pure material culture—focused only on the environmental or the architectural—ignores the socio-economic structures which impacted and continue to do harm in the city and upon its nature... humans very much included.

Considering post-industrial landscapes in terms of their built presences and absences helps to understand the recent attention, both scholarly and popular, paid to ruins and ruination. The widespread term "ruin porn" captures both the desire and violence in these legacies of American landscape aesthetics. In the long history of envisioning American ruins, the erasure of humans from view has been consistent. Post-industrial ruin imagery today excludes African-American and other minority urban dwellers as well as blue-collar white industrial workers in much the same way as Romantic ruins erased Native Americans, Chinese railroad builders, miners, factory and mill workers (DeSilvey and Edensor 2013). The effect of this imagery runs so deep that in the ongoing debate on DC statehood in June 2020, Senator Tom Cotton stated that,

[y]es, Wyoming is smaller than Washington in population, but it has three times as many workers in mining, logging, and construction, and ten times as many workers in manufacturing. In other words, Wyoming is a well-rounded working class state. A new state of Washington would not be.

Figure 27.4 Cornish Pumping Engine and Mining Museum, Iron Mountain, Michigan. Photo by Sarah Fayen Scarlett 2015.

His bias towards resource extraction and capitalist production erases the predominantly Black and working-class industrial legacy of DC in favor of a white-washed industrial narrative (Hutson 2015).

In addition to the problem of erasure, ruin imagery also ignores non-visual senses and haptic experience, which are vital to understanding post-industrial landscapes (Figure 27.4). Edensor emphasizes the importance of crunching glass underfoot, the icy chill emanating from concrete factory walls, the odors of oil and pigeon droppings, and the echo of one's voice when exploring abandoned industrial complexes (Edensor 2005, 837–838). This experience of discovery conjures up the workers of the past as well as the processes of natural reclaiming, decay, and personal (dis)investment that characterize post-industrial places (Edensor 2005; Millington 2013). Ruins make it too easy to forget that processes of deindustrialization include a general quieting down and cleaning up, in which passing trains, clanging machinery, and noxious smells all become out-of-place and out-of-time (Chatterjee and High 2017; Benjamin 2015). Likewise, heritage sites have to be "de-cluttered" for middle-class audiences (and to satisfy ADA and safety regulations). Removing unwanted bodies (human and more-than-human), however, also removes the haptic qualities that simultaneously evoke layers of overlapping memory that defy chronology and linear narrative (Edensor 2005; DeSilvey and Edensor 2013).

Industrial Heritage

The absent presence of post-industrial landscapes makes it nearly impossible for these places to fit neatly into the legal systems of historic preservation in the United States, a mismatch that perpetuates patterns of erasure and forgetting. Industrial places are always characterized by frequent change. In most cases, a factory, mill, or mine achieved economic success and longevity by updating its technology, revamping its physical plant, and rearranging its built assets to respond to new market or labor conditions. By contrast, the National Historic Preservation Act values unchanging landscapes and "integrity" in the built environment. Some efforts have appeared in the last two decades to acknowledge this problem (Arnold and Lafreniere 2017), but emphasis remains on the ability of a landscape to "convey the collective image of a historically significant industrial operation." ("Cultural Landscapes Inventory Professional Procedures Guide," 2009, 7–3 https://irma.nps.gov/DataStore/DownloadFile/513401).

This institutionalization of the absent presence of industrial heritage in the US runs in sharp contrast to European practice. From numerous UNESCO World Heritage Sites (WHS), such as the Blaenavon Industrial Landscape in Wales or the Cornwall and West Devon Mining Landscape, the UK in particular stands in contrast to the US's absence of any WHS either recognizing industrial heritage or utilizing the concept of landscape. Switzerland, Spain, Portugal, Austria, and many other states regularly frame heritage through the diverse layering of landscape, however the United States has not. The primary US organization advocating for industrial heritage on the international stage is the Society for Industrial Archeology (SIA). However, the SIA and its host, Michigan Technological University (formerly the Michigan School of Mines), have been slow to engage with the racialized and colonialized landscapes they have and continue to shape. Moreover, America is just beginning to embrace a supraregional industrial heritage infrastructure, albeit at a smaller scale than the European Route of Industrial Heritage, for instance. The National Park Service recently formed Blackstone River Valley National Historical Park, which supports existing Rhode Island and Massachusetts industrial sites in a "partnership park" model already in use at the Keweenaw National Historical Park and others. While there are less than 100 industrial sites out of the 2,600 National Historic Landmarks, five of the twenty-one new Landmarks announced in January 2021 were industrial in their focus.

Archaeologist Hilary Orange writes that the best kinds of memory work in post-industrial heritage landscapes can "reanimate" material detritus. Reanimation, in her argument, includes three "interwoven" strands: scholarly study, commemorative acts, and the approach to landscape as a "cultural and material construct" in which transformation happens through daily physical experiences (Orange 2014, 15). To "reanimate" themselves, American post-industrial heritage sites need to leverage what public historian Cathy Stanton (2005) calls the "porosity" between past and present through a much more self-reflexive model of interpretation, such as in the urban park model at Lowell National Historical Park, where "the city is the park and the park is the city" (Marion 2014; Stanton 2005; Figure 27.5). But the performative nature of today's placemaking and place-marketing, which since the 1980s increasingly has relied on bottom-line-driven public–private partnerships rather than governmental funding, can often replicate the very patterns of exclusion and inequality that heritage professionals are trying to expose in the past. They can also disappoint local communities who expect Federal investment but instead are left doing the heavy-lifting themselves (Liesch 2016).

Among America's industrial heritage sites with the most promising possibilities for successful reanimation addressing racial and class othering is Pullman National Monument in Chicago's far South Side. America's most famous company town, planned in the 1880s by railroad magnate George Pullman and called "the most beautiful town in America," may be able to acknowledge the role of its middle-class aesthetic in driving racial and class divisions on the South Side long after the company divested itself of the town in 1907 (Scarlett and Rouleau 2022). With the National Park Service partnering with the National A. Philip Randolph Pullman Porter Museum

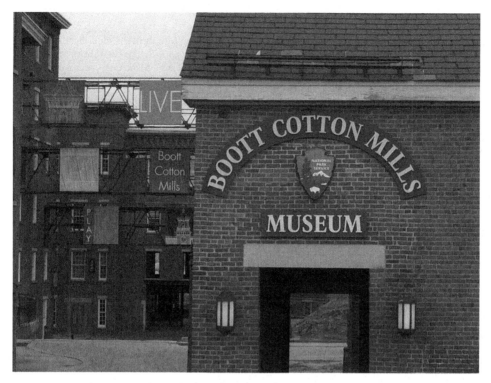

Figure 27.5 Entrance to the Boott Cotton Mills Museum, Lowell National Historical Park (right) alongside the independently owned Apartments at Boott Mills (left). Photo by Sarah Fayen Scarlett 2020.

and other Black-run community organizations, post-industrial heritage programming has the potential to capture the mutually constituted nature of landscape reality and imagination.

In some cases, digital public heritage projects can facilitate Orange's tripartite reanimation by using deep mapping and augmented reality. Researchers involved in creating the Keweenaw Time Traveler at Michigan Technological University argue that in post-industrial landscapes, deep maps—using historical GIS methods to facilitate the free exploration of data through time and space—help foster community–expert partnerships that simultaneously promote place-based heritage and economic regeneration (Trepal, Scarlett, and Lafreniere 2019). While digital reanimation runs the risk of disconnecting the audience from the materiality of place, it can often be worth the risk for post-industrial landscapes. Contemporary archaeologist Krysta Ryzewski argues that digital storytelling can be an important option in post-industrial landscapes where failing infrastructure and blight threatens physical loss and the continuity of memory (Ryzewski 2019).

Our aim has been to highlight the necessity of diversifying and de-romanticizing the post-industrial landscape through a generalized assessment of existing post-industrial landscape work. Many landscapes deemed *post*-industrial in name nonetheless remain dependent on industrial economies and the industrial communities which linger on and aim to thrive. At the same time, we have offered the presence of absence within these post-industrial spaces… of ruin, waste, and redevelopment juxtaposed with heritage, reinvention, and reanimation as a means of better understanding their form and function.

References

Addie, J.-P. D. and J. C. Fraser. 2019. "After Gentrification: Social Mix, Settler Colonialism, and Cruel Optimism in the Transformation of Neighbourhood Space." *Antipode* 51 (5): 1369–1394.

Andrzejewski, A. V. 2020. ""Selling Sunshine": The Mackle Company's Marketing Campaign to Build Retirement and Vacation Communities in South Florida, 1945–1975." *Buildings & Landscapes: Journal of the Vernacular Architecture Forum* 27 (2). University of Minnesota Press.

Arnold, J. D. M. and D. Lafreniere. 2017. "The Persistence of Time: Vernacular Preservation of the Postindustrial Landscape." *Change over Time* 7 (1): 114–33.

Bauman, Z. 2004. *Wasted Lives: Modernity and its Outcomes.* Cambridge: Polity.

Benjamin, J. 2015. "Listening to Industrial Silence: Sound as Artefact." In *Reanimating Industrial Spaces: Conducting Memory Work in Post-Industrial Societies*, edited by H. Orange, 108–124. Walnut Creek, CA: Left Coast Press.

Caignet, A. 2019. "Representing and Recording the Transformation of the Industrial Landscape in the North of England: A Reappraisal of Identity." *Polysèmes. Revue d'études intertextuelles et intermédiales* 22.

Chatterjee, P. and S. High. 2017. "The Deindustrialisation of Our Senses: Residual and Dominant Soundscapes in Montreal's Point Saint-Charles District." In *Telling Environmental Histories: Intersections of Memory, Narrative and Environment*, edited by K. Holmes and H. Goodall, 179–210. Palgrave Studies in World Environmental History. Cham: Springer. doi:10.1007/978-3-319-63772-3_8.

Coddington, K. 2011. "Spectral Geographies: Haunting and Everyday State Practices in Colonial and Present-Day Alaska." *Social & Cultural Geography* 12: 743–756.

Cowell, R. and H. Thomas. 2002. "Managing Nature and Narratives of Dispossession: Reclaiming Territory in Cardiff Bay." *Urban Studies* 39 (7): 1241–1260. doi-org.services.lib.mtu.edu/10.1080/00420980220135581.

Del Pozo, P. B. and P. A. Gonzalez. 2012. "Industrial Heritage and Place Identity in Spain: From Monuments to Landscapes." *Geographical Review* 102 (4): 446–464.

DeLyser, D. 1999. "Authenticity on the Ground: Engaging the Past in a California Ghost Town." *Annals of the Association of American Geographers* 89 (4): 602–632.

Derickson, K. D. 2017. "Urban Geography II: Urban Geography in the Age of Ferguson." *Progress in Human Geography* 41 (2): 230–244.

DeSilvey, C. 2017. *Curated Decay: Heritage Beyond Saving.* University of Minnesota Press.

DeSilvey, C. and T. Edensor. 2013. "Reckoning with Ruins." *Progress in Human Geography* 37 (4): 465–485. doi:10.1177/0309132512462271.

Dillon, L. 2014. "Race, Waste, and Space: Brownfield Redevelopment and Environmental Justice at the Hunters Point Shipyard." *Antipode* 46 (5): 1205–1221.

Edensor, T. 2005. "The Ghosts of Industrial Ruins: Ordering and Disordering Memory in Excessive Space." *Environment and Planning D: Society and Space* 23 (6): 829–849. doi:10.1068/d58j.

———. 2007. "Sensing the Ruin." *The Senses and Society* 2 (2): 217–232.

Fong, C. 2019. "Bergbau, Tagebau, Umbau: The Post-Industrial Landscape Aesthetics of Repurposed Coal Mines in Germany." University of Michigan. PhD diss.

Gobin, A. 2011. "Constructing a Picturesque Landscape: Picturing Sugar Plantations in the Eighteenth-Century British West Indies." *Hemisphere: Visual Cultures of the Americas* 4 (1): 42–66.

Gohman, S. M. 2013. "It's Not Time to Be Wasted: Identifying, Evaluating, and Appreciating Mine Wastes in Michigan's Copper Country." *The Journal of the Society for Industrial Archaeology* 39 (1–2): 5–22.

Hallock, G. 2019. "Mulberry Row: Telling the Story of Slavery at Monticello." *SiteLINES: A Journal of Place* 14 (2): 3–8.

Heatherington, C., A. Jorgensen, and S. Walker. 2019. "Understanding Landscape Change in a Former Brownfield Site." *Landscape Research* 44 (1): 19–34.

Herstad, K. 2017. "'Reclaiming' Detroit: Demolition and Deconstruction in the Motor City." *The Public Historian* 39 (4): 85–113. doi:10.1525/tph.2017.39.4.85.

High, S., L. MacKinnon, and A. Perchard. 2017. *The Deindustrialized World: Confronting Ruination in Postindustrial Places.* UBC Press.

High, S. 2003. *Industrial Sunset: The Making of North America's Rust Belt, 1969–1984.* University of Toronto Press.

Hill, L. 2013. "Archaeologies and Geographies of the Post-Industrial Past: Landscape, Memory and the Spectral." *cultural geographies* 20: 379–396.

Hutson, M. A. 2015. *The Urban Struggle for Economic, Environmental and Social Justice: Deepening Their Roots.* London and New York: Routledge.

Inwood, J. 2018. "'It is the Innocence which Constitutes the Crime': Political Geographies of White Supremacy, the Construction of White Innocence, and the Flint Water Crisis." *Geography Compass* 12 (3). doi:10.1111/gec3.12361.

Kern, L. 2015. "From Toxic Wreck to Crunchy Chic: Environmental Gentrification Through the Body." *Environment and Planning D: Society and Space* 33 (1): 67–83.

Knapp, C. E. 2018. *Constructing the Dynamo of Dixie: Race, Urban Planning, and Cosmopolitanism in Chattanooga, Tennessee.* UNC Press Books.

Liesch, M. 2016. "Creating Keweenaw: Parkmaking as Response to Post-Mining Economic Decline." *The Extractive Industries and Society* 3 (2): 527–538. doi:10.1016/j.exis.2015.12.009.

MacLennan, C. A. 2014. *Sovereign Sugar: Industry and Environment in Hawai'i.* University of Hawaii Press.

Marion, P. 2014. *Mill Power the Origin and Impact of Lowell National Historical Park.* Lanham, Md. Rowman & Littlefield.

Mathews, V. and R. M. Picton. 2014. "Intoxifying Gentrification: Brew Pubs and the Geography of Post-Industrial Heritage." *Urban Geography* 35 (3): 337–356.

McAtackney, L. and K. Ryzewski. 2017. *Contemporary Archaeology and the City: Creativity, Ruination, and Political Action.* Oxford University Press.

McKittrick, K. 2011. "On Plantations, Prisons, and a Black Sense of Place." *Social & Cultural Geography* 12 (8): 947–963.

Millington, N. 2013. "Post-Industrial Imaginaries: Nature, Representation and Ruin in Detroit, Michigan." *International Journal of Urban and Regional Research* 37 (1): 279–296.

Mitchell, D. 2013. "Labour's Geography and Geography's Labour: California as an (anti-) Revolutionary Landscape." *Geografiska Annaler: Series B, Human Geography* 95 (3): 219–233.

Nally, D. 2014. "Development." In J. Morrissey, et al. *Key Concepts in Historical Geography.* Sage.

Neumann, Tr. 2016. *Remaking the Rust Belt: The Postindustrial Transformation of North America.* University of Pennsylvania Press.

Orange, H., ed. 2014. *Reanimating Industrial Spaces: Conducting Memory Work in Post-Industrial Societies.* Walnut Creek, CA: Left Coast Press.

Del Pozo, P. B. and Pablo A. Gonzalez. 2012. "Industrial Heritage and Place Identity in Spain: From Monuments to Landscapes." *Geographical Review* 102 (4): 446–464.

Price, W. R. and M. Rhodes. 2020. "Coal Dust in the Wind: Interpreting the Industrial Past of South Wales." *Tourism Geographies.* doi:10.1080/14616688.2020.1801825.

Price, W. R. and C. L. Ronck. 2018. "Gushing about Black Gold: Oil and Natural Gas Tourism in Texas." *Journal of Heritage Tourism* 13 (5): 440–454.

Proulx, G. and N. J. Crane. 2020. "'To See Things in an Objective Light': The Dakota Access Pipeline and the Ongoing Construction of Settler Colonial Landscapes." *Journal of Cultural Geography* 37 (1): 46–66.

Quivik, F. L. 2013. "Nuisance, Source of Wealth, or Potentially Practical Material: Visions of Tailings in Idaho's Coeur d'Alene Mining District, 1888–2001." *The Journal of the Society for Industrial Archaeology* 39 (1–2): 41–64.

Ranganathan, M. 2016. "Thinking with Flint: Racial Liberalism and the Roots of an American Water Tragedy." *Capitalism Nature Socialism* 27 (3): 17–33.

Rapson, J. K. 2020. "Refining memory: Sugar, Oil and Plantation Tourism on Louisiana's River Road." *Memory Studies* 13 (4): 752–766.

Rhodes, M. 2020. "Memory." In *International Encyclopedia of Human Geography, Second Edition, Volume 9,* 49–52, A. Kobayashi, ed. Elsevier.

Rhodes, M., W. Price, and A. Walker, eds. 2021. *Geographies of Post-Industrial Place, Memory, and Heritage.* London and New York: Routledge.

Ryzewski, K. 2019. "Detroit 139: Archaeology and the Future-Making of a Post–Industrial City." *Journal of Contemporary Archaeology* 6 (1): 85–100. doi:10.1558/jca.33835.

Scarlett, S. F. 2013. "Crossing the Milwaukee River: A Case Study in Mapping Mobility and Class Geographies." In *Landscapes of Mobility: Culture, Politics, and Place-Making,* edited by A. Sen and J. Johung, 87–104. Surrey [England]; Burlington, VT: Ashgate Pub. Limited.

Scarlett, S. F. 2021. *Company Suburbs: Architecture, Power, and the Transformation of Michigan's Mining Frontier.* Knoxville: University of Tennessee Press.

Scarlett, S. F. and L. W. Rouleau. 2022. "Object Lesson: Architecture at Pullman National Monument as Both an Agent of Division and Collective Identity." *Buildings and Landscapes* 29 (2): 99–117.

Schein, R. H. 1997. "The Place of Landscape: A Conceptual Framework for Interpreting an American Scene." *Annals of the Association of American Geographers* 87 (4): 660–680.

Sharp, J. 2009. *Geographies of Postcolonialism*. Thousand Oaks, CA: Sage.

Smith, N. 1996. *The New Urban Frontier: Gentrification and the Revanchist City*. Psychology Press.

Stanton, C. 2005. "Serving Up Culture: Heritage and Its Discontents at an Industrial History Site." *International Journal of Heritage Studies* 11 (5): 415–431. doi:10.1080/13527250500337454.

———. 2007. "Performing the Postindustrial: The Limits of Radical History in Lowell, Massachusetts." *Radical History Review* 2007 (98): 81–96. doi:10.1215/01636545-2006-028.

Storm, A. 2016. *Post-Industrial Landscape Scars*. Basingstoke; New York: Palgrave Macmillan.

Sweitz, S. R. 2010. "The Production and Negotiation of Working-Class Space and Place at Central Aguirre, Puerto Rico." *IA. The Journal of the Society for Industrial Archeology*: 24–46.

Tengan, T. P. K. 2008. "Re-membering Panalā'au: Masculinities, Nation, and Empire in Hawai'i and the Pacific." *The Contemporary Pacific*: 27–53.

Thomas, H. 2004. "Identity Building and Cultural Projects in Butetown, Cardiff." *City* 8 (2): 274–278.

Thompson, C. F. 2021. "Industrial Heritage in an Era of Climate Catastrophe." In Rhodes, M., W. Price, and A. Walker (eds). *Geographies of Post-Industrial Place, Memory, and Heritage*, 172–185. London and New York: Routledge.

Trepal, D., S. F. Scarlett, and D. Lafreniere. 2019. "Heritage Making through Community Archaeology and the Spatial Humanities." *Journal of Community Archaeology & Heritage* 6 (4): 238–256. doi:10.1080/20518196.2019.1653516.

Weedon, C. and G. Jordan. 2011. "Special Section on Collective Memory." *Cultural Studies* 25 (6): 843–847.

Yates, M. 2011. "The Human-As-Waste, the Labor Theory of Value and Disposability in Contemporary Capitalism." *Antipode* 43 (5): 1679–1695.

28

LABOR AND THE CHANGING ECONOMIC LANDSCAPE OF THE UNITED STATES

Andrew Herod

The American economic landscape has looked very different throughout history, as the country has transitioned from an agrarian to an industrial and now a post-industrial society. Some important nineteenth-century economic centers are today economically depressed while others that are today leading hubs for production and innovation had not even been founded then. Yet other places were prominent both then and now. These changes have played significant roles in shaping the historical geography of US labor activity, opening up possibilities for workers to engage in some activities and forestalling others. This is because the way in which both capitalists and workers are idiosyncratically embedded in particular places but also differentially move or interact across the landscape means that how US capitalism's economic geography is structured has significant effects upon their behavior. Simultaneously, through their deeds such social actors can reshape the landscape's form. This recursive relationship between spatial structure and social behavior forms a sociospatial dialectic (see box). Drawing inspiration from this concept, the chapter examines links between the evolution of the United States's economic landscape and worker activity.

Early Days

Whereas the US remained largely agrarian for its first century of existence, nineteenth-century industrialization dramatically transformed its economic landscape. Several large industrial agglomerations developed, including steel making in Pittsburgh, coal mining in Appalachia, and meat packing in Chicago. Drawing upon the concept of environmental determinism (see box), many early geographers explained this industrial and agricultural geography as largely reflecting "the generosity of Nature"—some places, they argued, were simply favored with valuable raw materials, navigable rivers, fertile soil, or favorable climates while others were not (Smith 1984). In reality, however, many other factors were also important, including the immigration of large quantities of cheap labor from places like Europe and China and, of course, the removal of the Indigenous population.

Regardless of how these early economic geographies were explained, the concentration of various agricultural and industrial activities in particular places had important consequences for US labor politics. Across the southern United States, for instance, prior to the Civil War the agricultural economy was dominated by crops like cotton and tobacco largely grown by

DOI: 10.4324/9781003121800-34

slaves on plantations. Though they were obviously not free to form labor unions, neverthe-less many slaves did find ways to resist their owners' demands, engaging in everything from working slowly to minor acts of sabotage (what anthropologist James Scott [1985] has called the "small arms fire of the class war") to participating in full-blown revolts, like that led by Nat Turner in Virginia in 1831. The southern plantation landscape, in other words, was the product of struggle, even if plantation owners had much more power to shape it than did slaves. In the free labor states of the Northeast and Midwest, however, crop production was dominated by small-scale yeoman farmers who, although they sometimes hired farm hands, had quite different labor demands and so produced quite different landscapes, ones dominated by myriad smallholdings rather than large plantations. For much of the nineteenth century organizing by such farm laborers was little known, especially as often they were the farmer's own family members. However, as many farmers expanded their commercial operations and the hiring of non-familial labor spread, various labor and farmer organizations became active in the nation's agricultural regions, especially in the South where, by 1900, half of US farm laborers toiled (Jamieson 1945, 6). For instance, sugar workers in Louisiana joined the Knights of Labor in the 1880s while the National Farmers' Alliance and Industrial Union (more commonly called the Southern Farmers' Alliance) was formed in the 1870s to improve white farmers' eco-nomic conditions through creating cooperatives and engaging in political advocacy. Excluded from membership, African Americans formed the Colored Farmers' National Alliance and Cooperative Union (aka the "Colored Farmers' Alliance"), although there was tension between the two as many of the latter's members worked as laborers for members of the former—an 1891 strike for higher wages by African American cotton pickers in Lee County, Arkansas, for instance, was met by violence from a biracial posse (Holmes 1973). Later, in the 1930s, the Southern Tenant Farmers' Union—an integrated union in which women would play signifi-cant roles—was created to organize small-scale tenant farmers in the South (see Grubbs 1971).

In discussing the making of the US's agricultural landscapes, it is important to understand how their physical layouts shaped how workers could organize. For instance, the Knights of Labor's success with sugar workers was due, in part, to the fact that workers toiled together in gangs, with their physical proximity in the fields allowing them to strategize in ways that typically more dispersed workers, like cotton pickers, could not (Zinn 2005). Likewise, the transformation of California agriculture, from a landscape dominated in the 1860s by wheat farms to one increasingly structured by 1890 around smaller operations producing vegetables and citrus fruit, had significant implications for labor politics. In particular, as the transition unfolded, landowners needed a much larger supply of cheap seasonal and temporary workers than previously—wheat farming was largely mechanized but vegetable and fruit production is much more labor-intensive (Mitchell 2012). With a growing reliance upon migrant agricul-tural laborers, class conflicts between waged laborers and growers became more common and increasingly resembled conflicts in other sectors of the economy—as Jamieson (1945, 6–7) noted, "An industrial structure of operations when adapted to agriculture tended to bring a correspondingly industrialized pattern of labor relations."

Industrialization Comes

While patterns of crop production shaped labor landscapes in rural areas, the concentration of workers in the country's rapidly industrializing communities helped create various *milieux* in which powerful unions would develop, as workers' solidarities could crystalize out of their proximity to one another. One of the earliest examples of this involved mill women in Lowell, Massachusetts. After its incorporation in 1826, large numbers of young women moved to the city for work and Lowell quickly became an important center of the early industrial revolution, to the point where it became known as the "City of Spindles." The majority of these women lived in boardinghouses established by mill owners, who hoped that having the women live together would facilitate monitoring and controlling them. Significantly, though, the spatial interconnectedness of their domestic lives (they typically shared bedrooms and dined in a common area) allowed the women to develop important friendships out of their collective experiences, such that the boardinghouses quickly became informal centers of organizing activity—an outcome neither anticipated nor desired by the mill owners. Indeed, the mill workers organized several strikes in the 1830s and 1840s—a time before women could even vote—aimed at securing better working conditions. Although they suffered several defeats, the women were successful in establishing the Lowell Female Labor Reform Association, the first union of working women in the US. Thinking spatially, they also spread their movement geographically by organizing union branches in other mill towns in Massachusetts and New Hampshire, publishing various "Factory Tracts" to expose the miserable working conditions they faced. Similar developments occurred in other industries where workers were brought together in large numbers. It is, then, perhaps no coincidence that the Knights of Labor had some successes in organizing coal miners in Kentucky and Alabama in the 1870s (Garlock 1974, 150), where the fact that they worked underground, out of their employers' sight, allowed miners to come together to strategize concerning how to improve their conditions.

As the Gilded Age proceeded, the pattern of industrialization shaped the geography of worker power significantly. Between 1881 and 1894 about 90 percent of all strikes occurred within an area stretching from Maine to Maryland to Missouri to Minnesota, where industrialization was unfolding most dramatically and where most of the country's largest cities were located. There were, though, important variations (Bennett and Earle 1982). Thus, in communities with populations under about 85,000, labor's clout generally declined in the last decades of the nineteenth century as local capitalists often managed to create what were, essentially, company towns ("the textile town," "the furniture-making town," "the pottery-making town," and so forth), where the small number of employers could work in concert to exert control over labor by blacklisting "troublesome" workers who enjoyed few other employment options. In larger cities, on the other hand, workers' greater opportunities to find alternative work should they be fired facilitated strike activity, with disputes generally transmitted from place to place either across the urban hierarchy (larger cities were more prone than were smaller ones) or by contagion, as strikes spread from their starting points in big cities to adjacent, less-urbanized counties. At the same time, the spread of the distance-shrinking railroads also facilitated the expansion of worker power across the landscape. Hence, with its origin in a wage dispute in Martinsburg, West Virginia, the 1877 railroad strike was the country's first national work stoppage, quickly spreading down the tracks until some 100,000 workers—some as far away as Texas—became involved (Zinn 2005).

The first half of the twentieth century witnessed a dramatic remaking of the economic landscape as manufacturing activities both intensified in the traditional industrial core of the Northeast and Midwest but also spread to new places, like California and several Southern states. The growth of assembly-line production, spearheaded by firms like the Ford Motor Company, facilitated the transition of manufacturing away from the relatively small-scale, craft manufacturing previously common in urban centers. Given that these new forms of manufacturing were typically fairly labor-intensive, huge industrial agglomerations grew in places like Detroit, Youngstown (Ohio), Gary (Indiana), Chicago, and other parts of the industrial heartland. This was exacerbated by the dominant model of industrial organization at the time, that of the vertically-integrated firm wherein a single company owns every part of the production process. Determined to be independent of suppliers, Henry Ford, for instance, developed his Rouge River complex in Michigan as a virtually self-contained industrial city. Completed in 1928 and at one time employing over 100,000 workers, the Rouge contained not only vehicle assembly operations but also its own electricity plant, together with a glass factory and steel mill making components for the vehicles assembled there. It was, in fact, the largest integrated factory anywhere in the world. Such geographical concentrations of workers facilitated unionization, despite Ford's use of hired ruffians to attack labor organizers. Similar patterns of spatial propinquity abetting unionization were evident in many communities across the country.

At the same time that the Northeast and Midwest continued to industrialize, other places that had not been part of the US's initial industrialization began experiencing growth in manufacturing. This new geography of work had implications for organizing. For example, between 1880 and 1920 much of the textile industry relocated from New England to places like South Carolina as the implementation of automatic spinning and weaving machinery reduced the need for skilled workers, as Southern states offered Northern industrialists inducements to relocate to the South (e.g., Mississippi's "Balancing Agriculture With Industry" program, which provided tax breaks and cheap land), and as the industry transitioned away from water and steam power (which the geology of states like Massachusetts afforded in abundance) towards electricity. Other industries also moved southward, including chemicals, paper, and furniture manufacturing. The lack of industrial experience on the part of large numbers of Southern workers, with many only recently having left farms to look for industrial work, made them much more pliable than were their Northern contemporaries, where industrialization had occurred decades previously. Nevertheless, there were many violent labor struggles as Southern workers attempted to unionize—in 1934, for instance, two-thirds of Southern textile workers went on strike for better wages and conditions, and several were killed in clashes with strike-breakers and local law enforcement (Irons 2000).

As the landscape became more industrialized, and as the US became the manufacturing hub of the world economy, unionization rates grew. Part of this was the result of laws passed during the New Deal making it easier for workers to organize but part was the fact that tight labor markets in the early post-war era gave workers greater negotiating power. Consequently, union membership hit a high point of about 35% of the workforce in the late 1940s and remained above 30% until the early 1960s, when it began a long decline (though in terms of absolute numbers its highpoint was in 1979, when some 21 million workers belonged to unions). With deindustrialization increasingly transforming the US economic landscape during the 1970s and 1980s, rates of unionization and labor's ability to secure higher wages and better conditions began to diminish.

Box 28.1 Marxist Understandings of Landscape

Andrew Herod

Marxist geographers have generally argued that there have been three broad approaches to understanding the landscape. The first dominated in the late nineteenth/early twentieth centuries and was associated with "environmental determinism," an ideology which argued that nature determines patterns of social development. This perspective was often used to justify imperial efforts—the countries of Africa, environmental determinists argued, were poor because of harsh environments and not because of anything that imperialists had done to them and their peoples. This position weakened in the 1920s with the emergence of "environmental possibilism," which held that humans have some degree of behavioral freedom within particular environments but that, ultimately, they still can not overcome its constraints.

The second approach, arising in the 1970s, represented a Marxist response to what were perceived to be more conservative understandings of the landscape's making. This response averred that the landscape's configuration was, essentially, a spatial reflection of capitalism's internal workings. Accordingly, comprehending why the landscape looks the way it does, early Marxists argued, really only requires understanding how capital operates. However, in many ways this, too, was a highly deterministic view of landscape production.

The third approach emerged in the 1980s and regarded the landscape as the product of a sociospatial dialectic (Soja 1980). By this Marxist geographers meant that it both reflected the spatial practices of various social actors but that its structure also shaped how such actors exist in it and remake it through their actions. Adapting one of Marx's (1963 [1852]) famous maxims about people making history but not under the conditions of their own choosing, such geographers argued that people make their own landscapes but not under the conditions of their own choosing, as the landscapes of the past shape what people may do in the present. They also argued that there are links between the landscape's physical form and the ideological work it performs. French Marxist Henri Lefebvre (1991 [1973], 33–40) maintained that the interconnections between the material and the ideological are articulated through what he called the "production of space," which involves three deeply intertwined elements:

- *Spatial practice* (the activities which physically make landscapes);
- *Representations of space* (the images of landscape produced by planners, architects, artists, etc. in maps, models, and so forth that shape how people conceptualize the landscape); and
- *Spaces of representation* (the material spaces that are imbued with complex symbolic meanings in which everyday life is lived).

The landscape, then, is the fusion of three types of space—physical, mental, and social. However, while all spaces exhibit these three elements simultaneously, each is deeply contradictory. Consequently, there is not necessarily any coherence in how space is produced, imagined, and experienced. This presents opportunities for struggle over the landscape's form.

References

Lefebvre, H. 1991 [1973]. *The Production of Space*. Oxford: Blackwell.

Marx, K. 1963 [1852]. *The Eighteenth Brumaire of Louis Bonaparte*. New York: International Publishers.

Soja, E. 1980. "The Socio-spatial Dialectic." *Annals of the Association of American Geographers* 70 (2): 207–225. doi:10.1111/j.1467-8306.1980.tb01308.x.

Labor and More Recent Changes in the Landscape's Structure

During the past few decades changes in how the economic landscape is structured have in many ways made it more difficult for unions to organize workers. For one thing, automation and the shift away from vertical integration, with firms increasingly outsourcing activities to subcontractors located some distance away, has resulted in a decline in the size of the average manufacturing facility. The days of thousands of workers toiling in a single complex like Ford's Rouge River plant have largely gone—in 1967 the average manufacturing plant employed just over 60 people whereas by 2012 that had fallen to about 35 (Atkinson 2017). This lack of geographical concentration has reduced the influence over local communities that workers and their organizations often previously enjoyed. In addition, the physical distance between where many components are manufactured and where the final products (like automobiles) into which they are incorporated are assembled has increasingly spatially isolated the assorted workers involved in bringing products to market, thereby making it more difficult for them to exercise power collectively—workers assembling pick-up trucks in the US may know nothing about workers in, say, India or Mexico who make the components they use.

For its part, the massive suburbanization that occurred after World War II and which dramatically transformed the country's economic landscape has also meant that the type of community which developed in the late nineteenth and early twentieth centuries in many places, one where workers who labored in the same facility lived close to one another and often socialized during their off hours, thereby building social networks upon which they came to rely for building political power in the workplace, has gradually disappeared in many instances. As workers suburbanized, encouraged by Congress's passage of the 1949 Housing Act making it easier for working-class people to buy homes, they often increasingly lived farther from one another, which made the previously common kinds of socializing after work more difficult. This combined with a ruralization of much manufacturing as employers sought out non-union, malleable labor in previously non-industrialized regions. Consequently, whereas for much of the US's industrial history manufacturing had clustered in metropolitan areas, during the 1950s and 1960s rural counties began gaining manufacturing employment at a significantly higher rate than did metropolitan counties. For instance, between 1959 and 1969 manufacturing employment in Southern states increased by 44% overall but by 61% in counties located more than 50 miles from the nearest urban area (Hekman and Smith 1982, 8). These changes mean that the "United States has probably experienced more decentralization, and, in effect, non-metropolitan industrialization, than any other major industrial nation, capitalist or socialist" (Lonsdale and Seyler 1979, 5). Despite some ups and downs, this trend has largely continued. Thus, in 2013, of all of the counties where at least 20% of total earnings were derived from manufacturing, 68% were rural or micropolitan areas (Bond 2013).

At a broader spatial scale, other changes in much manufacturing's location have also impacted labor unions and workers' abilities to organize. Arguably, the main one has been the so-called Snowbelt–Sunbelt shift, as large numbers of people and manufacturing jobs have moved from old industrial states like Pennsylvania and Ohio to new ones like Georgia and South Carolina. This has been combined with significant new investment by overseas firms like BMW and Toyota coming to low-waged and lightly unionized Sunbelt states. Thus, between 1950 and 2000 Illinois, Indiana, Michigan, New York, Ohio, Pennsylvania, West Virginia, and Wisconsin saw the proportion of the country's manufacturing employment for which they accounted drop from 52% to 33% (Alder et al. 2014, 4). Although in 2013 the industrial heartland Midwestern states still had a higher share of their total employment in manufacturing than did any other region (12.3% of their jobs, compared to a national average of 8.8%), the South (at 8.1%) was more reliant upon manufacturing than was either the Northeast (7.5%) or the West (7.6%) (Scott 2015, Table 1).

This remaking of the US's industrial geography has helped create what Davis (1986, 129) has called a "new union-resistant geography of American industry." However, while many of the

Sunbelt states in which manufacturing has grown have long histories of anti-union and anti-worker legislation, unions have also faced difficulties in states with a reputation for being more pro-worker. In California's Silicon Valley, home to many companies seen as leading the emergence of the hi-tech "new economy," unions have found organizing challenging. Part of the reason for this is the extremely anti-union culture of firms like Google and Facebook, part of it is because a lot of work done for hi-tech firms is outsourced (meaning that Google, Facebook, etc. are not legally workers' direct employers), but part of it also is due to the fact that many tech workers are isolated spatially because they work remotely. Nevertheless, despite workers' separation in physical space, some labor organizations have started to make progress, often using cyberspace as their medium. For instance, a Tech Workers Coalition representing white-collar workers like coders, programmers, and designers in several countries has begun working to organize Silicon Valley workers, while the group Silicon Valley Rising, a coalition of labor, faith leaders, community-based organizations, and workers, has begun to organize the mostly Black and Latino subcontracted janitors, cafeteria workers, security officers, bus drivers, and other service workers who work on the big tech firms' campuses but are paid much less than direct employees.

The shift towards a service economy has also had implications for how unions go about organizing. One of the reasons for this is that the model of organizing many used during much of the twentieth century was largely developed in the manufacturing sector and made several assumptions about the workplace's micro-geography. For one thing, it supposed that workers are employed full-time in large, centralized workplaces with few entrances/exits and regular shift changes, thus allowing a handful of organizers to stand at the factory gates at the beginning/end of shifts and hand out leaflets (Green and Tilly 1987). However, service sector workplaces are often much smaller than, and laid out quite differently to, manufacturing workplaces. This can make them harder to organize. Hence, leafletting is often not as successful because distinguishing workers from managers through their clothing is more difficult, shifts are more varied, and smaller workforces mean that organizers may spend a lot of time leafletting yet reach few potential recruits. Also, whereas manufacturing workers often toil some distance from managers, shops' and offices' layouts mean that service sector workers frequently work closely with supervisors. This not only makes it difficult for organizers to access workers beyond their managers' gaze but also can lead workers to feel that joining the union would be a betrayal of their boss (Berman 1998). Furthermore, unlike manufacturing workers, service sector workers are frequently attached not to a single workplace but to several—janitors, for instance, might clean several buildings, owned by different firms, in disparate parts of a city in a single night. Consequently, trying to organize them on a workplace-by-workplace basis, as is common in manufacturing, would be virtually impossible. As a way to address these issues, in the 1990s the Service Employees International Union that represents janitors developed an innovative campaign in Los Angeles that focused upon organizing not individual buildings but whole office districts—that is, they reconfigured the geographical scale of their efforts to try to achieve union representation across each district and then to standardize wages and benefits within them. They also moved organizing from the private spaces of work to more public spaces by engaging in street theater to press building owners to encourage the janitorial companies with whom they contract to improve janitors' wages and conditions. Other workers and their supporters have used similar tactics. In 1998 the Los Angeles group Clergy and Laity United for Economic Justice (CLUE) developed a "Java for Justice" campaign wherein members would order coffee in hotels and then talk to guests about poor working conditions, with the hope that these guests would pressure hotel management to improve conditions when they provided feedback on their stays.

Although deindustrialization has undercut the power of many manufacturing workers and their organizations, it is important to recognize, however, that some changes in how production is organized socially and spatially have provided opportunities for workers to exert significant political power. For example, whereas in the past manufacturers would typically stockpile components in massive warehouses on site in readiness for use, as vertical disintegration has progressed many have come to rely upon a production model in which components arrive just before they are needed. These "just-in-time" models of production have allowed them to save on costs associated with storing parts. Not only is this transforming the economic landscape's structure—in the past components suppliers might be located far away but today are typically within fairly close reach—but it has also made firms more reliant upon timely delivery of parts and thus on the logistics industry to ensure their supply chains function smoothly. In turn, this has allowed workers to easily disrupt firms' procurement networks, as delaying components' arrival can bring production to a standstill. In 1998, for instance, strikes at just two General Motors components plants in Flint, Michigan, led the company to shutter 27 of its 29 North American assembly plants while 117 other supplier plants had to either close or cut back production, leading GM to post a $2.3 billion after-tax loss in the second and third quarters of 1998. This made it one of the most expensive strikes in US history (Herod 2000).

The recent emergence of a so-called "gig economy," in which workers are hired for very short periods of time with no guarantee of continued employment, is also affecting the geography of work across the economic landscape and thus the ability of workers to make a living. One aspect of the gig economy is associated with the rise of "platform capitalism," in which workers are hired through various computer platforms like Uber (Figure 28.1) or Amazon's MTurk. Hence, MTurk, which digitally connects employers with workers to perform activities ranging from simple data management to more complex ones like producing content for publications, can facilitate the movement of work from where it is generated (like corporate headquarters in downtown urban areas) to lower-cost labor in places like India or in rural communities within the US. This can have significant impacts upon cities—the more that routine office work is outsourced, the less demand there will likely be for office space in either cities' business districts or in the suburban back-office locations to which much such work migrated in the post-WWII era. These spatial changes can also impact the coterie of workers who might otherwise have serviced office workers located in central city areas, such as those working in sandwich shops, taxi drivers, newspaper sellers, and so forth. By way of contrast, Uber obviously cannot outsource passenger rides from US cities to drivers in India but must operate in the same geographical areas where its services are demanded. However, even here there is an important micro-geography at work. To be effective, Uber must ensure that its drivers are scattered across the urban landscape in roughly the same geographical distribution as its customers and cannot have them concentrated in just one part of the city, which would result in some areas being underserved while others have too many drivers competing for too few riders. Another type of gig work is that involving "zero hour" contracts, in which workers have to wait around all day (without getting paid) just to be available for a job that may only last a few hours or even minutes. This typically requires workers like delivery drivers, bicycle messengers, or even factory workers to remain in close proximity to employers, waiting to be called to work if needed. Their gig employment, then, limits the geographical range over which they can travel in any given workday. As these examples illustrate, the gig economy augurs new economic geographies with which workers must contend.

Figure 28.1   A car belonging to a driver for both Uber and Lyft ride share services. Photo by Raysonho and available via Public Domain and Wikimedia Commons: https://upload.wikimedia.org/wikipedia/commons/thumb/a/ae/LyftUberCar.jpg/2560px-LyftUberCar.jpg

Summary

The economic landscape has clearly undergone significant transformations over the past two centuries. This has had important implications for workers' abilities to organize to improve their economic positions. There are at least five important conceptual points to be made in this regard. First, how workers are embedded in particular places can shape the actions in which they may be involved—for instance, industrial development that concentrates workers in particular places can spur them to collective action while the relocation of work (as in the shift to the Sunbelt) may present new challenges (or, sometimes, opportunities). Second, the micro-geographies of the workplace have implications for workers' abilities to exert political and economic power—do they work alongside their bosses or are their bosses more remote, do they organize in the spaces of the workplace or in the public spaces beyond the factory gates and office walls, etc.? Third, workers can reshape the economic landscape through their actions, as when they disrupt supply chains operating according to just-in-time delivery principles, forcing manufacturers to reconfigure spatially their logistics operations. Fourth, it is important to recognize that economic landscapes have a degree of geographical path-dependence, that is to say that how they are structured may continue to shape social actors' behavior even after the social relations which led to these landscapes' initial creation have changed. Hence, because various institutions and traditions of organizing can become embedded in the local community, areas which once fostered strong worker organizations may continue to do so even after the industries which gave rise to them have closed. Finally, given that creating labor solidarity is a matter of developing linkages across space, workers must engage with capitalism's unevenly developed economic geography to be successful.

References

Alder, S., D. Lagatos, and L. Ohanian. 2014. *The Decline of the U.S. Rust Belt: A Macroeconomic Analysis.* Working Paper No. 14–5, Federal Reserve Bank of Atlanta Center for Quantitative Economic Research.

Atkinson, R. 2017. "U.S. Manufacturing: Past Trends, Future Prospects and Needed Policies." Keynote remarks to the National Council for Advanced Manufacturing Annual Conference, Hyatt Regency Crystal City Hotel, Arlington, VA, September 7. www.nacfam.org/2017-nacfam-annual-conference/.

Bennett, S. and C. Earle. 1982. "The Geography of Strikes in the United States, 1881–1894." *The Journal of Interdisciplinary History* 13 (1): 63–84, doi:10.2307/203836.

Berman, L. 1998. "In Your Face, In Your Space: Spatial Strategies in Organizing Clerical Workers at Yale." In *Organizing the Landscape: Geographical Perspectives on Labor Unionism*, edited by A. Herod, 203–224. Minneapolis: University of Minnesota Press.

Bond, B. 2013. *The Geographic Concentration of Manufacturing Across the United States.* U.S. Department of Commerce, Washington, DC.

Davis, M. 1986. *Prisoners of the American Dream: Politics and Economy in the History of the US Working Class.* New York: Verso.

Garlock, J. 1974. "A Structural Analysis of the Knights of Labor: A Prolegomenon to the History of the Producing Classes." PhD diss., Department of History, University of Rochester, NY.

Green, J. and C. Tilly. 1987. "Service Unionism: Directions for Organizing." *Labor Law Journal* 38 (8): 486–495.

Grubbs, D. 1971. *Cry from the Cotton: The Southern Tenant Farmers' Union and the New Deal.* Chapel Hill: University of North Carolina Press.

Hekman, J. and A. Smith. 1982. "Behind the Sunbelt's Growth: Industrial Decentralization," *Economic Review (Federal Reserve Bank of Atlanta)* March, 4–13.

Herod, A. 2000. "Implications of Just-in-Time Production for Union Strategy: Lessons from the 1998 General Motors-United Auto Workers Dispute." *Annals of the Association of American Geographers* 90 (3): 521–547, doi:10.1111/0004-5608.00207.

Holmes, W. 1973. "The Arkansas Cotton Pickers Strike of 1891 and the Demise of the Colored Farmers's Alliance." *The Arkansas Historical Quarterly* 32 (2): 107–119, doi:10.2307/40030730.

Irons, J. 2000. *Testing the New Deal: The General Textile Strike of 1934 in the American South.* Urbana: University of Illinois Press.

Jamieson, S. 1945. *Labor Unionism in American Agriculture.* Bulletin No. 836, United States Department of Labor, Washington, DC.

Lonsdale, R. and H. Seyler, eds. 1979. *Nonmetropolitan Industrialization* Washington, D.C.: V.H. Winston & Sons.

Mitchell, D. 2012. *They Saved the Crops: Labor, Landscape, and the Struggle Over Industrial Farming in Bracero-Era California.* Athens: University of Georgia Press.

Scott, J. 1985. *Weapons of the Weak: Everyday Forms of Peasant Resistance.* New Haven: Yale University Press.

Scott, R. 2015. *The Manufacturing Footprint and the Importance of U.S. Manufacturing Jobs.* Briefing Paper #388, Economic Policy Institute, Washington, DC.

Smith, N. 1984. *Uneven Development: Nature, Capital, and the Production of Space.* Oxford: Blackwell.

Zinn, H. 2005. *A People's History of the United States: 1492-Present.* New York: HarperCollins.

29

AFTERWORD

William Wyckoff

In my corner of southwest Montana, two warm spring days can turn a cottonwood green. In nearby Bozeman, where I live, the trendy lofts and sushi bars pop up almost as fast. Landscapes can change quickly in America and they seem to outrun our ability to rein them in and make sense of what we see. But they still manage to beguile and provoke us. Thankfully our three editors, Chris Post, Alyson Greiner, and Geoff Buckley, all leading figures in the study of the American landscape, have brought together this large assembly of talented scholars and in the process they offer us an opportunity to take a collective breath and assess the current state of the twenty-first century American landscape and how contemporary observers study it.

The topics tackled by our contributors run the gamut from traditional to cutting edge, but all of the authors are committed to taking a fresh look at how the American landscape has changed in the past quarter century and how the related scholarly literature has interpreted those changes. As the editors note, taking stock of where we are in the early 2020s seems particularly important in this moment of historically unpredictable, contentious change. No status quo for the American landscape or the powers that shape it. New virus variants appear one day, *Roe v. Wade* is overturned the next. Racially charged murders with assault weapons happen with numbing frequency, border hysteria seems to grip many Americans, and the human and economic consequences of climate change are items in the daily news, not prognostications for our grandchildren. As our contributors note, all of these titanic environmental, cultural, political, and economic shifts reverberate in interwoven ways across the everyday American landscape. And, of course, the landscapes we live in shape us as well. How do we make sense of these changes? As landscape observers, what stories do we tell? As landscape transformers, what changes do we advocate?

This ambitious compendium offers many examples and ideas for us to reflect upon. Readers will find it amazingly fertile ground in which to harvest fascinating reviews of the current landscape literature, much of it from far beyond the bounds of academic geography. The contributors also suggest timely and emergent conceptual frameworks to organize our thinking, whether they involve energy landscapes or the imagined landscapes of science fiction. We can also ponder the more prosaic landscapes of retailing in an era where "bricks and clicks" mingle unpredictably to reshape the ways we consume (Fekete). There is a plethora of empirical work. Landscapes after all take place (and make place) in real localities. Landscapes are stubbornly specific and they shape the lives of real people every day.

DOI: 10.4324/9781003121800-35

Particularly revealing are the case studies that explore how landscapes reflect and shape gender and ethnic identities in settings such as North Dakota's man camps (Dando) or in the South Asian communities of suburban Phoenix (Skop and Suh). Multiple authors also ponder the complex meanings of Black Lives Matter Plaza in Washington, DC and how it contributes to that city's legacy as a center of Black life and Black political agency. In related ways, landscapes of heritage are probed at the site of Judy Baca's artwork in Los Angeles (Doss), revealing how twenty-first century Americans experience the past and how memory—often in contentious ways—is socially produced in particular places. And what about the symbolic and political power captured in the different versions of "the Wall" as they were assembled along our southern border (Madsen)? Landscapes play central roles in all of these examples and these authors effectively reveal how these landscapes produce multiple, often contested meanings in these settings.

The book is divided into four parts. Each collection of longer essays (and a related set of shorter boxed features) recognizes distinctive impulses in our study of the American landscape. In "Environmental Landscapes" (Part I), several chapters reveal how the "environmental turn" more generally in historical geography has shaped related work focused on the American landscape. Whether we examine water (Colten), energy (Black), agriculture (Harrington and Laingen) or waste (Melosi), our contributors remind us that the Anthropocene remains an era in which the natural world continues to confound and complicate our human designs to transform it. This part of the book also highlights growing interest in environmental justice studies (Holifield) and how the longer-term impacts of climate change will affect the ways we manage resources such as our public lands (Wilson and Youngs).

In Part II of the collection ("Social, Cultural, and Popular Identities in the American Landscape"), the contributors probe how twenty-first century popular culture is produced, how it shapes our experiences and identities, and how landscapes—both real and imagined— often play pivotal roles in the process. Religion (Avery-Quinn), music (Keeling and Bell), sports (Lauermann), film (Lukinbeal), and futuristic science fiction (Davidson) are all given attention. Race and racial injustice are also featured. Aretina Hamilton's "Reading the White Unseen within the American Landscape" explores the country's long but often concealed history of white violence against people of color.

There is a special section between Parts II and III of the book. The essay in this section is entitled "Bridging Social and Political Landscapes." Authors Joshua Inwood and Derek Alderman ("American Landscapes Under Siege: A Provocation") provoke us to consider how landscapes are interwoven with social conflict and racial injustice. Importantly, the authors also suggest that scholars need to actively work to make these landscapes of racial violence and persisting inequality more visible to foster meaningful social change.

In Part III ("US Political Landscapes"), the essays span scales from the global to the local. Barney Warf's description of America's enduring imperialism casts the net broadly, appreciating how American power is asserted with military might, economic and corporate control, and more subtle cultural influences that diffuse our impacts globally. Other essays illustrate how political control is manifest in the landscape with examples that include surveying the public domain (Anderson), incarcerating our criminals (Mitchelson), naming our places (Brasher), or electing our public officials (Shelley and Hollen). Ultimately, we are reminded how all human landscapes are an expression of power shaping place.

The final group of essays (Part IV) is entitled "Urban and Economic Landscapes in the US." Our contributors look with fresh eyes at how the changing postindustrial American economy (Rhodes and Scarlett) continues to rework our everyday landscapes. We can ponder how automobility transforms how and where we live (Hostetter). Tourism is also revisited as a key driver of landscape change and Velvet Nelson laments how gentrification and "touristification" can displace poor residents in a city or how "overtourism" can negatively impact the local quality of life. Benton-Short and Keeley review the recent literature on urban sustainability, noting that creating more sustainable cities must include fostering both resilient physical environments and more just social and economic environments. Large-scale transformations refashioning America's rural landscapes (Husa) are considered as populations decline and once-bustling Main Streets attempt to find new identities with fewer residents nearby. Finally, Andrew Herod's essay on labor offers a fascinating narrative that describes how different landscapes of labor (from pre-industrial, industrial, and postindustrial eras) have offered strikingly different opportunities and challenges for unionizing workers.

Several key themes run through the entire book. First, many contributors highlight *the connections between technological change and landscape change*. Technology has redefined how we consume products and services, where we work, and how and where we play. Changing processes of production, distribution, and consumption are redefining every aspect of American life, whether we consider the energy economy or the creation and impact of popular culture (such as music). Second, *the neoliberal earthquake in late twentieth-century capitalism continues to reshape places, rework the labor force, transform landscapes, increase economic disparities, and provoke lively political responses and reactions*. The growing concentration of wealth in America along with the shifting geographies of economic marginalization continue to rework landscapes every day. In addition, *America's failure to deal with racial injustice and racial violence continues to play out across the landscape in myriad ways*. Many of our contributors explore this topic through the nation's contemporary cultural politics, implicating the landscape as a key part of the story and also suggesting how *landscapes play key roles in how we commemorate places, preserve the past, define our heritage, and create our national identity*. Finally, the essays demonstrate a broadly shared belief that *landscapes, while often used to obscure power relations and reinforce inequalities, can also be made more visible and truly transformative in addressing issues of environmental and social justice*.

Where do we go from here? This rich collection of essays on the American landscape suggests several productive directions in the years ahead. Most obviously, there is a need to *continue to diversify our approaches to the American landscape, delve into new areas of inquiry, and enrich our theoretical toolbox*. These essays demonstrate that a landscape perspective can be a powerful tool in our critical look at topics such as gender, race, heritage, popular culture, politics, and social justice. What will the next quarter century bring? No doubt more opportunities to dive even deeper into these areas of research and to explore new subjects yet unimagined. How can a landscape approach be useful as we contemplate topics such as a post-carbon world, automation of the labor force, artificial intelligence, and environmental justice in a century of accelerating climate change? In addition, the idea of "landscape" has proven useful in theorizing connections between culture and place as well as the social and political conflicts that often revolve around its evolution and control. As new theoretical frameworks come into focus, we need to explore how they can also be integrated into our thinking about landscapes, how they are made, and what they mean to us.

Second, *there remains a need for deeply nuanced, carefully crafted landscape histories of particular places*. I'm not imagining mere field studies of fence or barn types (although they can be useful), but rather well written, superbly illustrated narratives that fully flesh out how and why landscapes in a place have evolved and what that can tell us about key processes and people that have shaped a specific neighborhood, community, or region over time. Such stories, effectively rendered and contextualized, contribute to a better sense of larger forces at work as well as an effective way to connect with the American public. Thankfully, we have plenty of recent examples to suggest the possibilities. Consider Tim Cresswell's *Maxwell Street: Writing and Thinking Place* (2019),

Adam Mandelmann's *The Place With No Edge: An Intimate History of People, Technology, and the Mississippi River Delta* (2020), or John Harner's *Profiting from the Peak: Landscape and Liberty in Colorado Springs* (2021).

Finally, *we can do something to create more environmentally sustainable and socially just landscapes in the world we will leave behind.* This can mean many things. It might mean producing academic research that reveals how landscapes have normalized injustice and suggesting ways how revised interpretations can improve our shared understanding of a place and its past. It might mean participating directly in political acts or protests that address a particular issue focused on the landscape. The options are enormous, including everything from advocating for responsible management of public lands to lobbying for more affordable low-cost housing in an urban neighborhood. It can mean participating in how the rules are changed, becoming a constituent part of the process that we study. It can begin locally with a county planning board or a historic preservation initiative. It can extend globally to advocating for a carbon-neutral future and a world more resilient to the inevitable challenges climate change will bring.

Fortunately, the key participants going forward in such efforts have already been identified in this extraordinary assemblage of landscape expertise. The contributors to this fine volume will form a phalanx of energy and creativity moving forward, knowing they have produced an enduring collection of case studies and innovative ideas that will inform our work for many years to come.

References

Cresswell, T. 2019. *Maxwell Street: Writing and Thinking Place.* Chicago: University of University Press.

Harner, J. 2021. *Profiting from the Peak: Landscape and Liberty in Colorado Springs.* Louisville, CO: University Press of Colorado.

Mandelman, A. 2020. *The Place With No Edge: An Intimate History of People, Technology, and the Mississippi River Delta.* Baton Rouge: Louisiana State University Press.

INDEX

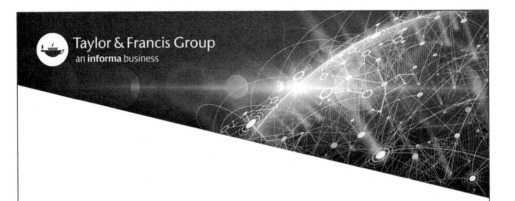

Printed in the United States
by Baker & Taylor Publisher Services